£17.95+

Community
impact evaluation

Community
impact evaluation

Nathaniel Lichfield

UCL
PRESS

First published in 1996 by UCL Press

UCL Press Limited
University College London
Gower Street
London WC1E 6BT

and

1900 Frost Road, Suite 101
Bristol
Pennsylvania 19007-1598

The name of University College London (UCL) is a registered
trade mark used by UCL Press with the consent of the owner.

British Library Cataloguing-in-Publication Data
A CIP catalogue record for this book is available from the British Library.

Library of Congress Cataloging-in-Publication Data are available

ISBNs: 1-85728-237-X HB
       1-85728-238-8 PB

Typeset in Times Roman and Optima.
Printed and bound in Great Britain by Biddles Ltd, Guildford and King's Lynn.

Contents

Preface

The central focus of this book is the application in the planning process (plan-making, implementation and review) of the theory, principles and practice of planning balance sheet analysis (PBSA), in its contemporary form of community impact analysis and evaluation (CIA/CIE). By definition, the main applications are *ex ante* in plan-making. But the approach and principles can be applied *ex post*, that is, in the subsequent monitoring, evaluation and review of the adopted plan, with the resultant lessons being applied in the next round of planning.

The concept, conceived early in my academic and professional life, has endured throughout in the sense that the fundamentals of the approach have remained the same. But this consummatory work records how the approach has been refined through its exposure to comment, discussion and experience, not least in its testing in a large number of case studies based on everyday practice.

Having launched the concept in the mid-1950s (Lichfield 1956) and learned how to apply it in practice in the 1960s (see Appendix), I then explored the role of *ex ante* evaluation in the urban and regional planning process (Lichfield et al. 1975). This started with the principles governing the use of evaluation in the process, went on to present case studies of other evaluation methods then being used, and finally presented the implications for planning procedures.

In the 1975 study, the authors justified their marked preference amongst the various evaluation methods for PBSA, but in that study illustrated the use of PBSA in only one particular case, the West Midlands region (Ch. 7). By contrast, although this book contains a brief comparative review of the various available methods of evaluation, it concentrates on the author's preferred method, which is essentially the adaptation of orthodox cost–benefit analysis (CBA) for making decisions in urban and regional planning (Lichfield 1960, 1964, 1968). As with cost–benefit analysis, it evaluates *economic impacts*, but also extends the scope of the analysis to *other* relevant impacts in urban and regional planning, primarily social, environmental and cultural. As such, in the 1960s it introduced the concept, barely understood then, of *impacts*. It was when the knowledge of them became more fully understood in the 1970s, as a result of American practice in impact assessment, that formal impact analysis became a second foundation to planning balance sheet, thus justifying the change of name to *community impact analysis and evaluation.*

But although the concept and the method have endured, the world in which they were conceived has not. Not only has the field of urban and regional planning become wider (for example, by now embracing community and environmental planning) but the number of fields that present a challenge for open evaluation in public decision making has considerably multiplied (for example, in education, health, environment, taxation). More and more, the call for open government becomes louder, and sensible and defensible decisions are demanded in the public interest. That demand becomes more strident as privatization grows, since there is the growing need to reconcile private-sector orientated decisions with the public interest.

But if the particular focus of the book is pursued, namely that decisions relating to action should be taken by comparing the potential impacts on the groups of people who are impacted (i.e. as though people mattered), then the method has much wider application. This stems from the realization that all decisions, however sectional, could have their public interest ramifications, and should therefore be considered by having regard to this wider interest in addition to the criteria of the particular decision-taker. Some examples will illustrate:

- The continuing battle by supporters and opponents of further involvement in Europe tends to be based in the main on a single issue: the degree to which interference by Brussels in national sovereignty can be accepted. Leaving aside the unnecessary confusion that arises from the failure to agree a definition of federalism, or its corollary of subsidiarity, perhaps more daylight could be thrown by comparing the predicted impact on the various groups in Britain of accepting or rejecting the *specifics* of European Union intervention, from the minor (food labelling) to the major (single currency). Who would benefit and who would lose, who would gain and who would pay, from the *specifics* of the additional or reduced intervention?

- Similarly, with privatization of former government enterprises in Britain: electricity, gas, water, air, rail. Instead of what appears to be the simple politically based ideology of sale to generate money for tax cuts, the following range of options could be evaluated for the various sectors and subsectors that would be impacted:
  - whether selling the family silver is better than keeping it in public ownership
  - whether the proceeds from selling the family silver should be invested in improving current infrastructure (so reducing public spending requirements) or instead used directly to reduce taxes
  - whether using family silver proceeds for augmenting current services, or providing new ones, is better than increasing disposable income in individual pockets by tax cuts.

- As a final example, there is the controversy over hospital closures. What seems to be the central issue is the need to reduce public expenditure, with the consequential managerial economies in the hospital service. What is not ventilated are the full repercussions on the way of life in the community in general of the options of retention or closure, by reference to all the impacted sectors, for example, health authorities, hospital functioning, transportation of patients and visitors, other social service spending, the economy.

These examples show that evaluation in the land use/development planning system, can be considered as an exemplar of the appropriate ways of taking decisions in the public interest in the other fields. The book thereby presents a challenge to all concerned with better and more accountable public-interest decisions outside the planning field.

In retrospect, the road traversed by this book has been a particularly long one and, in prospect, the journey end is far from reached. There is scope for improvement in the wide-ranging components of CIE in development planning; for extending the use of CIE to *ex post* evaluation; for linking the *ex ante* evaluation of CIE with the *ex post* monitoring and review; and for evolution of CIE for applications in new fields.

In these past journeys I have been fortified by this nugget from Pirke Avot, *The ethics of the fathers*, which, freely translated from the Hebrew, admonishes:

It is not upon you to complete the work but nor do you have the freedom to abstain from embarking on it.

The nugget continues to fortify me.

NL

Acknowledgements

Ruminating over the past forty years, in which I have been occupied and pre-occupied with the subject of this book, my thoughts return to all those from whom I have learned and benefited.

In Britain, to many individuals in central and local government; colleagues in my consultancies, both Nathaniel Lichfield and Associates/Partners and Dalia and Nathaniel Lichfield Associates; colleagues and students in University College London and many university departments; participants in many conferences, seminars, and workshops.

My thanks spread around the world. To Israel, where I collaborated with Professor Joseph Schweid and enjoyed fruitful discussions with Professors Morris Hill, Eric Cohen and Arie Shachar. To Italy, where I enjoyed support in Naples from Professors Roberto Di Stefano and Luigi Fusco Girard, and in Rome from Professors Franco Archibugi and Pasquale Scandozzi. To Australia, where the work was introduced and extended by Dr Ian Alexander and Professor John McMaster. To Sweden, where Bengt Holmberg and Professor Abdul Khakee made my work known. To Norway, where a research team consulted me on how to complete their use of PBSA in choosing Oslo's second airport. And to other countries via the International Committee on Economics of ICOMOS, where I chaired Professors William Hendon (USA), Christian Ost (Belgium), Peter Nijkamp (Holland) and Pietro Rostirolla and Almerico Realfonso (Italy).

Special mention must be made of the University of California at Berkeley. It was in 1959/60, as a visiting research economist in the Real Estate Research Program, that I came under the stimulating and enduring influence of Professors John Dyckman, Julius Margolis, Paul Wendt, William Wheaton and Melvin Webber. It was their encouragement during that year which helped me to find out how to make a planning balance sheet, and the generous opening of his office and files by Justin Herman, Director of the San Francisco Renewal Agency, which gave me the case studies. And it was there again in 1968, as visiting Mellon Professor, in the Department of City & Regional Planning, that I explored further in teaching some very active and collaborative students.

Next, I recall the many and diverse case studies in which the methods have been applied, firstly in research until around 1966 and then in consultancy commissions. My thanks are due to the clients who backed their views that evaluation would help them in decision-taking or in putting their case forward; to learned counsel who invited me to give evidence using evaluation on appeals; to the firms and individuals with whom we collaborated; to those who worked with me on the studies in my consultancy firms. Amongst these I must single out in particular for their significant contributions Professor Michael Beesley, David Campbell-Jackson, Honor Chapman, Peter Kettle, Dalia Lichfield and Michael Whitbread. The proof of the pudding was certainly in the eating.

I now turn to the specific help I have had in preparing this book. All the following gave generously of their time and helped me avoid solecisms in both content and style. Professor Lewis Keeble and John Delafons nobly read much of the text. Others made contributions to specific chapters on which I solicited their special knowledge: Professor Lyn Davies, Dr Norman Lee, Dalia Lichfield, Clare Mackney, Denzil Millichap, Stefano Moroni, Professor David Pearce, Dr Robert Rapoport, Geoffrey Smith, Roger Suddards and William Tyson. Any remaining errors are mine.

Finally we come to production. My thanks to my former secretary at NLP, Annabelle Disson, who word processed impeccably from tape to text, nobly followed by our present partnership secretary at DNLA, Tracy Buller, in the equally impeccable revisions of the text. To our planning assistant, Florence Herrero, who edited and master-minded the figures and tables to a seemly level. To my publishers, Roger Jones and his colleagues, who encouraged and disciplined me to this very fine production. To my family, Dalia, Gideon and Shulamit, who did not resent *too much*, those many, many preoccupied hours.

To all, and to others not specifically mentioned, my thanks for helping me to realize the ambition of this work.

I wish to thank the following for permission to use illustrative material or text:

British Airways for Figures 17.1 and 17.2
Built Environment for Figure 5.6
Cambridge University Press for Tables 7.9, 7.11, 7.14, 16.1, 16.2 and 16.3
Earthscan Publications for Figure 5.7
Glanford Borough Council for Figure 13.1
Heinemann Educational Books for Figure 1.4
Professor Lynn Davies for Tables 14.2, 14.3, 14.4 and 14.5
Her Majesty's Stationery Office for Figure 5.3
Routledge for Figures 5.1 and 5.2
Urban Studies for Tables 7.3, 7.4 and 7.12

To Dalia

for so ably and constructively participating in the excitement and struggle

Summaries

Choice, decision and action in urban and regional planning: Chapters 1–4

Although evaluation is only one feature in planning, it nonetheless pervades the whole of the process of plan-making, plan implementation and review. Accordingly, to place community impact analysis and evaluation in its appropriate context, it is necessary to have a broad introduction. Chapter 1 describes the generic nature of the urban and regional planning process. Since there are many models for such planning, and the evaluation process itself is influenced by the particular one in which it is made, it is necessary to focus on the model that is practised in Britain, namely development planning, as the framework for the application of the evaluation to practice in Chapters 12–19. Nonetheless, the evaluation described here is applicable to other models and other countries, as the case studies in the Appendix show.

Since *ex ante* evaluation is an aid to choice, Chapter 2 is placed in the context of choice, decision and action in everyday life. This review reflects a considerable body of theory and principles, upon which evaluation in planning can lean.

Chapter 3 starts with a brief description of the *tests* which are used in the planning process, many of which could legitimately be included in *evaluation* in the dictionary sense. Each test requires a specific method, with perhaps several dimensions, as in feasibility: financial and economic; accordance with the plans of the same, or higher- or lower-level planning agencies; or acceptability to the public. Within these different kinds of tests, this book concentrates on one, namely: given optional plans, how to compare them via community impact analysis (CIA) and evaluation (CIE) with a view to aiding the choice in the public interest of a particular option, as a preliminary to decision and action in implementation.

The results of this test will give the total net benefit (surplus over costs), as well as the distribution between sectors of the benefits and costs, thus showing how the plan would contribute to the *welfare* of the community in both efficiency and equity terms. It is this specific *assessment* or *appraisal* test to which we here confine the term *evaluation*, while accepting that in general discourse the term has a much wider connotation. We then proceed to a comparative review of available methods for *ex ante* evaluation, in order to distinguish between the roles they can perform, and to show the logic for our choice from them of CIE. This leads to the conclusion that, for evaluation as defined here (to assess a plan's contribution to welfare), the method itself needs to be founded on that form of economic analysis devised for the same purpose, namely social cost–benefit. Although this term popularly relates to a specific orthodox method of assessment/appraisal, Chapter 4 shows how it can be more usefully seen as a family of methods adopting a common approach to predicting and evaluating the costs and benefits of a proposal. Within that family there are particular members that are suited to answer particular questions which are of interest to particular kinds of decision-takers or others involved in the decision-making process, i.e. the stakeholders. Community impact evaluation is shown to be the member of the family most suitable for urban and regional planning.

Theory and principles of community impact evaluation: Chapters 5–11

The heart of the second part of the book is Chapter 7, which presents in detail the *generic* method of community impact evaluation applied to projects. The method is built around an algorithm, which has been devised with an eye to its computerization. At the heart of the method are the twin foundations of impact assessment, for predicting the impacts, and cost–benefit analysis, for evaluating the impacts. For that reason, chapters 5 and 6 respectively present the theory and principles of each, to enable their role to be better understood in the evaluation model. Since the results of the former is an *input* to the evaluation, the chapter aims to present a broad picture of this vast, wide ranging field. And since the latter is at the heart of the evaluation, which is essentially an adaptation of orthodox cost–benefit analysis for urban and regional planning, the coverage is in depth rather than width.

The generic method itself inevitably displays a large amount of material. In its presentation in Chapter 7, particular aspects are illustrated by reference to features of case studies, selected from the many that have been made in using the method since 1962. A complete list is given in the Appendix, amounting to some 60 in all, which show the diversity of application. Because of the large amount of material displayed in a CIE, its conclusions are not easy to comprehend. Chapter 8 is confined to this aspect, showing why conclusions cannot be simplified to a simple index, leading to guidelines as to how firm conclusions can nonetheless be drawn, with illustration from the case studies. Chapter 9 presents a simplification of the method, which can be used for those occasions when, for example, there is neither the time nor the financial resources to undertake the full generic method itself.

Having presented the theory and principles of planning, impact assessment and cost–benefit analysis, in Chapters 1, 5 and 6, and the application of CIE in Chapters 7–9, the scene is set for the theory and principles of community impact evaluation itself in Chapter 10. They are presented by reference to the sequential steps in the generic algorithm model. Then comes a discussion of the relation between fact and value in CIE, leading to a description of the relationship between *measurement* and *valuation*, and to how *valuation* leads on to *evaluation*. The chapter concludes with a description of the relationship between CIE and utilitarianism, this being the philosophy on which CIE is based.

This middle part of the book closes with discussion of an important use of community impact evaluation, namely the way in which its operation can advance and help the democratic process in planning. It first questions how democracy operates in the current planning process and then shows how it can be advanced, first in general and then by the use of CIE.

Community impact evaluation in practice: Chapters 12–20

Having described the generic method of community impact evaluation for projects and its theory and principles, in Chapters 12–19 we move towards its application in some discrete fields of British development planning, which are found in everyday practice.

Chapter 12 shows the adaptation of the generic method for projects to the planning process as a whole, namely plan-making for strategies, policies, proposals, programmes and projects; plan implementation; and plan monitoring and review. Of necessity this is somewhat generalized, since the field is very wide and the experience and application relatively limited compared with that on projects.

Chapter 13 presents one of the case studies on roads, by applying three evaluation methods to a particular proposal for an inner relief road to the town centre at Brigg, namely: the DoT's cost–benefit analysis (COBA); the SACTRA's assessment summary report; and community impact evaluation. It shows how CIE brought out a more meaningful conclusion for decisions in the public interest than the other two.

Chapter 14 goes into some depth in the application of CIE to the everyday practice of development control. It shows how this important field in urban and regional planning implementation, which currently lacks a clear discipline in decision-taking, can be enriched by the approach of community impact evaluation, and made more rigorous and meaningful. In particular it shows how all the planning considerations that are conventionally used can be translated into *impacts on people*, and thereby can be susceptible to CIE, and improved decision-taking "as though people really mattered".

The following chapters (15–18) deal with particular issues in development control, and in turn shows how community impact evaluation will enlighten and enrich the application to them, viz: planning gain/obligations; conservation of the cultural built heritage; integration of environmental assessment with planning assessment; balancing and weighing up the subtly conflicting policy directives relating to planning applications in green belts.

Whereas the preceding four chapters are concerned with regulation of the market, Chapter 19 turns to the practice which grew up in the 1980s, and is being extended in the 1990s, relating to control over the introduction of *new* regulations. Government guidance was to consider in advance the impacts of any new regulations, and their benefits and costs, prior to acceptance. Although the appraisal is welcomed as consistent with the approach to CIE, it is criticized for its limitation in considering primarily the business and economic sectors and not others which would necessarily be impacted. The chapter then goes on to show how the approach should be extended to the equally significant practice of *deregulation* and *decommissioning*, in the interests of the community as a whole.

The book ends by describing the management of CIE studies. Given the diverse array of applications of the generic model, it formulates an approach, based on the author's years of consultancy commissions in the field, to managing a study in practice. The management approach is to be effective in terms of the client objectives and the analyst's available time and resources.

Abbreviations and acronyms

ACTRA   Advisory Committee on Trunk Road Analysis
AIDA   analysis of interconnected decision areas
AONB   Area of Outstanding Natural Beauty
BART   Bay Area Rapid Transit [system]
BATNEEC   best available techniques not excluding excessive cost
BPEO   best practicable environmental option
BPM   best practicable means
CAA   cost compliance assessment
CAM   community analysis model
CBA   cost–benefit analysis
CBH   cultural built heritage
CCA   assessment of cost compliance
CE   cost effectiveness
CIE   community impact evaluation
CM   cost minimization
COBA   cost–benefit analysis
CRA   cost–revenue analysis
CV   contingent valuation
DM   do minimum
DNH   Department of National Heritage
DoE   Department of the Environment
DoT   Department of Transport
DTI   Department of Trade and Industry
ECIA   economic impact assessment
EDU   Enterprise and Deregulation Unit
EEC   European Economic Community
EIA   environmental impact assessment
EIS   environmental impact statement
EMA   East Midlands Airport
EP   environmental protection
EPA   Environment Protection Agency
FA   financial analysis
GA   goals achievement
GDO   General Development Order

GMPTE Greater Manchester Passenger Transport Executive
GNP gross national product
HIP housing investment programme
HMIP Her Majesty's Inspectorate of Pollution
HP hedonic price
HTPA Housing and Town Planning Act
IA impact assessment
ICOMOS International Council of Monuments and Sites
IPC integrated pollution control
IRR Inner Relief Road (Brigg)
LIA Location of Industry Act
LPA local planning authority
MAA multi-attribute analysis
MADM multi-attribute decision-making
MAJAG Manchester Airport's Joint Action Group
MAM multi-attribute method
MAUT multi-attribute utility theory
MCDM multi-criteria decision-making
MCE multi-criteria evaluation
MHLG Ministry of Housing and Local Government
MODM multi-objective decision-making
MOLGP Ministry of Local Government and Planning
MOTCP Ministry of Town and Country Planning
NEPA National Environment Protection Act (USA)
NNP net national product
NPV net present value
NT New Towns Act
OECD Organisation for Economic Cooperation and Development
PAG Planning Advisory Group
PBSA planning balance sheet analysis
PCA Planning and Compensation Act
PCT public choice theory
PE preventive expenditure
PLBCA Planning (Listed Buildings and Conservation Areas) Act
PPB planning, programming, budgeting
PPG Planning Policy Guidance [Note]
PPP "polluter pays" principle
PVB present value of benefits
PVC present value of cost
RC replacement costs
RTPI Royal Town Planning Institute
SACTRA Standing Advisory Committee on Trunk Road Assessment
SCBA social cost–benefit analysis
SEA strategic environmental assessment

SFA   social financial analysis
SFW   social welfare function
SIA   social impact assessment
SSSI   Site of Special Scientific Interest
TC   travel costs
TCP   Town and Country Planning [Act]
TCPA   Town and Country Planning Association
TCPIDA   Town and Country Planning (Interim Development) Act
TPP   transport policies and programmes
TR   technical report
UCIA/UIA   urban and community impact analysis
UDP   unitary district plan
UNESCO   United Nations Economic, Social and Cultural Organisation
WA   Water Act
WCA   Wildlife and Countryside Act
WTA   willingness to accept
WTP   willingness to pay

CHAPTER ONE

The nature of
urban and regional planning

1.1 A concept of the urban and regional system[1]

No human settlement, except perhaps the remotest jungle-bound or desert village, is
self-contained in that only residents use it. In contrast, the typical town is used in part
by residents and in part by others who visit for various activities (work, education,
recreation); and some of the town's residents will travel outside for some of their
activities (work, recreation). Any town can thus be defined in relation to this functional criss-crossing of "urban activities". This definition in itself must bring in the
hinterland in which the town functions, i.e. its subregion. Together they make up the
"urban and regional system".

Any town or region thus comprises a diverse array of physical elements (buildings
of all kinds and spaces between them, parks, roads) and of diverse human activities
(shopping, production, recreation). In the diversity there is some order, for otherwise
people would not get to work on time, would not have fresh milk available each
morning, and would not meet in groups (religion, culture). Figure 1.1a,b give one
version of this order (see also Reif 1973, Wilson 1974, Batty 1976).

Within any community in the urban and regional system, from village to metrop-
olis, each individual will participate in activities that will vary according to his/her
stage in the life-cycle (Fig. 1.1a). Some of the activities will be purely individual, others as members of families, and others as members of wider-ranging groups (clubs,
associations, youth organizations, etc.). In sum, the activities make up the way of life
in the community, be it limited or full.

People, however, do not have full freedom in choosing their way of life; they are
subject to external constraints over which they have little control, such as the economy or national policies. A concept of how a particular community achieves a particular way of life is shown in Figure 1.1b, to which the numbers in parentheses refer.

At the top of the Figure are the people (1), not at any one moment in time but as
they change over their life-cycle. Their activities (2) require institutions (organizations and style of management) (5), ranging from highly centralized direction to considerable freedom for initiative, innovation or self-management.

---

1. Based on Lichfield (1988c: §1.1–2).

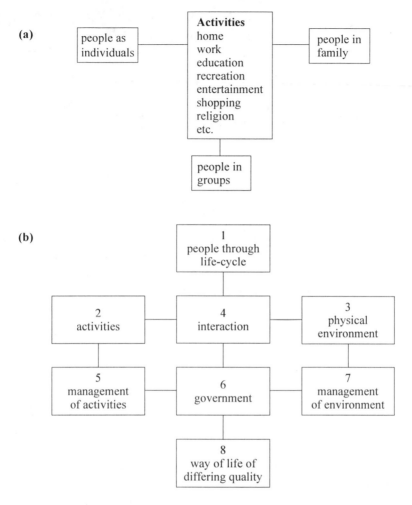

Figure 1.1 (a) Activity of people in a community through their life-cycle. (b) Way of life in community.

People engaged in these activities have a physical environment, both natural and man-made (3), with institutions for their organization and management (7).

The physical environment (3) and activities (2) will interact (4). Correspondingly, the management of the activities (5) and environment (7) will interact with government (6). Good housing will help good family living. The absence of schools and community centres will stultify education and recreation.

All these influences will affect the way of life (8) and thereby people's perception of the quality of that life (Perloff 1969). They are concerned not simply with what they do but how they do it. In this a critical factor is the way in which that life is managed (5 and 7) and governed (6): the greater the degree of self-management, the greater the likelihood of people responding quickly to external changes and adopting solutions that suit their own perception of their needs and values. A high standard of life under a dictatorship is quite different in quality from a poor standard of life with freedom in law and self-management in an open democracy. A high standard of housing and landscaping can be coupled with a low level of personal fulfilment and poor-quality social relationships.

A high quality of life gives people, whether as individuals, families or groups, the opportunity to fulfil themselves as human beings. For this they need not only an appropriate material standard of life but also appropriate management of their environment in all spheres (social, economic, institutional, cultural, physical) and appropriate administration by government.

1.2 Modelling the system

Of the many ways of modelling the urban and regional system, one that is well suited to our purpose is in Figure 1.2 showing Chadwick's framework ". . . within which the central relationship of Man and Nature can be seen clearly" (Chadwick 1978: 19). The Figure shows how Man is part of the total ecosystem. Within this he differs from other animals in having a value system that influences the activities that make up individual and social life, and the capacity to adapt Nature and space to his needs, to both good and bad effect.

The systems in the Figure are ". . . a set of objects together with relationships between the objects and between their attributes", with the objects being parts or components of the system, the attributes being the properties of the objects, and the relationships tying the system together (Chadwick 1978: 36).

But, while attempting to describe the real world, the system ". . . is *not* the real world, but a way of looking at it. Definitions of systems therefore depend in part on the purposes and objectives for which they are to be used" (McLoughlin 1969: 79).

However, defining the system we wish to study is by no means easy since ". . .if

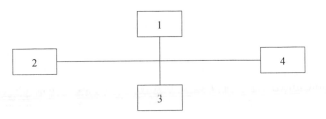

Figure 1.2   Relationship of man and nature
1. The ecosystem: Nature, including Man and the natural landscape of the Earth and its flora and fauna.
2. Man's value system: values, goals, objectives.
3. Man's system of activities: activities, flows, abstract spaces.
4. Man's system of adapted spaces: adapted physical spaces, channels.
Source: Chadwick (1978).

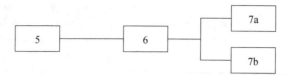

Figure 1.3   Change in the urban and regional system through a project
5. The urban project.
6. The system of activities and adapted spaces (Fig. 1.2, Box 3 and 4).
7. Post-project system of activities and adapted spaces:
   (a) without government intervention;
   (b) with government intervention.

we wish to consider the interactions affecting one simply entity, then we shall have to define that entity of the system. The system we choose to define is a system because it contains interrelated parts, and is in some sense the complete whole in itself. But the entity we are considering will certainly be part of a number of systems, each of which is a subsystem of a series of larger systems" (Beer 1966: ch. II).

Accordingly we need here to offer our own concept of the urban and regional system to suit our own *purposes and objectives* (McLoughlin 1969: 79), bearing in mind that ". . . the recognition of a system does not necessitate or imply its complete description" (Chadwick 1978: 9). In essence, our concept derives from our need in this book to explore the process summarized in Figure 1.3 to show how injections into the system (5) by projects (6) will bring changes into that system as it is evolving at any moment in time without planning intervention (7a); or how the changes can be shaped in accordance with such intervention (7b). It is the difference in the changes in the system represented by 7a and 7b whose impacts we are in effect aiming to evaluate.

1.3 A model of the urban and regional system[2]

Figure 1.1b shows the interaction between people's activity and their physical environment in a community. In this section we amplify that interaction.

Table 1.1 shows a representation of the urban and regional system that is suitable for our purpose. It is made up on the one hand of the supply of physical stock (col. 2) and, on the other, of the human population in their activities as producers and consumers in the socio-economic system (work, distribution, recreation, education) that exercise the demand on that stock (col. 3).

The physical stock can be divided into: natural resources and the man-made built environment, containing fabric and movables. The former comprise the human population inhabiting the land, water, minerals and so on, the product of the land (fauna

---

2. Based on Lichfield (1988c: §1.2, §1.6.3); there could be many other models, for example Wilson (1974), Batty (1976).

Table 1.1 Stock in the urban and regional system, past, present and future.

1. Physical elements	2. Supply: physical stock			3. Demand: flow of activities	
	Natural resources	Man-made fabric*	Man-made movables*	Producers	Consumers
1. Human population					
men	x			x	x
women	x			x	x
children	x			x	x
2. Natural resources					
land, water, minerals, etc.	x			x	x
fauna	x			x	x
flora	x			x	x
air	x			x	x
sun	x			x	x
3. Utilities infrastructure					
gas	x	x		x	x
electricity	x	x		x	x
water, sewerage, etc.	x	x		x	x
4. Transportation infrastructure					
train	x	x		x	x
bus	x	x		x	x
car	x	x		x	x
cycle	x	x		x	x
foot	x	x		x	x
parking (various)	x	x		x	x
5. Telecommunications					
telephone	x	x		x	x
radio	x	x		x	x
television	x	x		x	x
cable	x	x		x	x
6. Urban fabric					
buildings and sites	x	x		x	x
residential	x	x		x	x
industrial	x	x		x	x
commercial	x	x		x	x
shopping	x	x		x	x
administrative	x	x		x	x
recreation and leisure	x	x		x	x
educational	x	x		x	x
7. Open spaces					
city-wide	x	x		x	x
district	x	x		x	x
neighbourhood	x	x		x	x

* Built environment.
Source: adapted from Lichfield (1988: Table 1.1).

and flora) and the life-giving air, sun, rain. The latter comprises both the infrastructure on which the utilization of the buildings and places depend: utility services (water, sewerage, gas, electricity); the means of transportation between buildings used for these various purposes (private automobiles, buses, trains); and also in substitute, the means of telecommunication between them (telephone, radio, television); and the buildings and open places used for residential purposes (homes, hotels, barracks) and for social activities (work, education, leisure, recreation).

The activities take place within the framework of Figure 1.1a,b. There is a flow of people to their activities, from within the town and surrounding area, who use the physical stock as producers and consumers: for part of the time they live in their dwellings (homes, hotels, etc.) and for part of the time occupy the buildings and places designed for social activities (work, recreation, leisure, education, etc.). As linkages between the activities there are the means of transportation and communication.

These activities take place within two physical frameworks which provide Man's total environment: Nature's ecosystem and Man's system of adapting spaces in the built environment (Fig. 1.2 above). His activities make their impacts upon this total environment, which in turn makes its impact on Man in an interacting impact chain (Ch. 7). It is the human impact on the natural environment which gives rise to human concern about the erosion of natural resources and environmental pollution, which in turn affect mankind in its daily life (smog, pollution of water for consumption and recreation).

Table 1.3 shows the situation at any moment in time past, present or future. Changes can be brought about by the interaction between the physical stock and human demands on that stock. Decline in the numbers of population and their requirements for work (Fig. 1.1b) will give rise to growth or decline in the numbers of dwellings, work places, etc. which are needed. These changes will come about not only from forces within an urban area but also from outside, in the particular region supporting a particular urban area, or in the wider system of urban areas of which the particular urban area is part (Rodwin 1970). Conversely, the availability or non-availability of a stock of man-made fabric of various kinds will attract to itself or deny human activities, as a stimulant or a constraint to socio-economic life.

It is through this interaction of the man-made fabric and human activities that cities, towns and villages change, grow and decline, through what is known as the development process. To this we now turn.

1.4 The adaptation of space through development process without government intervention[3]

The built environment typically comes into existence on open land which is used for some form of agriculture, or perhaps some transition between agricultural and urban use (for nursery gardens, storage of materials, etc.). The earlier use is displaced so

---

3. Based on Lichfield (1988c: §1.6.3, §4.7).

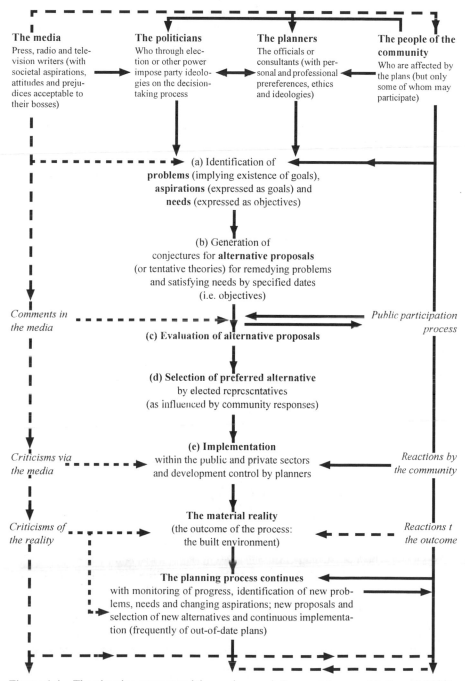

Figure 1.4   The planning process and those who may influence it (*source:* McConnell 1981).

that the vacant land becomes a development site. During construction, natural or man-made resources are destroyed, and environmental nuisance will arise (in noise, water, air). On completion of construction, the first life-cycle of the built fabric starts, with a use associated with the purpose for which it was designed (dwellings, manufacturing, retailing, etc.).

In the nature of the durable material typically used for human settlement, its life tends to be long, exceptions being the cane huts of an African village, tins and board of a Latin American squatter barrio, or the demountable tents of the nomadic Arab. This apart, the life will vary: relatively short (the ten years design life of "temporary housing"), or lasting over centuries (the monumental palazza of medieval Italy). By contrast, human activities which have to be accommodated within the fabric change much more rapidly (e.g. the size of families, modes of production, abandonment of cinemas for television, dislocation through war or earthquake). Thus, changes in human activities will tend in the first instance to be accommodated within the existing fabric, adapted as appropriate through refurbishment or other kinds of renewal, perhaps for a brief time and perhaps over a long period. At that time the physical stock reflects the then-current demands on it. But it may not do so later, following changes in location of activities or in means of accessibility of people to the physical stock.

Eventually the interaction cannot be quantitatively and/or qualitatively accommodated in the existing or adapted urban fabric, giving rise to the need for new physical stock on open land, either infill within the urban fabric or on its edge, which we call new urbanization. The adaptation of the current stock and the new development are in competition with each other in satisfying the common need for the matching of the fabric to contemporary requirements. The competition is not even, for the provision of new stock on open land is generally easier than renewal, in time, complexity and more profitable use of resources.

Over this life, the use and the conditions of the fabric, as a whole or in its separate parts, or within parts, do not remain constant. Maintenance and renovation lengthens physical life, but after a certain point, before it reaches exhaustion, the fabric becomes obsolescent. Then some form of renewal (in the form of rehabilitation or remodelling) is carried out, enabling the fabric to enter a new stage of life. This process will be repeated, once or more, before the degree of obsolescence is such that reconstruction, redevelopment or abandonment takes place. This is the beginning of a second life-cycle on the original site.

But it could be that, instead of renewal taking place at the stage indicated, there is resistance because of the objectives of conservation of the fabric. In this case the process of renewal, and the uses that would flow from the works, would be different; conservation is a special kind of renewal.

All the adaptation of space over the life-cycle of the built environment, as just described, takes place within what is generally recognized as the *urban development process* (Lichfield 1956: ch. 1; Chadwick 1978: 6–11; Ratcliffe 1978: chs 5, 6; Baum 1982; Baum & Tolbert 1985; Cadman & Austin-Crowe 1993: chs 1–8). The process starts with the gleam in the eye of anyone who sees the potential for change and finishes when the completed construction is handed over for the occupation for which it was designed.

Such development can be seen as an economic process carried out by the *development industry,* whereby an entrepreneur/undertaker/agency will bring together the various factors of production (land, labour and materials in the construction industry, finance) to produce the finished product to meet the economic demand which has been predicted. For this production/consumption process to be undertaken it is clearly necessary for the *land* itself to be appropriated (become property). As a result, there is a market in land which enables it to become part of the production/consumption process (Lichfield & Darin-Drabkin 1980).

This development process can be carried out either completely by the private sector (building houses on privately owned land) or completely by the public sector (the Town Hall on publicly owned land) or by various combinations of the two. And even if not participating directly in the development, the public sector will be involved directly in the provision of infrastructure or indirectly through, for example, providing subsidies to potential consumers of the finished development (occupiers of the dwellings) so that in combination they can express an adequate level of collective demand. Although this process represents that of the typical firm in the textbooks in economics, or of the typical public agency in the textbooks on public-sector management, it has distinct characteristics. The greatest is its complex and long drawn-out nature compared with, for example, production in agriculture.

From this description of the physical development process it is apparent that there are many parties involved: there might be the original landowner or any subsequent purchaser of the land; the public or private providers of infrastructure; the private or public developer of various kinds who undertakes the process; the building industry, including firms, labour, those supplying the material, and those producing them; the institutions lending finance; the ultimate occupiers, whatever the tenure; and the ultimate consumers of the services provided, e.g. shoppers via retailers. And all these parties are supported by an array of professionals: the town planners, economists, surveyors, valuers, financial analysts, architects, engineers, quantity surveyors; lawyers, estate agents, and so on (Esher & Llewelyn-Davies 1968, Fabrick & O'Rourke 1982, Knox 1989).

From this general description of the development process, it is apparent that its execution can be complex indeed. It involves many sides of the development industry; the contributions of many different professional skills; the need to predict conditions over a long time ahead since the product's development will be long-lived; and the need to keep an eye during the development process on variations in the assumptions and facts, to be built into the complex decisions required in the process.

1.5 Change in the system with government intervention

1.5.1 Why intervene?

Changes in the urban and regional system as just described come about through, in essence, the "regulated market process", namely that interplay between entrepreneurial agencies (private and public) in the market subject to governmental regulation. This market can be said to *plan* through the ". . . disposal of available means towards reaching an end or goal" and implies ". . . an automatic direction of economic life towards an inherent goal, that is of non-plan planning" (Myrdal 1960: 3–4). However, the result of such "non-planned planning" leaves much to be desired in terms of the ends of society, as opposed to those of operators prevailing in the unregulated market. It is the socially unsatisfactory outcome of this process which has given rise to the perceived need of intervention by government, be it at the simplest level (co-ordination of utility infrastructure) or at the more complex (urban and regional planning).

The case for this intervention reflects a variety of interests. To the economist there is "market failure" for various well known reasons: preoccupation with efficiency and not equity – the efficiency criterion being judged in the main by private and not social costs and benefits; difficulty in providing public goods; imperfections in the market resulting in waste (Bator 1958). To the environmentalist, there is the need to limit the pollution of Nature, which comes from market-decisions which ignore externalities (O'Riordan 1976). To the conservationist there is the need to protect what exists in order to preserve or to secure "sustainable development" – that is, the need to manage natural and man-made resources so as not to prejudice the needs of future generations (Pearce et al. 1989: 60–81). To the lawyer there is the need to prevent un-neighbourly conduct which is only cumbersomely provided for in the law of nuisance, tort and contract (McAuslan 1980).

The pressures from these various sources have resulted over the years in the introduction of different kinds of government regulation and control. For example laws relating to protection of public health, provision of utility services, elimination of slums, provision of transportation, protection of the natural environment, conservation of natural resources, preservation and conservation of the cultural built heritage (Ashworth 1954, Cullingworth 1994). The timing of such measures has varied from country to country, depending on its stage of socio-economic development, the degree of the problems that have emerged and its political philosophy. We now present the informed view of Friedmann as to how this philosophy has evolved.

1.5.2 Origins of societal planning

In the modern sense the logic for planning has ideological roots in the early 19th century, in the work of Henri de Saint Simon and August Comte, in which the vision first took shape of science working in the service of humanity (Friedmann 1987: 21). From these beginnings in ideological development, a theory of planning evolved in response to many different influences (Friedmann 1987: ch. 2):

The need for planning, for integration of different policies and programmes, and for government intervention, could all be related to the use and development of land. Thus, the concept of land-use and development planning evolved and provided a framework for specific intervention. This concept was first implemented in Britain in the HTP Act (1909). But this did not result in the early integration of all the various types of government intervention into the planning system. There were various reasons for this. The planning concept was relatively underdeveloped; the statutory mechanisms were slow to evolve; the planning system could not handle the problems as effectively as specialist legislation; and there were departmental and administrative jealousies to be overcome.

But consolidation within the planning system is gaining ground. In Britain, for example, many streams of intervention which had distinct origins have become incorporated as land-use planning objectives and their implementation has become more effective by incorporation into the planning system. Major examples are the protection of agricultural land, nature reserves, water-gathering grounds for reservoirs, flood plains, architectural heritage, natural resources, natural environment (Cullingworth 1994).

The struggle for this consolidation continues at two levels. The first is one of logic: how can you plan land use and development and yet respect the natural environment? The second is institutional: when government enacts legislation that crosses departmental boundaries. A contemporary example in Britain is the recent introduction of unified control of environmental pollution (EP Act 1990), replacing early legislation dating from the nineteenth century, and its integration with the development planning system itself (see Ch. 17).

Thus, the reasons for government intervention in the urban and regional system are very widespread. But it is not practicable within the scope of this book to follow them all. Rather we concentrate on land-use/development planning, including the other types of intervention which have become incorporated within it.

1.5.3 How intervene?

The case for government intervention through planning can be robustly made. But since the market does plan, and does deliver goods and services with great efficiency despite the recognized limitations (Myrdal 1960) and since Man in the 20th century finds difficulty in planning (Osborn 1959) it does not follow that *any* plan is better than *no* plan. Thus, to ensure an improvement over what otherwise would happen, the critical test is to devise forms of planning intervention that can be clearly justified in terms of the need for intervention, the cost of the planning, the impact on individual freedoms, and the bureaucratic implications. This is the nub of the dispute over planning, as well put by Hayek (1944: 26):

All economic activity, in particular, is planning decisions about the use of resources for all the competing ends. It would therefore seem particularly absurd for an economist to oppose "planning" in this most general sense of the word . . . a dispute between the modern planners and their opponents . . . is not a dispute on whether we ought to choose intelligently between the various possible organizations of society . . . it is a dispute about what is the best way of so doing. And these disputes which have accompanied attempts in this century to control the market have by no means resolved the issues, and are central to the debates between political parties (social market economy, etc.).

The reasons for Man's difficulties in planning are varied. The art and science in the generic sense, and its application to any particular field, is in its infancy; it must be where medicine and economics found themselves perhaps one hundred years ago. Thus, whereas the case can be made *for* planning as described above (§1.5.1) the very introduction of the planning produces "failures" which can be as unsettling as "market failure"; there is "government failure" in the inabilities of bureaucracies to deliver; "political failure" in the need to trade off decisions requiring a long view against those reflecting short-term electoral advantage; apprehension about the erosion of individual freedom that arises when the economic individualism of the market, where in theory at least the consumer is sovereign, is modified by attempts to increase "political freedom" in the redressing of inequities and maldistribution (Buchanan & Tollison 1972, Lichfield 1980).

How the change is to be managed, in a way that demonstrably improves upon the product of the market, is a central theme of debate between theorists and practitioners, between politicians and managers, and between professionals and the public. This book is seen as a contribution to that continuing interchange.

1.6 Planning in the evolution of the urban and regional system

1.6.1 The generic planning process applied to various fields

Planning is a human characteristic: that process of forethought about the future and then choice, decision and action based upon that forethought (Ch. 2). As such, the process of planning is generic, in that it can be applied to many human activities: from a holiday to a national economy (Archibugi 1993). But while the generic approach is common to many possible fields of human forethought (business, defence, the economy, manpower, etc.) its application must vary necessarily with the field itself, in terms of plan-making form and content, implementation, monitoring and review. Planning an educational system requires methods different from planning for defence.

In this book our concern could be with any of the systems described in Figure 1.2. But our prime concern is with one field only: namely where government, central and/ or local, has the opportunity or duty to plan the use and development of land. Planning in this field is applied the world over, under different adjectival descriptions: land, land use, physical/development, spatial, territorial, town, town & country, environmental, urban and regional, settlement, community. For our purpose the adjectives are synonymous. For others, the adjective would have significance in showing their emphasis: geographers might opt for "spatial", environmental scientists for "environmental", and sociologists for "community". In order to avoid confusion we adopt "land use" or "development" unless otherwise indicated.

1.6.2 The land-use/development planning process

As described above, land-use/development planning in essence applies the generic planning process to the urban and regional system through government intervention applied to the adaptation of space through the market. This has the twin components brought out above (§1.1). First is the current and future scenario of the present and future users of the plan area, that is those resident in or making use of it in their everyday socio-economic activities: work, shopping, education, recreation. Second is the provision for them of the appropriate mix of land uses for their activities (industry, shopping centres, education) which are associated with the appropriate physical development for the uses.

It is the latter component which has earned for the process the title "physical planning", which conveys that land-use/development planning is concerned more with the physical aspect than with the related socio-economic activity. This is not so. The two components are inextricably intermixed in practice, since "land use" implies the socio-economic activity for that use. Thus, the term "physical planning" is not altogether accurate for covering both components, and is often misconstrued as meaning only the first.

The planning process for this purpose is illustrated in Figure 1.4 (McConnell 1981). Within this can be seen its three interrelated phases, namely: plan-making, plan implementation and plan review.

This three-phase process is the prime preoccupation of the planning professionals: the general planners and the many others who contribute to the process from applying their individual disciplines (economics, geography, sociology, urban design, surveying, transportation, law, finance, etc.). But they do not act in isolation. As can be seen from Figure 1.5, there are many others outside professional circles who can influence the planning process throughout its run: the media, the politicians and the people of the community who are affected.

We now amplify each of the planning phases in turn.

1.6.3 Plan making

The actual plan-making process differs from country to country, and for particular plans within countries. This stems from three interdependent sources.

(a) The current and emerging philosophy behind State intervention in the market, in the country in question. This is to be found, for example, in the degree to which there is tightening or loosening of the State over enterprise, public control over private land ownership and the degree of dirigisme in the urban and regional planning system;

(b) Stemming from (a), the legal and institutional framework within which the statutory planning process is carried on. Here there are wide variations. For example, in much of western Europe the local plan when adopted has the force of local law, so much so that development has to take place in accordance with that plan, or it will be refused a planning permit (Davies et al. 1986a,b). This system is also found around the world as a residue of the inter-war system

Development planning	Common elements	Operational planning
Long history, strong role since 1947		Relatively new; a matter of guidance to authorities not of statute
10 years, plus perspective beyond	1. Planning use of resources to achieve aims of local community – with constraints	3–5 years, plus perspective beyond
Land use and location emphasis	2. Identification of problems information 3. Assembly of	Local government operations emphasis
Long-run social cost-effectiveness constrained by overall resource availability	4. Evaluation of alternatives; packages of programmes (plans) and individual programmes	Short-run social cost effectiveness constrained by local government finance
Uses operational planning programmes of district spending, plus intervention in market, plus representations	5. Implementation programmes	Uses spending programmes predominantly
	6. Monitoring effectiveness which returns to (2) and (3)	Additional emphasis on efficiency monitoring

10+ year structure plan being updated by county, extended development planning and local plans by district

Constituent programmes in the plan. Most of all, these need to be incorporated in operational planning as time for implementation approaches.

programmes comprising operational planning

Some longer-range perspectives (e.g. education, social services)

Figure 1.5 Relationships between development plans and operational plans (*source:* Lichfield 1988c: diagram 2.1).

which Britain left behind as a colonial power (New Zealand, Australia, Israel). By contrast, in Britain itself the pre-war planning schemes were abolished in 1943 and replaced by the development plan system in 1947 (TCP Act 1943, 1947). In this the plans do not have the force of law but have to be taken into account, alongside other material considerations, in making decisions on any applications for development (Ch. 14). The post-1947 Act system is thus more flexible and carries with it considerable discretion in the granting of permits, for both professionals and politicians, at local and central government levels.

(c) The art and science of the planning, which are found in the particular country, reflecting the conditions under (a) and (b). It follows that there are variations here also.

The influence of (a), (b) and (c) in the various countries around the world has produced an array of planning styles or models. The following categorization is based on Alexander (1986: ch. 4).

- substantive or sectoral function: related to the object of concern such as physical, economic, transportation, health, environment, resource planning
- instrumental: relating to the tools that are deployed, such as co-ordination, regulation, allocation of resources, development planning or indicative planning
- contextual: relating to the time, social institutions and value ideological premises, such as comprehensive, social, advocacy or bureaucratic planning.

Interest in the possible diversity of planning models has grown of recent years with the formation of the European Community. For one thing, there is diversity in different European countries in the light of the pressure for a level playing field in the different planning systems. For another there is the need to reconsider any individual planning system, including that of Britain, in the light of the distinctive type of planning model being generated from Brussels for application on a community-wide basis (Davies et al. 1994).

Although these models are varied, running through them there is a relatively standardized thread (the so-called "rational model") which is adapted with varying degrees of emphasis. More than that, since planning is a generic human activity which is adapted to many different fields, of which land-use planning is one, it is to be expected that the rational model in plan-making can be found throughout other fields. That this was so was found in an examination of ". . . about 100 prescriptive and normative studies of planning procedures, selected from the academic literature of the past decade (that is 1965–75) and largely originated in England and the United States" (Lichfield 1975: §2.1). From these "33 distinct models of planning procedures, both normative and descriptive, were identified and more fully compared". Although, as it might be expected, the models were reminiscent of the rational process, they were not identical: "From these comparisons a classification was made of planning activities from which the structure of a general model of an urban and regional planning process was constructed" (ibid.).

It was this generalized, *synthetic* model, which was then used for the purpose of comparing methods of evaluation in urban and regional planning. It was intended:

"To provide a checklist of all main activities or operations which would characterise the planning process from the time that some 'problem' is identified and a study is initiated through to implementation and review of the adopted course of action". Following are its main steps:

1. Preliminary recognition and definition of problems
2. Decision to act and definition of the planning task
3. Data collection, analysis, and forecasting
4. Determination of constraints and objectives
5. Formulation of operational criteria for design
6. Plan design
7. Testing of alternative plans
8. Plan evaluation
9. Decision taking
10. Plan implementation
11. Review of planned developments through time.

1.6.4 Plan implementation[4]

Not all plans are prepared with an eye to implementation, as for example the pre- Second World War planning schemes. But the post-war development plan was. Accordingly, from the plan-making process a programme for implementation could emerge. This could comprise a broad approach in terms of policies, proposals, strategy or plan, together with specific elements of a programme of projects (buildings, environmental areas, conservation areas). The essence of the programme would be its phasing by priorities and time, with a clear indication of the means whereby the programme is to be implemented by the implementation agencies (developers, local authority departments, governmental departments) and the resources they will need.

Within the framework of the plan, the planning authority will steer, influence or control the actions of the implementation agencies (development, renewal, conservation, and renewal) who are responsible for implementing their individual sectors (housing, shopping, industry, roads, etc.). Sometimes the power of implementation of the plan is not with the planning authority, but with another which could be termed the "implementation authority".

The nature of the implementation process will vary from country to country and from locality to locality, according to the implementation measures, economic resources, political interest and will, and so on. Within the process a variety of implementation measures will be available. From what was described above (§1.1), on the interaction between the physical fabric and human activity in the urban and regional system, it is apparent that the implementation can be directed at either the fabric or the activities, or both. The former involves measures ranging from the indirect (control over the actions of others) to the direct (positive action taken by authorities, as in the provision of new roads or new buildings). The latter involves the exercise of influence by indirect means (general persuasion, taxation or subsidy). This leads to

---

4. Based on Lichfield & Darin-Drabkin 1980: ch. 2; Lichfield 1956: chs 20, 21.

the following categorization of implementation measures, presented in terms of ascending order of social control, and the degree to which their effects can be predicted:

- General influence: such as education of planners and the planned, advocacy planning, the role of the planner as change agent, indicative planning (i.e. of what could happen as a guide to developers, without intention to implement).
- Intelligence and information: for example, improve upon the supply of data, information on proposed development projects, intentions and programmes of the planning authority.
- Organization for the management of change, for example: co-ordination between public and private implementation agencies, different central and local government departments, statutory bodies, etc.; rationalization of land release and infrastructure investment, etc.
- Fiscal measures: such as taxation, both general and specific, and financial incentives, such as subsidies, which influence the economics of development projects. Where they are not project specific, as in subsidies for car-parking or housing, the impacts are generalized; indeed they have been called the "urban impacts of non-urban policies" (Glickman 1980), as for example taxation relief for expenditure on commuting by public transport or rehabilitation of historic buildings; or subsidies via development for welfare and education services.
- Direct implementation measures: the most familiar in development planning are (negative) in development control and positive (land assembly, direct public investment). The former will considerably influence development projects, since they are associated with the grant of planning permits. The latter will crystallize into projects whose effect is reasonably clear.
- Direction of people and family: while happily of no relevance in non-totalitarian countries, it is mentioned as an example of one implementation measure which could be most effective.

1.6.5 Monitoring and review (ex post evaluation)

Implementation will continue, typically starting during the plan-making process itself. Over the years of implementation there will be continuing change, some of which will affect the basis and assumptions on which the plan-making was founded. As time elapses, the programme for implementation will need extending beyond that initially prepared.

Thus, a process of monitoring change is needed. This will facilitate a consequential review of the plan, its implementation and programme, leading to changes brought out by experience. While such monitoring and review is necessarily periodic, in practice the plan and its implementation could be continually under continuing review. As such, the review is closely related in practice to the continuing ex post assessment or audit of programmes (health, education, welfare, etc.) which are pursued, with or without the benefit of a prior plan (Rossi et al. 1979: 21–9).

1.7 Planning and management

1.7.1 Managing urban change

From the preceding it is apparent why the planning process for the urban and regional system is more than just the making of plans, with which it is commonly associated. A continuing process, with implementation and monitoring/review, is better suited to the more recent label which is being applied, namely "managing urban change" (Rose 1974). This recognizes that the development planning process can be fused with the urban change process (Bruton & Nicholson 1987: §2.4). Indeed the two are being seen as seamless.

1.7.2 Urban management[5]

"Managing urban change" is quite distinct from "urban management". This recognizes the urban system as a resource, capable of being used to satisfy human needs, wants and desires, and relates to its continuing administration. Given this, the objective of the management is taking conscious decisions, with an eye to the future, about continuing operations or the use of assets, or both in combination, within a structured organization. The price of not so doing, of bad management, can mean that the fullest use is not made of the operation or asset.

In practice, such urban management is not holistic in the same way as urban planning or managing urban change. While these should be applied to the whole urban area in its regional system, outside a "company town" (where ownership, administration and the operation of the main activities are concentrated in one organization), urban management tends to be fragmented. Within it perhaps five different agencies can be detected. These relate to:
- land units of fragmented property owners, both private and public
- larger estates in one ownership
- merging of land ownership and comprehensive development, as in building a new town
- corporate management for all local authority and other public sector activities
- environmental management in the regulatory sense.

1.7.3 Relationship of urban management and planning

The different agencies in urban management each "plans" by a model similar to that used in planning, namely the rational process (§1.6.3). Thus, not only are management and planning closely interrelated in functional terms, but also in their approach to the task in hand. This in practice makes for strength in the planning and management process, since each can be taken as complementary, and each uses a "planning process" reminiscent of the other.

Urban management and planning are thus interrelated. Managers see planning as an input to help tactical management. Planners see management as an output in tactical plan implementation. Both overlap. Their critical difference is in the agencies

---

5. Based on Lichfield (1988c: ch. 2).

responsible for each, and the kind of skills and professions they typically rely upon for advice.

Management looks in the main to managers, accountants and administrators. The tactics of their day-to-day management are clearly helped by having regard to objectives, and these become meaningful if translated into policies, strategies and plans which have regard to the longer and broader view.

Land-use/development planning on the other hand stems from the responsibility of government to take the wide and long view over the future of the urban and regional area. In this it relies for implementation upon the agencies concerned with the management, conservation and development of the urban and regional resources. It looks for advice to development planners working with a variety of other skills: in development (architects, engineers, surveyors, landscapers); in conservation (soil scientists, ecologists); in social sciences (sociologists, economists). It thus provides plans and policies to illuminate critical management decisions in relation to resources of land and property.

It is apparent that there are potential conflicts between the objective of management and planning, not least in the fact that each is studied, taught and practised by different academic and professional skills; in universities the management and planning schools tend to be distinct, with distinct academic strengths. But the conflicts can be readily reconciled in practice. This was achieved in one study, jointly carried out by planners and managers, where the relationship between development planning and operational planning was presented (Fig. 1.5).

1.8 Development planning as the focus of this book

1.8.1 Need to select a particular style of planning

Our concern in this book is with the application of the generic planning process to one particular field, that of land use and development. But while the practice has sufficient similarities around the world to facilitate international discussion, it is differentiated as between countries simply because of the differences in their kind of statutory and non-statutory plan-making.

The particular mode of planning adopted by government in a country heavily influences the form and content of the planning carried out within the official system (statutory planning) and also, to a lesser extent, outside that system (advisory planning). Accordingly, when exploring the practice of land-use planning (or any of its synonyms) it is necessary to pick out one particular form and content from the generality.

In this book we focus on one particular approach, namely *development planning* as initiated in Britain following Second World War under the Town and Country Planning Acts (Cullingworth 1994).

1.8.2 Macro and micro development planning

The adoption in post-war land-use planning of the terms *development* and *development plan* resulted in some confusion, since the same terms were also used in a quite distinct field. This was the national and regional economic planning at the macro level, which had been initiated in the attempt to give some direction and coherence to the economies of those countries moving from capitalism to socialism, and from colonialism to independence (Waterston 1966, Mabogunje 1980). This would not have mattered so much if the fields had been quite distinct. But clearly they were not, particularly at the regional level where the two have most chance of meeting (Lichfield 1967). Planning at the micro level (literally from the ground up, in terms of human settlement) clearly had some relationship with planning from the top down, for the country as a whole and its regions. But although in the latter, development planning quickly amassed its own theories, principles and practice (Lewis 1955), the interface between the two has not been adequately considered. In general, the macroeconomic models have had too little to say about physical and spatial planning (Mabogunje 1980). And for the latter, even when the theory became a matter for considerable attention, there has been little use of that introduced in national and regional development planning (Faludi 1973).

This gap is significant not simply because of lack of attention to theory but rather because the two streams generally fail to meet in practice. At the physical, spatial level, there is uneasiness at the introduction of the socio-economic; at the economic level, there has been the failure to recognize that the planning of physical development on the ground can make a significant contribution in the implementation of macro national and regional economic plans.

Having drawn attention to this dichotomy it is not our intention in this book to pursue the two fields of development planning. Rather it is to concentrate on development planning as it has evolved in relation to land use and settlement, i.e. *local development planning,* while bringing in as appropriate some cross-reference to socio-economic planning at the national and regional level.

1.8.3 Evolution of local development planning in Britain[6]

It is still a matter of wonder how the Coalition Government in Britain, during a war requiring every effort to avert conquest by Hitler's Germany, could find the spirit and fortitude to prepare revolutionary policies and plans for its post-war reconstruction in so many different fields: education, welfare, health, urban and regional planning. In the last, the field of our concern, the revolution was outlined in the three classic reports: Barlow (1940), Scott (1942) and Uthwatt (1942). These were followed through in a series of remarkable statutes: three by the Coalition Government before 1945 (MoTCP Act 1943, TCPID Act 1943, TCP Act 1944) and three by the new Labour Government (LIA Act 1946, NT Act 1946, TCP Act 1947), which reflected much of Coalition Government[7] policy.

---

6. For example, *The control of land use* (London: HMSO, 1944).
7. For a full review of the evolution in relation to London, see Simmie (1994: chs 6 and 7).

The 1947 Act introduced two key innovations: a new development planning system to replace that of the 1932 Act planning schemes, which were in effect local acts setting out regulations for the use and development of land (for the 1932 Act, see Roberts 1946); and financial provisions aimed at a solution of the compensation and betterment problem which had bedevilled the 1932 Act schemes, through fear of compensation yet inability to collect betterment (Lichfield 1956: ch 23).

The first, the development plans, expressed a new paradigm in planning. Because of the fear of compensation liability, the planning schemes zoned far more land for development than could possibly be needed in the reasonable future with, generally speaking, no indication as to when the development would take place, how, by whom, and at what cost. By contrast, the new development plans set out to predict the development that was likely to be needed and to make provision for it (Collins 1950; for a later account, see Morgan & Knott 1988: ch. 4). It could do so because of another innovation; the financial provisions, which ensured that any curtailment of potential development value on land not allocated for development would not attract compensation (Cullingworth 1994). The nature of this war-time revolution becomes all the more significant when it is recognized that many countries of post-war dynamic change (USA, Australia, Israel) continue to operate under a regulatory type of planning, as do many countries that needed to reconstruct after the war (France, Germany, Holland) (Delafons 1969 and Wakeford 1990 for the USA, Davies et al. 1986b for western Europe).

The term *development plan*, refers to the group of plans for a particular locality which have been carried out under repeatedly modified statutory requirements since 1947. It was conceived as a ". . . framework or pattern of proposed land use, in the form of a coherent set of proposals for the use of land, against which day-to-day development can be considered" (Lichfield 1956: 12, citing MoLGP 1951: 23). The coherent set of proposals were presented in an hierarchical set of plans that moved from a general outline for counties (county map), more detailed proposals for towns (town map and inset), a programme map setting out the changes visualized in the initial five years and the remaining fifteen years of the plan, and more detailed proposals for areas of immediate change (comprehensive development-area maps and supplementary town maps).

The working of the new planning system was articulated in Ministerial Circulars and Advisory Booklets. Here we do not attempt to describe the system as a whole (Lichfield & Darin-Drabkin 1980: ch. 4). Rather we aim to explore one specific aspect of local development planning of particular interest for this book, namely the relationship of development and the development plan.

In the initial plans of the 1947 Act, as put by this then-contemporary observer, the development plan (Lichfield 1956: ch. 20):

. . . is made because it is a way of formulating a policy which can act as the basis for guiding, controlling or initiating development; for deciding, for example, whether development should take place at all on a particular piece of land, or which of two or more kinds of development should be preferred on it, or whether development should take the form proposed by a developer. It is expected that as a result of making the plan, and of steering development in accordance with it, the ill-effects of unco-ordinated development will be avoided, an area must be better to live and work in than it might have been without planning control, and some contribution will thereby be made to national wealth and well-being.

The logic of this approach in the initial development plans implied *planned development*. For this to be carried out meant the local development planning process had the following features:

- In making the plan there should be the "bottom up" approach which ". . . leads the planning authority as part, but part only, of the preparation of its plan to try and comprehend what development would be likely to take place in the area if there were no planning control: in other words, what are the development programmes in the area" (Lichfield 1956: 27 and ch. 21).
- Since public authorities were responsible for some 90% of the development, because of the slackness of the private sector after the Second World War, there was a means of introducing public sector project-led proposals into the planning framework, which became known as "positive planning". These included joint venture partnerships between local authorities and developers in war damage reconstruction in city centres.
- When the development programmes had been screened through the planning policies and objectives in the development plan, the result was summarized in the form of a "programme map", for the ensuing five and fifteen years. This "programming" facilitated a clear understanding of the planned changes which were envisaged in the locality.
- Since there was a decided limitation of resources, in both the public and private sectors, it was necessary for the prospective development programmes to be tested for feasibility in terms of resources, including in respect of central and local government budgets.
- The fact that the potential development programmes themselves came from the development agencies rather than the planning office helped considerably in ensuring that the development plan and programme were in themselves practicable in terms of implementation. This aspect was emphasized in the Schuster Report (1950: 19, quoted in Lichfield 1956: 36):

> Now that the plan is no longer a series of broad restrictions, and has become a programme for development, every local authority must consider whether its proposals are feasible; within what period they can be carried out; whether, if carried out as planned, they will prove economically justifiable and how the burden of the cost is to the borne. The cost in fact must dominate positive planning, since there is no end to the improvements in the environment which are ideally desirable. This cost sets the limit, and when cost is ignored – as indeed appears to have been the case in some of the contemporary planning proposals – then planning is just so much waste paper.

- This hard-nosed approach to the feasibility and implementation of plans sharpened the setting of priorities and sequencing of the development programme in the plans.

These features visualized the development plan as a positive and feasible means for the transformation of the urban and regional system. Such an approach was clearly in the mood of the times, which saw such transformation as one of the benefits to be derived after an exhausting and impoverishing war. But the mood changed with the difficulties of achieving the transformation in the traumatic conditions of post-war Britain, and was expressed in the rejection by the electorate in 1951 of the Labour Administration. And with the change came a different view of the purpose of the development plans. There were many strands (Committee of Enquiry 1986: ch. 1; Hall et al. 1973: vol. 2; Bruton 1984: chs 1, 5, 9; Ravetz 1986; Cullingworth 1994). Development plans were seen as instruments more of development control and less of purposive planning. The abandonment in 1952 of the compensation and betterment system of the TCP Act 1947 initiated the pressures from the market on the planning system, in the quest for the spoils to the private sector resulting from planning permission. The formulation of development plan programmes became much more difficult when the private sector increased its share in the potential development.

Recognizing that change to the planning system was needed, the then Minister appointed a Planning Advisory Group (PAG) which reported in 1965 (PAG 1965). Their recommendations were reflected in the ensuing White Paper of 1967 (MoHLG 1967). This recognized the valuable contribution of the 1947 Act system (consolidated in the TCP Act 1962) and added "that the time has now come to profit from twenty years' experience and to make the changes required by new circumstances, new policies and new advances in planning techniques". In this the Ministry addressed itself to three major defects which had appeared in the current system:

First, it has become overloaded and subject to delays and cumbersome procedures. Second, there has been inadequate participation by the individual citizens in the planning process, and insufficient regard to his interests. Third, the system has been better as a negative control on undesirable development than as a positive stimulus to the creation of a good environment.

These are the main defects which the revision of the system must tackle and the Government propose to remedy them. To combine the safeguarding of individual interests with quicker decisions means streamlining; to emphasize the positive environmental approach requires a concentration of effort on what is vital and less central control of the details; and, in considering the changes necessary, we must recognize that planning now is operating in a very different context from that immediately after the war.

This approach culminated in the new style development planning of the TCP Act, 1968, which became absorbed in the consolidating TCP Act of 1971. Amongst the many changes, our concern here is just with the form of the "new style development plan". In essence this comprised (MoHLG 1970): structure plan (county and urban) and local plans (district, action area and subject). In this the hierarchy of planning of the 1947 Act was continued, but a new concept was introduced:

> The term structure is used here to mean the social, economic and physical systems of an area, so far as they are subject to planning control or influence. The structure is, in effect, the planning framework for an area and includes such matters as distribution of the population, the activities and the relationships between them, the patterns of land use and the development the activities give rise to, together with the network of communications and the systems of utility services. (MoHLG 1970: para. 3.6)

This change, the PAG argued (PAG 1965: para. 1.22), was not so much a progression from the 1947 Act system but an attempt to get back to the original concept of the development plan, as seen in the first draft of the Bill leading to the 1947 Act. Here a development plan was defined as "A plan indicating the general principles upon which development in [the]area will be promoted and controlled. This is very much the kind of emphasis which we would like to see reintroduced into the planning system. The main defects of the present system flow from the abandonment of this concept". This original concept provided for both an outline/basic plan giving policy, which would influence the programme map and comprehensive development area map for the early areas of prospective change (Lichfield 1956: ch. 2), a concept that was echoed in the PAG proposals for structure plans and local plans (Solesbury 1974). In proposing the change PAG seemed to have been more concerned with obtaining a *form* and *content* of development plans which would overcome the largely *procedural* defects of the 1947 Act system, which were highlighted in the 1967 White Paper. These in themselves were not inherent in the initial concept, but were the result of the subsequent plan-making, approval and review machinery becoming bogged down in the creaking machine of the statutory planning system. Attempts at improvement in this regard did not result in a return to the original concept of development plans, as sought by PAG. The "programmatic" nature of the original development plan was not in evidence.

The change of mood and economic environment certainly justified the reformulation of the approach to development planning in the 1960s, with its weakness in relation to physical–socio-economic planning, the role of policy in decision-making, and so on. But it hardly seems to have justified what is virtually the abandonment of development planning philosophy *per se*, namely to see the plan as an instrument for development. This hardly appears in the "new style" development plans following the 1968 Act. That this was unfortunate became apparent in the late 1970s and early 1980s when continuous economic growth since the war began to falter and Britain was seen to be in need of a development strategy, plan and programme to cope with its structural economic deficiencies, and decline of the regions based on traditional industry, and so on. And it is not surprising that the concurrent defects of too much reliance on policy planning, and not enough on development prospects and potential, is seen as a fault (Lichfield 1990b).

This dilution was all the more surprising since in the 1970s certain sectoral departments introduced the concept of sectoral investment programmes, such as transport policies and programmes (TPP) of 1975, housing investment programmes (HIP) of 1978, and inner area programmes of 1978. As Bruton & Nicholson point out (1987: 221), "All of these devices have a strong expenditure component: each consists typically of a plan or strategy statement explicitly linked with a budget, representing an attempt in this particular policy area to relate plan proposals to their cost and to the availability of resources".

But the full integration of these programmes with the development plan itself has many difficulties, despite the DoE advice that (Bruton & Nicholson 1987: 225) "structure and local plans provide a useful framework for those aspects of investment programmes which have a bearing on the development and use of land: the preparation of development plans and other plans and programmes should stem from a common planning process so that assumptions, for example in HIPs, are consistent and the policies and proposals are compatible". However, in practice, there are instances where (Bruton & Nicholson 1987) ". . . the development plan system is left out in the cold, the main implication being an explicit separation of structure and local plan proposals from the public sector resources which are partly necessary for their implementation".

It is perhaps no coincidence that such programmes were used in a manner reminiscent of the post-war years, as a means of central government control over local government expenditure. But the reasons were different. Post-war it was necessary to ensure that the authorities were resource-minded in their programmes; in the 1970s and 1980s the perceived need was to control public financial expenditure in terms of macroeconomic policy.

We now come to the 1980s, the Thatcher years. It might have been expected that the radical move towards the market, clarification of the ideology of the Right, and the impatience with centralized control, would have a made a fundamental impact on that most decided arena of social control over the market which is represented by urban and regional planning. But although there was a decided shift towards the importance of the market in decisions taken at local and central level (Thornley 1988) the planning system emerged basically as before. This is evidenced in the consolidating TCPA Act of 1990, which echoed in its basic principles that of 1971. And even the concerted attempt at deregulation spearheaded by Lord Young left in the end only relatively minor influences (see Ch. 19).

Indeed, in certain respects the planning system became more and not less planning orientated. For instance, the practice of planning gain/obligations, which was started in the 1970s and consolidated in the 1980s, grew in strength and marked a decided shift in the traditional frontier between private and public funding of infrastructure (see Ch. 15). Secondly, whereas there were tendencies in making planning decisions in the 1980s to play down the weight of the development plan against other material considerations, the Planning and Compensation Act of 1991 saw a decided break with the 1947 Act tradition of giving each of these considerations an equal weight. Planning decisions were now to be planning led. And while the shift in weight may not in practice be as significant as first thought, it has nonetheless changed the scene considerably (Ch. 14). Thirdly, whatever the change in weight, the shift took on a particular importance in that there was also a big drive towards covering the whole country with local plans, in particular urban development plans. With these in place, and the disinclination of the Secretary of State to interfere in local decisions unless called for, planning led decisions must rise in number and area.

However, a contrary trend has been seen in relation to strategic planning. The metropolitan counties have been abandoned, starting with the Greater London Council. After considerable vacillation, the planning life of the counties has been undermined *vis-à-vis* the district authorities, both because of the creation of the unitary district plan (UDP) in the TCP Act 1990, and in the threats to the counties as distinct entities by the Local Government Commission set up in 1992 (Cullingworth 1994: ch. 2). The location of central government regional bodies will go some way to compensate, but at central not local government level. But perhaps the greatest change of all, in relation to environmental considerations, came not so much from the Conservative Government policy or ideology but from pressures from outside. Notably, there was the recognition in the 1980s that the previous concern for the environment had been inadequate, and growing concern was essential. Thus, as regards practice in Britain it was the EEC rather than the government which made the running, in the process, without doubt increasing the need to have regard for the environment in planning at all levels. This led to the overhaul of environmental controls which had been inaugurated in the 19th century, and their consolidation in the Environmental Protection Act 1990. But the integration of planning and environmental controls has been slow to mature; to this we return below (Chs 5 and 17).

CHAPTER TWO

Choice, decision and action in everyday life

2.1 The boundaries of choice

An inexorable fate shared by every arrival in the world, and a great leveller, is the running of time from before birth, at conception. On each tick of the clock, what is past is sunk, what lies ahead is yet to come. This everyday fact leads to momentous consequences. The past cannot be undone. The future can be shaped, to a degree that is limited by what is outside control, both in nature (e.g. typhoons) and in society (e.g. other individuals and organizations). Within this limitation are made choices between alternative futures. This is found even in bacteria, in which all organisms show "... evidence that they act with a direction or aim, namely to ensure that life continues ... first of all that of the individual and then of the species" (Gregory 1987). As for humans, "As organisms have become more complicated the number of possible lines of action open to them has increased: in human life it is very great indeed" and "choice implies a capacity to select from a set of possible actions that are likely to achieve some end, given the circumstances" (ibid.). If no choice be made, that is also a choice. And any particular choice precludes others at that moment in time.

Within the limits noted above, of external constraint circumscribing the area of choice available, there is the long-standing question: are our choices predetermined or do we have free will? Is our apparently free choice dictated by the deities, by the stars via astrology, or by fate, as in the appointment with death in Samaria (Eilon 1971: ch. 7)? Or can we choose freely within the limits stipulated by constraints? The world largely assumes that we can, as we do here.

2.2 Choice and decision

In popular language, making a choice, or choosing, could be thought equivalent to making a decision or deciding (Mackenzie 1975). But a distinction is helpful. Whereas they both convey the selection from options, decision also conveys action upon the choice, as for example commitment to allocate resources, which implies a power to do so (Faludi 1986: 2). No decision, in a situation in which a decision is required, is also a decision not to decide. And any decision that carries with it a commitment to resources and action could preclude some other commitment. By the same token, a decision which does not become operational (in the sense of commitment to resources and action) is a non-decision (Jenkins 1978).

There is another richer meaning to the non-decision: the decision-taker decides, in the full sense described above, but is frustrated by the power of others involved in the problem at hand, as for example where the commitment of certain resources required for action is denied by their owners, so aborting the initial decision itself (Parry & Morris 1974). This distinction brings out a relationship to which we return below: of the prime decision-*taker* and others who are affected by the decision, the *stakehold-ers* (Friend & Jessop 1969: part I; Friend & Hickling 1987: 267). The former is the one empowered to take the critical decision under consideration on the issue at hand. But within this, the prime issue, there may be others also involved in deciding on related issues, which could be subsidiary to the prime issue, on wider or higher levels.

2.3 Decision and action

Where a decision does not become a non-decision, and leads to action, then the deci-sion is *implemented*. This could be under the control of the decision-taker (such as an individual using his own time or money for his own activities), or of a different agency (such as where the father gives money to the son to carry out a decision already reached). In this there might have been no relationship between the two in the process of decision-taking, or there might have been involvement of the son as stake-holder (e.g. willingness to spend the money in the manner indicated).

2.4 Action and review

The use of resources to implement a decision can have limited short-term implica-tions (walking to the pillar box to post a letter today rather than take it to the office for that purpose tomorrow), or be very significant and wide-ranging (to build a par-ticular new model of motor-car or inaugurate a manned spaceship journey to the Moon).

Whatever the implications, the action results in experience, from making the ini-tial choice to living with the ultimate consequences. Such experience is clearly a source of learning by all who are affected by the process. It can therefore be a topic for monitoring (audit) and review; perhaps in the consequence of the decision, or in the approach used to reach the decision and action, as a guide to future essays in the decision process.

This review will clearly help to enlighten the question: Was it a good or bad deci-sion? The answer can relate to a variety of aspects: to the *mode* of the decision-taking having regard to all its ramifications; whether the decision itself resulted in the *out-come* that was visualized; if it did not, was the reason in the decision-taking process (and therefore remediable) or the product of certain circumstances unknown at the time, over which the decision-taker had no control (in terms of, say, freak weather, the strategies of competitors, change in consumer tastes, faulty technology in implementation).

2.5 Hard and soft choices/decisions

From the few examples presented above it is apparent that choices/decisions vary enormously in complexity, from those that are easy to make (soft) to those that are difficult (hard). The complexity does not necessarily relate to the problem with which the decision-taker is faced, as the following instances show.

A soft choice could be regarded as one that involves the minimum use of resources; but a walk to the pillar box in the irretrievable posting of the letter could carry with it significant potential repercussions (accepting a job abroad or breaking up an engagement in marriage). Equally a seemingly hard choice, of the educational channels to a career, becomes soft when the decision criteria have already been made in terms of tradition: that the second son should join the Church or the Army, as have generations before. The distinction is not simple, because of the many considerations involved in the choice/decision process, as shown in the following four attempts at categorization, each using variations in the adjectival description.

To Von Winterfeldt & Edwards (1986) the range relates to the *mode* in which the decision-taker wishes to resolve the problem. A sedentary academic, who remained intellectually active until the age of 78, had an arteriosclerotic condition, of which his father had died. Should he continue with his conservative medical treatment or have carotid artery surgery? He chose by simple reference to the medical advice (soft/trivial). But his choice could have been formalized from decision theory into decision analysis (hard).

To Cohen & Ben-Arie (1993) the "hard choice" relates to the need for trading off *values* that are, by definition, *incommensurable* and not rationally calculable: "We define a person adhering to a plurality of values, lacking a general standard of evaluation, as facing a "hard choice" that is, one which cannot in the philosophical sense be resolved rationally". Such hard choices are "neither rational or irrational but a-rational" and need to be made by intuition, or perhaps following a social consensus concerning the mode of trade-off between differing values.

To Rittel & Webber (1973; for critical comment, see Friedmann 1987: 165) it is the nature of the *problem* that makes the distinction between "wicked problems" which are malignant, vicious, tricky or aggressive, and "non-wicked" which are tame, and benign. In the latter the mission is clear, as is the conclusion of whether the problems can be or have been solved. By contrast, wicked problems, being nearly all issues of public policy, have neither of these clarifying traits.

To Hardin (1984), the difficulty arises where individual or group interests *compete* with those of the whole society or collectively. These are "no technical solution problems".

2.6 Decision theory

2.6.1 The field

Much of the above can be encapsulated in what is termed "decision theory". This can be descriptive, of how decisions were *in fact* made (Schaffer & Corbett 1966); or prescriptive or normative (that is how decisions *should be* made if they are to be improved in effectiveness (Braybrooke & Lindblom 1963, Friend & Jessop 1969).

Decision analysis stems from decision theory. It would be useful here to introduce some simple and compact lead into this topic. This is not practicable, for not only are libraries filled by the literature, but this stems from different sources, in each case with a considerable degree of complexity, as a tribute to its importance in human affairs. There have been contributions from economics (Von Neumann & Morgenstern 1964), political science and policy science (Dror 1971), operations research (Ackoff 1962), management science (Beer 1966), social science (Von Winterfeldt & Edwards 1986), psychology (Edwards & Tversky 1967) and mathematics (Keeney & Raiffa 1976, Luce & Raiffa 1957).

Thus, in order to avoid either an over-complex statement or a void, it is necessary to be selective. To suit our purpose below (Ch. 7) three concepts pertinent to the rational approach in land-use planning are put forward (§2.6.2–4).

2.6.2 Space/framework for choice and decision

While choice and decision runs through the whole of everyday life, each choice/ decision arises within its unique framework. Accordingly if a sensible approach to the decision issue is to be offered then the choice/decision environment (space/ framework) must be clearly articulated (McGrew & Wilson 1982). For this, several questions need to be answered, such as:

- *Choice context:* What is the problem requiring decision, the options being formulated, the role of the choice in the consequent sequence of activity? Is it strategic or tactical? Does it relate to ends or means or both?
- *Decision-taker:* Is this an individual, a family, an informal group or formal group (local or central government); and if a group, are those making the decisions delegates (with freedom of action) or representatives (needing to act within the constraints dictated by the group)?
- *Stakeholders:* How are they approaching their task? How does this bear on the approach and freedom to choose of the decision-taker?
- *The constraints on the freedom of generation and choice of options imposed on the decision-takers and stakeholders:* Are these constraints born from external compulsion (e.g. role in a hierarchy of governmental agencies); or are they internal (guidelines for choice)?
- *Criteria for choice/decision:* Have these been specially formulated by the decision-takers and the stakeholders or are they implied? Will they be by tradition or custom (chief of tribe in an accepted position of political power) or open to negotiation and bargaining (between the decision-taker and stakeholder); would they be implementable without consensus?

- *Constraints on resources for the evaluation process:* What time and resources are available for the deliberations needed for the decisions? Will there be professional analysis available or not?
- *Decision mode:* Will the decision-taking be closed (that is internal to those responsible); or open (deliberations to be accessible or made public with the rationale for decisions to be explained and justified); or command (a dictator);
- *Decision communication:* How will the result of the decision/choice be communicated to the relevant parties and the public;
- *Use of decision:* How is it intended to use the conclusion? Will it lead to commitment and action, or not?

2.6.3 Decision analysis

Keeney & Raiffa suggest one "simple" paradigm of decision analysis (Keeney & Raiffa 1976: 5–26):
- Pre-analysis: corresponding roughly with the decision space/framework above (2.6.2);
- Structural analysis (Von Winterfeldt & Edwards 1986: ch. 3): by a decision tree, structuring the qualitative anatomy of the problem, the choices that can be made now or deferred, the information that can be gathered purposefully, etc. Following the start this has nodes that are either under the control of the decision-taker or not: decision nodes or chance nodes respectively;
- Uncertainty analysis (ibid.: ch. 2): using value trees, event trees and inference trees, assign probabilities to the branches emanating from the decision nodes, based on past empirical data, assumptions, results from modelling, expert testimony or subjective judgements;
- Utility or value analysis (ibid.: ch. 7): using value trees for the assignment of utility values on consequences associated with the path through the trees;
- Optimization analysis (ibid.: ch. 8): from the preceding, the calculation of the optimum strategy, that which maximizes expected utility. It is here that multi-attribute utility theory (MAUT) was developed.

2.6.4 Decision communication

The consequences of the process of choice and decision need to be formulated for communication with those who are affected. Within the process itself there are several links: material for the decision-taker as an aid to choice; his choice or decision; the action to be taken as a result, including the commitment of resources; the liaison with the various stakeholders who have the power to frustrate the decision into a non-decision; the people who are affected by the decision itself and the action to be taken; the public at large in an issue of general interest.

In a simple situation the communication can be simple and perhaps even oral as between all the parties. But with growing complexity there is need to rely upon aids to communication that can convey precisely the conclusions reached, in a form intelligible to those who receive the communication and need to consider or act on it and keep it under review. From this it follows that a particular mode of communication between certain parties in the process, for example decision-takers and stakeholders, will necessarily be different from communication to the public at large, with a possible distinction between those who are affected and those who should be informed. And in any of these links the mode can take various form: print, visual, audio-visual, seminar, and so on.

Where the required decisions are particularly complex, and the consequences will be experienced in different groups, recourse can be made to "decision conferencing" which draws on experience and research from three disciplines: decision theory, group processes and information technology (Phillips 1987). Typically, in this process, a sequence of steps is followed in a group discussion of those concerned, in which the nature of the problem is clarified and an *ad hoc* computer model is constructed which is used to assist in the group discussion. In essence the process is not a decision-making model per se so much as a decision support system (Janssen 1990).

2.7 Context for decisions in practice

The decision theory just described is applied in practice to real life situations. These affect the manner of application. At the one extreme, the decision-taker acts as an individual having regard to his own environment over which he has complete control, as for example the soft choice of walking to the mail-box to post a letter. But in a large number, and perhaps majority, of cases his decision-taking environment is complex, in that it is dependent upon decisions that fall to be made by others. Even the soft choice of a career in terms of family tradition becomes harder when others need to make decisions in parallel about, for example, the financial resources needed or qualifications for entry. This leads to a helpful distinction between decision-*takers* on the one hand and decision-*makers* concerned with the dependent decisions on the other. The latter are referred to here as *stakeholders*. These clearly abound in the field of this book.

CHAPTER THREE

Evaluation for choice in land-use planning

3.1 The role of tests in planning

As shown above (Ch. 1), urban and regional planning has evolved in the real world in response to problems, opportunities and constraints in the development and use of land which government has legislated to tackle. To assist government in this a large array of professionals has grown up, not only the generalist urban and regional planner proper, but also specialist planners (economics, transportation, etc.; see §1.6.2). The professionals see their contribution in understanding the issues with which the planning authorities, development industry and public are confronted, and their objectives and values, in order to make the appropriate analysis and studies with a view to offering advice and recommendations. Thus, in accord with all professional practice, they do what they can in the light of the then current technology of the profession and its ethics.

In the early days of planning in Britain, before the Second World War, this professional practice was dominated by the limited array of professional skills that were directly involved (primarily architects, engineers, surveyors and lawyers). Accordingly, the planning essays were marked by the technology available to these professions, which were strong on design (the architect typically being the "leader of the team") and on the legal framework. This limited approach became enriched when the 1950s and 1960s saw the involvement of those from social science, operational research, systems analysis, management science and policy analysis, who made a significant contribution to theories and practice, primarily *in* and *for* planning, that helped so much to mature the process (Batey & Breheny 1978).

With this enrichment came the recognition that the urban and regional planning process could be helped by the contribution of the scientific method in indicating how planning *should* be done. From this came the introduction of the rational method in planning (Breheny & Hooper 1985), and its use of models, as put forward in the systems view (McLoughlin 1969), as the following statement shows (Chadwick 1971: 63):

By creating a conceptual system independent of, but corresponding to, the real world system, we can seek to understand the phenomena of process and

change, then to anticipate them, and finally to evaluate them; to concern our-
selves with the optimisation of real world system by seeking optimisation of
the conceptual system.

However, in the 1980s came a reaction against this approach (Healey et al. 1982,
Alexander 1984, Yewlett 1985, Checkofway 1986). In general, it asked: While systems analysis was clearly relevant for understanding nature and certain kinds of
human organization, was it really applicable to human settlements? Could it be
extended beyond the spatial to the social and economic aspects? Was the optimization inherent in the rational model capable of being achieved, or was it necessary to
settle for satisficing, for the simple reason that neither the human mind nor human
skills were up to the requirements? Would not the conclusions from the model
accordingly be too limited for the human issues involved? Should not planners content themselves with disjointed incremental decisions rather than ones flowing from
the rational models (Braybrooke & Lindblom 1963)?

This tendency, to abandon the use of rationality in planning because of its complexity, would appear to flow from the unsatisfactory attempts in practice to apply
rigorously a rational model of the kind just described. If so, the best was taken to be
the enemy of the good. The implied assumption was that the use of a rational model
is *in itself* the application of the scientific method capable of producing positive
answers, so overlooking the message from Karl Popper: that the scientific method
lies in recognizing that any proposition (theory, hypothesis, plan, etc.) is a conjecture
whose validity cannot be proven, however derived. It is only by refutation that it can
be disproved, and until disproved can only be assumed valid (for relevance to planning theory, see McConnell 1981, Faludi 1986).

The refutation can conveniently take the form of *tests*. Examples are a test for
internal consistency: is all the activity dealt with in the plan (employment, schooling,
car ownership, traffic generation, etc.) accurately related to predictions of population
(stratified by age, sex, family formation and groupings, life-cycle, etc.)? Or for feasibility, namely the practicability, of the plans being implemented? Here there would
be many subtests, such as the economic demand for the products of the plan, the
resources available to meet that demand, the availability of land for the development
proposals, and so on (Lichfield & Darin-Drabkin 1980: ch. 2).

Such an approach has been suggested by one innovator who is identified with the
use of models to enhance rationality in planning. While advocating systems analysis,
Chadwick states (1978: 260):

The earlier discussion of scientific method, and those methods deriving from
it, has emphasised the central function of testing in the planning process: testing which derives in the first place from innate behavioural sources, and which
becomes externalised and formalised in the external, formal process of planning. Planning has been described as a list of tests which have to be carried out:
tests which determine whether or not fundamental criteria are satisfied.

The tendency to abandon the use of rationality would appear also to fail to recognize that in a field as difficult, evolving and immature as planning, *satisficing* is acceptable as opposed to *optimizing*, and that procedure by trial and error is inevitable. On this basis what has been achieved since the 1950s in planning methodology (although not without blemish) was a decided improvement on previous practice and, with continuous improvement from experience, better and better planning decisions will be achieved.

Thus, *whatever* the method used for reaching solutions to problems, be it entirely intuitive without data (the creative image of the traditional architect) or the output of the most sophisticated systematic modelling exercises (as in transportation planning), the scientific method requires that conclusions be *tested* to see whether they satisfy the criteria that need to the met.

The nature of the test in mind has been described as follows (Lichfield et al. 1975: 151–2):

> The simplest way of visualising what is here meant by the testing (in the broad sense) of a plan or project is to imagine that somebody is intent on making a critical examination of it: for example, the planners themselves, commissioning body, a Ministry examining the plan of the local planning authority, an inspector or commissioner at a public hearing, other planners or a writer reviewing the plan for a journal. The review could take many forms.

Such tests can be applied in a piecemeal random manner, such as, for example, the listing of a series of critical questions (criteria) with a view to finding the answers that indicate whether, on the criteria posed, the plan is satisfactory or not, or whether the local authorities in considering the adoption of a plan would find it feasible to finance the local government services out of its financial resources. But the conclusions from the tests are more readily usable if they are seen as interrelated elements of the planning process, for then the relevance of the tests, criteria and conclusions, can be put in a better perspective. From this viewpoint it is logical to adopt a *rational model* as the context for the tests, for the three interrelated aspects of plan making, plan implementation and plan review. This is not to suggest that the plan making *ought* to be carried out in that way, or that it has been, but rather that the model is a *rational heuristic framework* for the carrying out of the tests. If the framework is *irrational* in itself, in the sense of *illogical,* then the results of the testing itself could not be relied upon.

While an array of tests such as those indicated would have been carried out as a preliminary to generating the alternative plans, it does not follow that the alternatives that have been produced will necessarily be fully *satisfactory*. This might be so if the "design" were carried out by a computerized model where the output would necessarily follow the inputs, as in a computerized optimization model using linear programming (Ben Shahar et al. 1969). But there, because of the inherent limitations of the model for design purposes, the inputs were necessarily cruder than those that would be adopted in a manual planning process. And indeed that will, for the foreseeable future, apply to all such models.

Chadwick, in his similar approach (Chadwick 1978: 260) referred to *all* the tests as "evaluation". Here we apply that term (and its synonyms, "appraisal" and "assessment") to just *one* in the array of tests, namely that used to compare the inputs and outputs of a plan or project options, which is the focus of this book. This, as the other tests just described, are ex ante; that is, in advance of choosing. There is also ex post testing/evaluation, termed "programme review" (Suchman 1967, Weiss 1972, Rossi 1979). This is a feature of many public sector programmes (health, education, welfare, etc.). The process also features in urban and regional planning under the description of "monitoring and review" of plan making and plan implementation in the light of later information and experience. This "ex post evaluation" (appraisal, assessment) and "ex ante evaluation" are interrelated, as the following shows.

The purposes behind such review (and by definition the kind of monitoring which is set up) are not standardized but vary according to the prime interest of the review body. The following examples will illustrate. If the planning process is predicated upon initial goals/objectives, there will be interest in whether these have been achieved or fulfilled. If objectives were set during the plan making (be they derived as instrumental goals or in other ways) then there will be a monitoring of performance in relation to such goals/objectives. If, however, the planning process were problem orientated then the performance monitoring would relate to the degree of resolution of the problems, that is the closing of the gap between the situation current when beginning the plan and the situation aimed for under normative goals. The interest could relate to the monitoring and review of outputs in terms of impacts, that is the results that flow from the implementation, as opposed to the goals and objectives which were set up in devising implementation (see Ch. 12). Where there was in fact in the plan-making process a stage of ex ante evaluation of options, then there would be interest in monitoring the variables that enter into the evaluation process. The aim here would be to check and improve on the future performance in the evaluation process itself. From this array of purposes in plan monitoring can be seen a further array of tests, which are geared to the review.

3.2 Ex ante evaluation as a particular test

To define "evaluation" we find accord with Suchman who, while dealing primarily with ex post or programme evaluation, had much to say that is relevant to the ex ante process (Suchman 1967: 31). To him (ibid.: 29), *evaluation* (which is interchangeable with assessment, appraisal or judgement) is: ". . . the general process of judging worthwhileness of some activity regardless of the method employed".

In judging worthwhileness we rely in this book on the second of the foundations of community impact analysis, namely of the economics concept in cost–benefit analysis. If this be the goal, the means is *evaluative research* (Suchman 1967: 7), ". . . the application of scientific research methods and techniques for the purpose of making an evaluation".

In essence, this compares the plan options, which have successfully come through the array of tests, as an aid to choosing that which should be recommended for adoption by the decision-takers. For this purpose there is an array of methods which can all legitimately be included in our sense of the term "evaluation", as amplified below (Ch. 3.6). The differences between them relate to the questions that they inherently pose on behalf of the decision-taker, as for example which plan would best meet pre-set objectives or alternatively maximization of net benefits. With this in mind, we now review the available methods.

3.3 An earlier review of ex ante evaluation methods

Given the wide array of potential questions from decision-takers, with the growth of plan evaluation practice, it was necessary to search for a method appropriate to the evaluation of urban and regional plans (Lichfield 1970; Lichfield et al. 1975: ch. 4). Such a search in 1970 identified some 8 families comprising 23 distinct methods, of which some were in the cost–benefit family. About one half of the methods had been devised and applied within planning practice, whereas others had been introduced from other fields. It was this variation in the parent field of the method and the diversity of disciplines amongst the analysts that gave rise to the considerable variety in approach.

From the review it was seen that, apart from the planning balance sheet analysis of 1956, the earliest known method (checklist of criteria) was used in 1960, thus indicating the comparatively recent growth of plan evaluation. Since then, other evaluation methods have come into the field. Some already existed but had escaped the 1970/75 review net (e.g. multi-attribute evaluation, Winterfeldt & Edwards 1986; strategic choice, Friend & Jessop 1969), whereas others have been developed since (requisite decision-making, Phillips 1987; multi-criteria evaluation, Voogd 1983).

3.4 The changing basis for choice between methods

The early interest in the plan evaluation method, in theory and practice, was from combatants, in showing superiority for plan evaluation of one method against the others. In this spirit, in the 1970 review just noted, the comparison of the 23 methods was made in relation to ten criteria to bring out what, "plan evaluation methodology should satisfy to discharge its full function" (Lichfield 1970: 154). In this analysis the Lichfield planning balance sheet showed best, not attributable to immodesty (Self 1985) but because it had been specifically devised in departure from established methods with plan evaluation in mind. This combat in specific methods was pursued with some vigour: in attacking planning balance-sheet analysis (Hill 1968, 1990), in returning the attack (Lichfield et al. 1975: ch. 5), in taking proponents at their face value and knocking their heads together (McAllister 1980). But the interest moved on from this limited kind of comparison of methods, on at least three fronts.

First, since the methods had been designed, implicitly or explicitly, to answer different kinds of questions or hypotheses, there could be horses for courses. Indeed, the use of different methods on the same problem would by definition indicate different choices between options, so that: "In essence therefore it is the choice of method which decides the choice of option to be implemented. Therefore the choice of method must reflect the constraints and criteria which the decision-takers wish to raise, for otherwise they are not getting the choice they would really favour" (Lichfield 1977: 150–53; Lichfield 1985: 53). This obvious conclusion was demonstrated on a specific study relating to public transport in Manchester, where a different ranking among ten options was derived from using ten different methods (Table 10.3; Lichfield 1987). Secondly, what is the planning/decision framework with which the analyst is faced, and for which therefore he should be indicating the appropriate method (Lichfield 1977: 154; 1985: 52)? Hill pursued this issue in order to facilitate the choice of an evaluation method which would be more context responsive. He scanned a matrix of 12 evaluation variables which may be affected by the decision-making context, against five different kinds of planning and decision-making modes or styles, and then explored the relationship between the decision-making context variables and evaluation variables (Hill 1990). Nijkamp made a similar exploration by providing a matrix of 29 evaluation methods (mostly derived from multi-criteria sources) for comparison with 17 requirements in relation to environmental problems (Nijkamp & Spronk 1982, Janssen ct al. 1984). Thirdly, this recognition of the qualities of the different methods led to another emphasis: that there could very well be the need to make a synthesis of different methods, if by so doing the analytical tool were better suited to the planning/decision-making purpose (Lichfield 1985: 54).

3.5 Basis for current review

It is our intention here to retrace the steps of the 1970 review, adding the methods that have subsequently emerged or come to notice, with the aim of bringing out those which can be included in the term "plan evaluation" as used here, i.e. offering a comparison of outputs and inputs. In doing so our approach is as follows. We have in mind all kinds of plan making, namely for projects, plans, programmes of projects, policies, strategies, and we are referring here to ex ante evaluation but not the *process* of decision-making itself. Our interest here is not in the method as such, nor its application in practice, but simply in the way in which the evaluation is used as a "test" (§3.3).

3.6 The review

The methods now presented in four groups (A, B, C, D) are those described above (Lichfield 1970, Lichfield et al. 1975, Voogd 1983, Hill 1985, Nijkamp 1985). Whereas group D is introduced descriptively, groups A, B, C are presented in a matrix in Table 3.1 in order to bring out the essential characteristics under the following four

Table 3.1 Methods of evaluation, showing general characteristics.

		Input		Output		Criterion		Sectors	
		Q	M	Q	M	N	M	S	M
		1	2	3	4	5	6	7	8
A Outputs only									
1. Checklist of criteria	CC			x		x			x
2. Quality of service	QS			x		x		x	x
3. Norms and standards	NS			x		x			x
4. Goals/objectives	GO			x		x		x	
5. Linear programming	LP			x		x			
6. Impact assessment	IA								
Fiscal/financial	IA/F				x		x	x	
Environmental	IA/EN			x		x			x
Social	IA/S			x		x			x
7. Urban/community impact	IA/UC			x	x	x			x
8. Multi-attribute	MA			x		x			x
9. Multi-criteria	MC			x		x			x
10. Multi-criteria decision-making	MCDM			x		x			x
B Inputs only									
1. Unit costs	UC			x	x		x	x	
2. Threshold analysis	TA	x	x			x		x	
3. Costs in use	CU	x	x	x		x		x	
C Both input and output									
1. Financial analysis	FA		x		x	x	x		
2. Social financial analysis	SFA		x		x		x		x
3. Cost revenue analysis	CR		x		x		x	x	x
4. Planning and programme budgeting	PPB		x			x	x	x	
5. Cost–benefit analysis: single objective	CBA		x		x		x		
Cost effectiveness		x	x						
Cost minimization		x	x						x
6. Social cost–benefit analysis: multiple objective	SCBA	x	x				x	x	
7. Framework appraisal /	FA				x	x	x		x
assessment summary	AS				x	x	x		
8. Optimization	O				x		x		
9. Cost–benefit matrix	CBM								
10. Planning balance sheet analysis	PBSA	x	x	x	x		x		
11. Community impact analysis	CIA	xx	x	x	x	x	x		x
12. Social audit	SA		x	x	x	x	x		x
						x	x		x

Notes: Input/output – Q = quantity, M = money value; criterion – N = number, M = number reflecting money value; sectors – S = single, M = multiple.

criteria, indicated by a cross in the appropriate cell. Since there could be considerable variation in the use of a particular method in practice, the indications in the table are only a typical or general primary contribution in the assessment/appraisal process. Thus, the table does not claim to do justice to all the essays under each method.

The four criteria are:
- Do they relate to inputs and/or outputs? (cols 1–2, 3–4)
- Are the inputs and outputs measured in quantity or money? (cols 1–4)
- Is the criteria for choice a number, or a number reflecting money value? (cols 5–6)
- Do they relate to single or more (multiple) sectors of the community? (cols 7–8)

We now summarize each method in turn.

A. Output (value, benefit) in the main

Checklist of criteria (Kitching 1963, 1969, Matthew et al. 1967, Llewellyn Davies et al. 1970)   Performance criteria, typically somewhat random, are listed by the analyst, on his own assessment or perhaps in consultation with the client, public, etc. The criteria relate specifically to the services which would be generated by the plan, and the activities that flow from them (e.g. reliability, comfort, cleanliness). They could apply to more than one sector.

Standards and norms (Harrison 1977, Monopolies and Mergers Commission 1982) Standards are criteria indicating numerical requirements for particular plan elements, e.g. open space per 1,000 population, or carparks per floorspace in offices. They may be minima or maxima. Where desirable standards are non-feasible for application, the *norm* maybe followed instead, that which already generally obtains.

Goals/objectives achievement (Hill 1966, Shankland Cox & Associates 1966, Buchanan 1966, Kreditor 1967, Hill 1968, Schlager 1968, Tyson & Cochrane 1977) The criteria are in the form of pre-set, pre-weighted (in accordance with relative importance) goals/objectives. The derivation of the goals and weights can be those formulated by the planner or analyst or the decision-taker, or in the more sophisticated models derived from widespread consultation with the decision-makers or public. Then the outcome is related by quantitative measures to the goals/objectives, to reflect the extent to which these have been achieved. An aggregate index of achievement is then calculated. The aggregation procedures have been improved in an extension of the method, in multi-dimensional scalogram analysis (Hill & Tzamir 1972).

Linear programming (Ben Shahar et al. 1969)   The aim is to optimize the output, an objective function, subject to a number of well defined linear inequality constraints. Although very successful in the typical operational research problems, it has encountered difficulties in application to land-use planning largely because the list of constraints is very long and some can be translated into money prices whereas others can not. The advantage of the method is in its computerization and capacity for searching out options.

Impact assessment (Carley 1980)   The concentration is on the impacts that would flow from the project or plan, etc. Since these could vary greatly in character (fiscal, economic, social, natural environment, etc.) the generic method is applied in specialisms such as environmental (EIA) for the natural environment, social (SIA) for the socioeconomic, and economic (ECIA) for the economy. Each could relate to a large array of sectors.

Urban and community impact analysis (Glickman 1980)   President Jimmy Carter introduced this method in his National Urban Policy of 1978, which called, *inter alia*, on Federal agencies to carry our "urban and community impact analysis (UIA) on the spatial effects of new programmes and policies" in order to gauge in *advance* the possible *unintended* consequences of the programmes. The variables to be measured were largely economic: employment, fiscal, population. In the literature the term "community" is dropped and the method is known as "urban impact analysis". This is not to be confused with the "community analysis model" (CAM) which was developed for urban spatial modelling (Birch 1977).

Multi-attribute method or analysis (MAM/MAA) (Edwards & Newman 1982, Von Winterfeldt & Edwards 1986)   The output is disaggregated into its many attributes and values, and the scoring performances are then ranked. Because of complexity, the method is typically attributable to single sectors.

Multi-criteria evaluation (MCE) (Nijkamp 1975, Nijkamp & Spronk 1981, Voogd 1983)   This term is popularly ascribed to any of the methods which use more than the one criterion, e.g. goals/objectives (A4), impact assessment (A6), planning balance sheet (C10), and community impact analysis (C11). But more strictly it is attached to the one particular school described in the references A8–10. The function here is wider than that relating to the multi-attribute method. At least four steps are suggested: descriptive analysis of the spatial system; selection of options, account for a proposed line of action or policy line, and test the likely appropriateness of a certain policy (Voogd 1983: 35). As such, the method has richness and mathematical dexterity which has a wide relevance in planning, outside evaluation as defined here.

Multi-criteria decision-making (MCDM) (Cochrane & Zeleny 1973, Massam 1988)   This method is even wider. It emphasizes decision-making, within which four closely related techniques are used for particular functions: multi-attribute decision-making (MADM), multi-attribute theory (MAUT), multi-objective decision-making (MODM), and public choice theory (PCT). MADM sets out to evaluate a given feasible set of alternatives to select the best one; whereas MODM also defines the set of alternatives as a preliminary. MAUT introduces the choice by the highest expected utility value. PCT examines the general problem of finding appropriate ways of incorporating the views and opinions of individuals towards the satisfaction of the collectivity. Taking MCDM problems as a whole, there are three major components (the alternative plans; the criteria used to evaluate the plans; and the interest groups) that are combined as a set of matrices.

B. Input cost

Here there is a concentration on the economic value of the resources required for the options, normally expressed in financial terms (Lichfield et al. 1975: 55–8). It has subdivisions:

Unit cost (Monopolies and Mergers Commission 1982)   The total cost of the input is divided by the total quantity of the output in a performance measure (vehicle miles, vehicle hours, passenger miles) to give costs per unit of output. Passenger miles per pound sterling has been used by London Transport as a useful approximation for cost–benefit analysis (Beesley et al. 1983).

Threshold analysis (Malisz 1966, Kozlowski & Hughes 1967, Kozlowski 1968)   The thresholds are the points/times at which new capital infrastructure costs need to be incurred to accommodate growth, for example through topography in sewerage schemes. This is a preliminary to estimating the numbers of new units (e.g. dwellings) which cannot be constructed and serviced at their previous unit cost levels without substantial additional outlay.

Costs in use (Stone 1980)   The aggregate unit costs of the input are estimated in respect of both capital and subsequent operating/running costs, it being recognized that variations in either will affect the other.

C. Both output (value, benefit) and input (cost).

Financial analysis (Merrett & Sykes 1973, Darlow 1982, 1988)   As the name implies, output and input are compared in financial terms (both capital and annual) and typically relate to one sector, that of the promoter or investor in the project.

Social financial analysis (Lichfield 1988)   In this case the financial costs and returns are explored not only for the promoter but for all stakeholders who are directly involved in the project.

Cost revenue analysis (Mace 1961)   This is similar to the preceding two methods, in that it deals also with financial costs and returns. But the nature of the costs and returns differ, since the costs relate to the municipal services required for the development, and the revenues to the changes in taxation receipts which will be obtained as a result of the development.

Planning, programming, budgeting (PPB) (Lyden & Miller 1968) The methods and techniques of cost–benefit analysis, in this field also called cost–utility analysis, were introduced into the process of *planning, programme* and *budgeting* via systems analysis and management science, which was applied to United States budgeting in the 1960s. Initially the application was in the Defence Department under Secretary of State McNamara, and then in 1965 in all departments and most agencies in the US

Government. Each department was asked to: ". . . develop its objectives and goals, precisely and carefully; evaluate each of its programmes to meet these objectives, weighing the benefits against the costs; examine, in every case, alternative means of achieving these objectives; shape its budget request on the basis of this analysis, and justify that request in the context of the long range programme and financial plan". The planning lies in systematic analyses of the alternative ways of providing outputs, both in the short and long term.

Cost–benefit analysis: single objective (McKean 1958, Dasgupta et al. 1972, Mishan 1982)   Whereas financial appraisal shows the money costs and returns to the under-taker/operator, it does not answer the question typically raised in public sector projects where the output is not sold in the market: What are the costs to the community in terms of real (economic) as opposed to financial resources, and what are the benefits to the users of the system as a surrogate for benefits to the undertaker/operator? It was to answer such questions that cost–benefit analysis was derived and practised extensively (see Ch. 6). In certain projects either costs or benefits are specified as given, with the other being the variable. In the first, the method is known as cost effectiveness and in the second as cost minimization.[1]

Social cost–benefit analysis: multiple objectives (Beesley & Foster 1965, Beesley & Gist 1983)   Whereas cost benefit analysis proper has regard to a single sector, social cost benefit analysis will consider more than one sector. In a public transport rail project, for example, were added the related benefits to users of the roads who would be relieved from congestion because of the diversion of other road travellers to rail.

Framework appraisal (ACTRA 1977, SACTRA 1979, 1986, 1992, DoT 1986, 1992) Specifically designed for roads, the framework is a tabular presentation of data summarizing the main likely direct and indirect impacts on people of the alternative options for a proposed highway scheme. The method has three elements: cost–benefit analysis on the conventional narrow basis of highway costs and benefits; identification of impacts (via environmental assessment) on three groups who would experience the effects of the traffic (travellers, occupiers of properties and users of facilities); and the way in which the proposals would affect policies of the authorities for conserving and enhancing the area and for development and transport (see Ch. 13).

Optimization (Broniewski & Jastrzebski 1970)   The extension of threshold theory where the attempt is not simply to minimize investment cost but also to maximize the effects that are the outputs from the investment. As such, the method comes close to cost–benefit analysis.

Cost–benefit matrix (Urban Motorways Committee 1972, Urban Motorways Project

---

1. Somewhat differing interpretations are given in the literature. See, for example, Harrison & Mackie 1973, Sugden & Williams 1978.

Team 1973)   The matrix was a contribution by consultants (Travers Morgan & Partners) for the work of the Urban Motorways Committee. In the columns are the bear-
ers of the costs and benefits (DOT, developer, compulsory mover, environmental and
non-environmental movers, local households, visitors and road users, and owners of
local businesses and public buildings); and in the rows the kinds of discounted cost
or benefit, whether quantified or not (road contraction, re-housing due to redevelopment).

Planning balance-sheet analysis (Lichfield 1956, 1960, 1964, 1968)   The extension
of social cost–benefit analysis to address multiple sectors/objectives in urban and
regional planning. It takes account of the total relevant costs and benefits, including
externalities, on a community, and brings out the incidence and distribution of such
costs and benefits on the various community sectors.

Community impact analysis (Lichfield 1988c)   Extends the approach of planning
balance-sheet analysis, by modifying to reflect the incorporation of impact analysis
(see Ch. 4).

Social audit/accounting (Medawar 1978, Harte 1986, Geddes 1988, Haughton
1988)   While definitions are not standard, social audit (generally a one-off exercise)
and social accounting (a continuing process) are distinguished from *corporate social
accounting* in being conducted externally rather than internally. The latter describes
a company's own costed reports on progress in social, safety, environmental and other
matters pertaining to their workforce and local community. In this it goes beyond conventional accounting principles and practice to take in broader financial, economic
and social perspectives. The reason for their inclusion here is because, while accounting for the past, they are also seen as "systematic attempts to incorporate into project
appraisal all major dimensions of both economic and social impact, extending beyond
the usual concern with internal commercial project viability to look at community
costs, benefits and opportunities" (Haughton 1988). As such it echoes community
impact analysis.

Anti-disaster methodology   The review of evaluation methodology within the mainstream is presented by Peter Hall in his devastating review of notable planning disasters (Hall 1981). Following a review of the disasters themselves (London's Third
Airport, London's motorways, Anglo–French Concorde, San Francisco BART System, Sydney's Opera House), Hall concludes with a chapter on "Towards prescription"; in essence, how can the disasters be avoided in future? As a conclusion Hall
suggests that the desired improvements fall logically into two main areas: in forecasting; and then in answers to the questions such as: On what criteria should we make
our choice? How do we measure individuals' or groups' gains against losses? How
rank gains and losses in the near against distant future?
    Following a helpful discussion, Hall presents an approach to a method which ". . .
assumes that judgement will be needed every step of the way. In this it departs from

the classic fences of cost–benefit analysis, and approaches closely to the planning balance-sheet of Lichfield and the British Leitch Committee."

Comprehensive weighing (Attfield 1989)   Although outside the main stream of evaluation literature, and originating from the fields of philosophy, law and environment, this approach has echoes of evaluation having regard to community interests. It recognizes the role of cost–benefit analysis in the process but is critical of its limitations. One of these is the concentration on the interest of human beings, to which animals are subservient, without including the direct impacts of "sentient organisms" in fauna and flora. Another is the placing of *values* into the forefront of the comparison between the options. From this standpoint it makes a most useful contributions to the debate, including on the role of basic human values and environmental ethics.

D. Both output and input in greater width

Evaluation in structure planning (DoE 1972, Wannop 1985)   The DoE introduced no new technique for evaluation in structure planning but instead a synthesis of all the then available techniques, namely cost–benefit analysis, cost effectiveness, cost minimization, goals achievement matrix, planning balance-sheet analysis. The synthesis was not so much to generate a new method but to employ the individual methods as appropriate to address particular issues in the evaluation, such as the following: effectiveness in achieving aims, resource implications, distribution, uncertainty, and decision-making.

Evaluation in inner cities (DoE 1986a,b,c, 1987a, Robson 1994)   Broadly the same approach has been adopted in these wide-ranging evaluation studies of particular aspects of inner-city programmes, such as environmental projects, industrial and commercial improvement areas and employment effects. In each case there was a wide-ranging set of questions on particular issues of relevance to the programme. And in each case an array of evaluation methods/techniques was adopted for the purpose. Some examples are:
- *Environmental projects:* cost effectiveness, being the extent to which individual projects within each category of expenditure have met project-specific objectives; value for money, referring to comparisons of the cost effectiveness of different types of project;
- *Industrial and commercial improvement areas:* effectiveness of measures pursued in relation to community development objectives (jobs created or retained, increase in turnover or investment by firms, effect on private sector development and confidence);
- *Employment effects:* the economic impacts of projects on firms and employment. This involves a wide range of criteria such as: objectives and priorities set by the local authorities, benefits accruing to the firms, urban programmes and departmental objectives, timescale, cost of job.

Strategic choice (Friend & Jessop 1977, Friend & Hickling 1987)   This is not so much a way of formal evaluation between plan options as a way of handling complex decision situations in the planning process via a "technology of choice under conditions of uncertainty". It does so by structuring them by an analysis of interconnected decision areas (AIDA) where the choice options selected in the different areas may be incompatible. In this array of options it is possible to introduce robustness analysis.

Postscript

It can be seen that, although the evaluation criterion is simply "worthwhileness", it can still have many dimensions, from the very narrow (output only) to the very broad (DoE structure planning and inner cities). The former is too narrow for planning evaluation; the latter addresses itself to varying questions and varying criteria, each of which could require different evaluation methods. We now proceed to crystallize from the array a more specific approach to "worthwhileness".

CHAPTER 4

The cost–benefit family in plan and project evaluation

4.1 Value for money as a common objective in economic life

Man lives several different lives concurrently: family, community, political, spiritual, cultural, economic. All are mutually interpenetrating: the quality of family life impinges on the working day; low rewards in economic life undermine family life. However unattractive the idea in appearing to advance materialism, one particular life, the economic, penetrates all the others. This arises simply because, whatever the nature of the several lives, they require the use of economic resources in their fulfilment: even the contemplation of nature in the countryside requires time for the contemplation and perhaps money for the trip.

This leads to the well known paradox. In these varied lives, man shows an almost limitless tendency to want to increase his "standard of life" (material needs) and then "quality of life" (all other needs) by the consumption of goods and services, yet his resources for pursuing that consumption are limited in relation to his demands. He therefore needs to trade off between different ways of spending those resources on the goods and services he desires in his varied lives. Given the scarcity of the resources, he is led to "economize", namely to seek the best "value for money": that bundle of purchases which will achieve the greatest value to himself from whatever expenditure of resources he is able to make. Put more formally (Robbins 1952: 14),

> [But] when time and the means for achieving ends are limited and capable of alternative application, and the ends are capable of being distinguished in order of importance, then behaviour necessarily assumes the form of choice. Every act that involves time and scarce means for the achievement of one end involves the relinquishment of their use for the achievement of another. It has an economic aspect.

Since "economizing" in this sense pervades all our mutually penetrating lives, it is natural that economics should have evolved techniques of analysis for identifying the value for money obtained from using resources which, however abundant in supply, are scarce in relation to human needs. In essence the techniques aim to answer three questions:[1]

- Should the purchase or investment be carried out *at all*, having regard to the alternative possible use of the resources? In brief, is the expenditure *worthwhile*?
- If it be *worthwhile*, should the purchase or investment be made in the particular *way* being considered, or are there other ways (options) to produce better results in terms of value for money?
- Given the answer in the affirmative, should the purchase/investment be made at that moment *in time*, or would better value for money be obtained by deferment?

Inherent in these questions is the recognition that, given the *scarcity* of resources, if any are taken up in a particular project, they cannot be used for another. Thus, in the search for "value for money", the value derived from a particular purchase/investment, needs to be compared with the value that could be obtained from the use of the given resources in another way, under each of the three questions posed. The latter value is the "opportunity cost" of the particular resources used.

It is the answers to these seemingly simple questions that have generated the vast theory, practice and literature on investment or project appraisal or assessment. In popular usage, these have become subsumed in the term "cost–benefit analysis", the prime tool in economic analysis used for the *appraisal, assessment* or *evaluation* of policies, projects and programmes (see Ch. 6).

4.2 The cost–benefit family of methods

The review in §3.6 was aimed at bringing out the distinction between the various categories of evaluation techniques by reference to their handling of inputs and outputs.

In (A), where the concern is primarily with output, benefit or value, with a *stated output* there can in certain methods be a search for value for money through *cost minimization*: in finding the minimum cost/resources to achieve that output.

In (B), where the primary concern is cost or resources, with *stated input* there can in certain methods be a search for value for money in cost *effectiveness*: in finding the maximum output/benefit/value for the *stated input*.

In (C), where neither the input nor output are fixed, it is in the relationship of the variables on *both* sides that conditions the search for value for money, whether this be in terms of the *optimal* (the best relationship of all possible options), or *satisfactory* (that which is acceptable). It is this last category of appraisal, assessment or evaluation, which is most closely identified with the method of cost–benefit analysis just described. But since the methods in (C) are varied, we have chosen to identify them as members of a "cost–benefit family".

But although using the popular and understandable criterion of "best value for money" for this family it does not follow that all measures of benefit or cost need necessarily be in money terms. For example, in (A) it is rare for outputs in any of the methods to be measured in money, but rather by performance levels, norms, standards, or other value attributes. And in (B), the cost could be put in terms of unpriced

---

1. These questions are generalized from the cost–benefit literature.

inputs, such as time, or additions to an infrastructure system. And, it follows, in (C) either side could be unpriced.

But clearly, in (C), where neither side of the equation is fixed, the translation into money as a common measuring rod (that is *valuation*) of either or both costs and benefits is of help for the purpose of comparison: not only do costs and benefits need to be compared but also marginal variations within each (Lichfield 1988b). But as we shall see below (Ch. 7), although a desirable aim, this is not of itself essential in the search, and moreover should not be followed where the search itself can be confused by obtaining false valuations because of the absence of acceptable measurement/valuation techniques, particularly in relation to certain aspects of life that cannot lend themselves to valuation. In brief, the limitations in *valuation* should not condition the *evaluation*. To this we turn below (Ch. 9).

For *evaluation*, in the strict sense of comparing outputs and inputs, we need to turn to category (C), i.e. the family of methods that are related under the name of cost–benefit. Other methods, in categories (A) and (B), concerned with output or input only, have their role as *tests* in plans. For example, the testing of a plan's input in terms of the pre-set goals and objectives (A.4) will answer questions on comparative acceptability to those who framed the objectives. And the testing of a plan's input in terms of unit costs or costs in use (B.1 and 3) will be helpful for financial feasibility.

4.3 Some differences in objectives within the cost–benefit family

Within the cost–benefit family there are several offspring who adapt the family method according to the subject matter of the analysis and the circumstances. These vary with the criteria for choice that are set, implicitly or explicitly, by the decision-taker or analyst in question, in relation to, for example:

(a) *Whose costs and benefits are to be taken into account?* The individual purchaser would think of his financial costs and benefits, or those of his family, and not others, unless he were altruistic. A private company would also think of financial costs that the company had to meet and benefits for which they could charge, so excluding the others, the "externalities" (Ch. 6). Such a company might be involved in "ethical investment", which brings in ethical constraints against certain choices, for example in industries involving drugs, armaments or apartheid. A local authority might be concerned only with the cost of all its corporate services and the benefits to the residents they serve.

(b) *Which costs and benefits in geographical terms?* The private individual or company would tend to think of the costs and benefits accruing to the household or project with which they are immediately concerned, and not those falling elsewhere. But a local authority, faced with the need to provide for the offsite impacts of a development project (traffic, water, sewerage, etc.) would also consider these cost implications.

(c) *Should the decision relate simply to efficiency or also equity and social justice?* Comparison of the direct benefits or costs is a measure of *efficiency* in terms of value for money, be these the costs and benefits to be experienced by the decision-taker (the individual or family) or the wider community (municipal services). However, the individual or company might not be concerned with the distributive consequences of the purchase or investment (unless they take an ethical stand) and would therefore not concurrently take into account criteria of *equity* and *social justice*, as between those to whom the product of the purchase or investment is distributed.

By contrast, a municipality that represents its electorate and is also concerned with prospects for return at the next election, would take account of distribution, in prospective votes (Downs 1967). A planning authority, choosing between optional plans for a community, would also take account of distribution, if only because of the pressures by the public that forces them to do so. Even if they do not wish to consider social justice/equity (on the proposition that only the creation of wealth makes possible its distribution), they would certainly be sensitive to the distributive aspects if only, if equity and social justice are ignored, to be warned of the opposition they are likely to encounter.

This review of varying criteria adopted by decision-takers, in answer to the three questions, present further instances of cost–benefit analysis aimed at multiple objectives, which we termed above "social cost–benefit analysis".

4.4 Specialization in the different methods

From these examples it can be seen that there is potential for considerable diversity in the criteria adopted for choice by decision-takers. For this reason, the members of the cost–benefit family have specialized, either expressly or implicitly.

To some extent this specialization has already been brought out in the differentiation above between the members of the family in terms of their four characteristics (§3.6). Here we proceed to differentiate further in terms of the method of project appraisal within the cost–benefit family that would be used by different types of decision-takers.

Table 4.1 gives a general impression of the differentiation. It shows in the rows the kinds of costs and benefits/disbenefits that can arise in relation to a particular project, and in the columns which of these a particular kind of decision-taker or his analyst would call upon in pursuing the choice in that particular instance. At the foot is shown the member of the cost–benefit family that could typically be used to make the analysis.

Table 4.1 gives a *general* impression, only because the precise methods are not standardized and particular studies call for a combination of aspects, without change in nomenclature. For example Schofield (1987) subdivides cost–benefit analysis into economic, relating to efficiency, and social, relating to distribution. In our approach the economic *can* include distribution whereas the socioeconomic *always* does, and it also embraces outputs that are *social*, in that they include aspects outside the conventional boundaries of *economic life*. Bearing this in mind, we now amplify in respect of each kind of decision-taker.

Table 4.1 Methods of project appraisal used by different decision-takers and stakeholders.

Kind of cost and benefits/disbenefits	Decision-taker or stakeholder							
	Developer/ entrepreneur/ financier		Business or industrial		Government		Planning authority	
	Private	Public	Private	Public	Municipal	Central	Central	Local
1	2	3	4	5	6	7	8	9
Costs/resources								
Financial	x	x	x	x	x	x	x	x
Economic					x	x	x	x
Benefits/disbenefits								
Financial	x	x	x	x	x	x	x	x
Fiscal					x	x	x	x
Economic					x	x	x	x
Social						x	x	x
Health						x	x	x
Cultural							x	x
Natural environment					x		x	x
Traffic					x		x	x
Possible method of project appraisal	FA SFA	FA SFA	FA SFA IA	FA SFA IA	FA SFA IA CBA SCBA CRA PPB	FA SFA IA CBA SCBA CRA PPB	CIA as nest for others	CIA as nest for others

Notes: FA = financial appraisal, SFA = social financial appraisal, CBA = cost–benefit analysis, SCBA = social cost–benefit analysis, CRA = cost–revenue analysis, CIA = community impact analysis, IA = impact assessment, PPB = planning programming budgeting.
Source: adapted from Lichfield (1988).

The developer/entrepreneur/financier, be s/he private or public, would be concerned with the financial costs s/he has to bear and the financial benefits for which s/he can charge, and so ask for a financial analysis (FA). Were s/he also interested in the repercussions on others directly involved in the project, for example consumers on site, s/he might ask also for a social financial analysis (SFA), as would developers/entrepreneurs (who have to assess what the market would bear, or have yet to raise finance) or a financier considering whether to lend money.

Where the decision-taker is in business or industry, s/he would typically be concerned only with the financial outcome to her/himself, and thereby use financial appraisal (FA). But in addition s/he might need to consider the impact of her/his activities on those outside the project (e.g. on the natural environment or traffic flows), either or both as a means of preparing an environmental assessment as part of her/his application for planning permission, or finding measures of amelioration to meet the requirements of environmental protection (Ch. 5). S/he would accordingly wish to use impact assessment (IA).

Municipal or central governments are clearly faced with decisions on a variety of issues and will therefore need to have a wider array of choice criteria leading to different methods. These could include the methods already discussed (FA, SFA, IA). But in addition they might need to employ cost–revenue analysis (CRA), in order to assess the tax burden implications of municipal or governmental projects; planning, programming, budgeting (PPB) for their municipal services; cost–benefit analysis (CBA) in assessing non-market projects (such as transportation); social cost–benefit analysis (SCBA), as where assessing non-market projects that have repercussions on other governmental activities (such as the closing of a hospital in terms of implications for patient welfare).

Where a planning authority needs to take into account the total array of costs and benefits that might arise in a community, it could use community impact analysis. Since this has the widest treatment (in embracing offsite impacts and all relevant costs/benefits), it can be set up in such a way to embrace all the other analyses; these, as it were, can *nest* within the CIA (see Ch. 7). Then, conclusions can be drawn from the CIA relating to the decision-takers and stakeholders, consistently with the overall analysis.

Because the different decision-takers/stakeholders just noted will, as clients, employ different kinds of professional advisers, it is inevitable (academic and professional divisions being what they are) that the different skills would have devised methods of their own in isolation from others. A clear example here is the use by accountants, surveyors and engineers of financial costs and returns in making financial analyses; and the use by economists of economic as opposed to financial measures in cost–benefit analysis.[2]

For all members of the cost–benefit family the measurements need to be reconcilable, even if the differences between them are necessary for the particular analysis. For this to occur, on any particular project, the different skills need to adopt terminology, definitions, rules of measurement, use of criteria and so on, that have greater similarity than they traditionally do. The different questions, which are justifiably asked by the array of decision-makers/stakeholders, can then be reconciled in the sense that differences and similarities in the answers can be better understood. This is particularly important in recognizing the distinction between decision-takers and stakeholders; unless those involved in particular aspects of the project can see clearly how their cost–benefit calculations agree with or differ from those of the others, there is room, without good reason, for confusion, heated debate and delay.

4.5 The limitations of CBA in urban and regional planning

From this review of the cost–benefit family of methods, we now return to the question posed above for this family and other methods (Ch. 4): their role in evaluation of plan-

---

2. See Harrison & Mackie (1973) and Sugden & Williams (1978), which contrasts the treatment of financial and the economic measures.

ning options. In this we are not considering the role in relation just to *projects* that are encountered in the field of land-use planning, such as in urban renewal, transportation, recreation, local economic development.[3] Rather we have in mind such projects seen within the *context* of plans for towns and regions as a whole. The answer to this question is not self-evident. It has required considerable exploration over time by the writer since its original formulation (Lichfield 1956: chs 18–19), and is accordingly presented in those terms. In this exploration there was the concurrent search for the principles and theory of the method, and for its application in case studies.[4]

The preceding section shows that the cost–benefit approach can be regarded as constituting a family with distinct members who, although closely related, have adapted themselves to particular kinds of decision situations. This differentiation enables us to consider the well known criticisms of evaluation theorists to the limitations of traditional cost–benefit analysis proper in its application to land-use planning.[5] It can now be seen that the criticism applies to *particular members* of the family, in particular cost–benefit analysis proper, and furthermore that the CBA being criticized tended to relate to the early forms, of the 1950s and 1960s, which we described above as having only a single objective (§3.6). In relation to that particular model, this was justifiable, and indeed presented the writer's own point of departure from this method (Lichfield 1960). But since the 1960s the theory, principles and practice of CBA have been changing in response to demands on it, and at a greater pace over the past twenty years with the advent of natural resource and environmental economics (Pearce 1976, Pearce et al. 1989, 1993, Winpenny 1991). In brief, many of the criticisms levelled by this writer and others have been taken on board.[6] But, confusingly, economists have kept the same name for both the traditional narrow and also the more recent widened approach. Accordingly, criticisms that have been accepted to the evolving traditional model are rejected as not applicable to contemporary cost–benefit analysis as currently practised. Some economists therefore tend to reject the suggestion that planning balance sheet analysis/community impact analysis is a model different from contemporary cost–benefit analysis.

Furthermore, the cumulative modifications to the traditional CBA technique are spread widely over the literature, so there is no clear statement of the satisfactory application of contemporary cost–benefit analysis to land-use planning. Thus, the only way to grasp the relevant modifications to the traditional CBA, that have been introduced by PBSA/CIA, is to record the differences as historically perceived by this writer in the evolution of community impact analysis. To that we now turn.

Financial analysis has the limitation that it is applied only to the *financial* outcome of projects for the particular promoter; and, although *social financial analysis* goes beyond the particular promoter, it still has the limitation of being appropriate to programmes that belong to the private rather than the public sector, so ignoring, for

---

3. These are explored in Schofield (1987: pt II).
4. For a list of the case studies, see the Appendix.
5. e.g. Lichfield, N. (1960) leading to planning balance-sheet analysis; Hill (1968) leading to goals achievement matrix; Voogd (1983) leading to multi-criteria analysis.
6. See for example HM Treasury 1991.

example, economic as opposed to financial accounting, and also externalities. Cost–benefit analysis overcame the *financial* limitation by applying itself to *economic* costs and benefits.

Traditional CBA continued the limitation of application to particular promoters, even though on the benefit side (e.g. in transportation) it was indirect rather than direct outputs that were taken into account. But it failed to recognize the diffused nature of the costs and benefits when seen from the community viewpoint; highway cost–benefit analysis concentrates on benefits from using the road itself, rather than from the trip that is made (Lichfield 1987). And, although social cost–benefit analysis did recognize the role of more than one promoter (as, for example, investment in rail transport taking account of repercussions on other modes; Beesley & Foster 1965), it was still limited in the wider community sense.

Being *primarily* economic analysis, traditional CBA was reluctant to spread to non-economic inputs and outputs, and it tended to push aside those coming under the category of, for example, social, environmental and cultural. It was thus over-discriminating in terms of *which* costs and benefits.

This discrimination was reinforced by the proclivity amongst economists to deal only with phenomena that could be valued (i.e. measured in terms of actual or simulated price) in order to introduce the undoubted richness that economics can offer: the capacity for sophisticated handling of apples and pears in the one set of equations. This led to the *intangibles* and *incommensurables* being treated separately in the analysis, which relegated them to a subsidiary place as *secondary* benefits and costs, so tending to leave them ignored or underplayed in the decision-taking process. And, although this proclivity has had the most valuable effect of stimulating investigation of more and more ingenious methods of estimating the value of benefits that are not recorded in the market process (Sinden & Worrell 1979, Nijkamp et al. 1986, Pearce et al. 1989, Pearce et al. 1991), economists nonetheless still find it difficult to handle the intangibles, incommensurables and unpriced values alongside money prices in the cost–benefit analysis itself.

Another limitation of traditional CBA was the proclivity of economists to concentrate on *efficiency*, which is in the mainstream of economic analysis and is capable of elegant exposition, to the neglect of *distribution*, resulting from the allocation of output inherent in any economic decision. Some economists are loath to regard the problems as one of economics, as opposed to politics or sociology, so departing from the earlier traditions of political economy (Friedman 1953, Lipsey 1963). And even those not so disposed found that their tools in so doing did not measure up in any way to those available for efficiency. Thus, distribution tended to be ignored. And, as indicated above, the issues in land-use planning insist that it cannot be.

In brief, although there must be agreement with those who maintain that traditional cost–benefit analysis as a tool has severe limitations for land-use planning (as opposed to public sector projects), the *approach* of such analysis is highly relevant, just because it recognizes the fundamental approach *of economics in the enhancement of welfare* (Lichfield 1960). But the approach had to be *adapted* to the contemporary requirements, not only in land-use planning but also to issues raised in natural resource and environmental economics. Indeed, it was in order to meet such criticisms that, although adopting the cost–benefit *approach* for plan evaluation, planning balance sheet/community impact analysis was devised and promulgated by the adaptation of the approach (Lichfield 1964, Lichfield 1968b). To this we now turn.

4.6 The role of the cost–benefit family in plan evaluation

4.6.1 Origins

The initial concept matured in the early 1950s, as part of a study of the economics in the development and the planning context (Lichfield 1956: chs 18, 19). In essence, the argument was as follows. Any development has many and varied implications for the community, some of which will be a benefit and some a cost. Since the community is not homogeneous, the benefits and costs will not be uniform for different groups. In making his decisions on a project, the developer would draw up a "development balance sheet" that would take account of those costs and benefits of direct concern to her/himself, so excluding others. By contrast, in making its own decisions on the same project, the planning office would have regard to not only the direct costs and benefits of concern to the developer but also the indirect costs and benefits falling on others. This distinction between the direct and indirect costs and benefits echoes that of private and social costs (Pigou 1948). In essence, private costs are those the developer *has to bear*, and private benefits those for which he *can charge*. In practice the dividing line is an institutional one, defined in the law and practice surrounding development, that antedates the planning system (e.g. what contribution must the developer make towards the cost of offsite roads and sewers?).

From this concept was developed the "planning balance sheet" method of plan evaluation, aimed at questions that concern the planning office in reaching decisions in the public interest, within the framework of planning law and practice (see Ch. 1). For example, how to predict and measure the direct and indirect social costs and benefits to the greatest possible extent, in order to weigh up the incidence on the various groups, with a view to drawing conclusions on both the efficient use of resources and the distribution of the product in accordance with current social conscience. How to explore the costs and gains of planning? What would be the distribution of the costs and benefits on the developer and the various groups in the community, with the planning intervention and without? And who pays the costs and who gets the benefits?

On this foundation the method was progressed in two mutually penetrating ways: in theory, and by application of the evolving method in case studies (see the Appendix). Each is presented in turn.

4.6.2 Theory

Land-use planning is carried out for the better attainment of certain community goals and objectives, that are aimed at enhancing the welfare of the people affected. Since the planning process hardly lends itself to proceeding rationally and logically to this end, there is the need to carry out welfare tests on planning proposals and options alongside the many other tests that are introduced (see Ch. 3). Since the planning authority in practice achieves the advance of that welfare by acting as a *supra-investment authority* (to guide, divert, influence, modify, alter, suppress or stimulate public and private sector development projects), the appropriate welfare tests should be in the cost–benefit family (Lichfield 1960).

The adoption of this family meant there was available a powerful tool rooted in well founded economic theory, which was then vigorously evolving in the United States (e.g. Eckstein 1958, McKean 1958). However, cost–benefit analysis, as then put forward, was designed primarily for non-market public sector projects and was therefore not readily usable for the different purpose of land-use planning, where the authority was a "supra-investment" and not an "investment" agency (Lichfield 1964, 1968b, building on Lichfield 1956, 1960). Some reasons were that the plan was in effect a series of projects, which were visualized as being interdependent in space and time; that the social costs and benefits (the externalities) were critically important; that many were not measurable and certainly not in money terms, yet nonetheless needed to be taken fully into account; that in the decision-taking equity was as important as efficiency; that value for money needed to be interpreted with regard to "social" and not "market" value; and also to "whose value" and "whose money". Thus, some adaptation was needed of the then conventional cost–benefit analysis, which led to planning balance-sheet analysis.

In retrospect, various reasons can be seen for the times being ripe for the evolution of PBSA. First, as seen above (§3.3), when the method was initiated in the middle 1950s, plans were not evaluated beyond generalities and, indeed, options were not generally put forward for testing. But by the early 1960s, when the method was made operational (see the Appendix), the need for options and their comparative evaluation was being recognized (Kitching 1963, Shankland Cox 1966, Kreditor 1967). By 1970 it was possible to present a considerable array of methods for comparison (Lichfield 1970; Ch. 3 above). The maturing of PBSA therefore took place over a period of dramatic change in the use of evaluation of options in planning practice.

Secondly, came another pertinent change in the 1960s. Economists were becoming more involved in the planning process (see Ch. 1). They were thus able to make their particular contribution to evaluation in that the cost–benefit family was founded and nurtured on the principles and practice of economics, which could best be applied by economists. As such it needed to be seen as part of the process whereby economics made its contribution in town planning (Lichfield 1956, 1968b). But more than that, since town planning was making demands on economics which went beyond the current theory and practice in the field, it was necessary for economics to extend its contribution (Peters 1973, Schofield 1987).

Thirdly, with the growth in evaluation practice, it was necessary to see evaluation not as a discrete step in the planning process but as integrated with it throughout (Lichfield et al. 1975).

Fourthly, although the generic method of PBSA remained reasonably constant in the 1960s it was subject to a major new influence in the 1970s. This resulted from the explosion into the urban & regional development field of impact assessment, which in its modern dress was initiated in the USA around 1970 (NEP Act 1969). From its beginning in the natural environment, the assessment exploded over the whole field of human environment (Clarke et al. 1980). This led to the writer's realization that, in the planning balance sheet studies as indeed in cost–benefit studies also, it was not so much the costs and benefits *per se* that were being predicted but rather the costs and benefits flowing from *impacts*, physical, social, economic, and so on (Lichfield & Marinov 1977). From being somewhat buried in the "black box" of the PBSA, it was now brought into the open, with considerable help from the work of impact assessment.

In essence, following earlier hesitation as to whether PBSA was best described as "cost–benefit analysis in planning", which carried with it the disadvantage of association with the then form of cost–benefit analysis which had been rejected for planning, PBSA emphasized the concern with *impacts*, and that it was (Lichfield 1985: 59) ". . . the whole array of impacts on the whole community which are under consideration and not simply particular impacts (economic, social, etc.) on particular sectors (e.g. transportation), or only those which are measured in money". PBSA was accordingly adapted and renamed *community impact analysis* (CIA), both in order to show that it is more comprehensive than other kinds of impact analysis (e.g. energy, transport, economic, social) and also to show that it is not simply the impact as *output* which is important (as in impact assessment proper) but the effect of that output on *people*, i.e. on a community. Furthermore, since the end-purpose of impact *analysis* in planning is not just *assessment* (as in IA proper) but also *evaluation* as an aid to choice, CIA is seen as a step towards aiding choice in alternatives, and so becomes *community impact evaluation* (CIE) (Lichfield 1985). In passing it might be noted that, although this distinction between *analysis* and *evaluation* is explicit in CIA/CIE, it is *implicit* in CBA, SCBA and PBSA, where it is implied that the *analysis* is the *evaluation*.[7]

4.6.3 Application by case study

Clearly, for application in practice it was necessary to do more than just describe the planning balance-sheet/community-impact approach; it was necessary to show how it should be used. This was the second of the two areas of study introduced (see the Appendix for case studies). From the initial exploration in 1956 (Lichfield 1956: ch. 19), and an initial exercise of 1959, several foundation case studies were published in 1962–6. From this research base, starting in 1966 the method was developed via a series of case studies in consultancy commissions, for aiding choice by decision-takers between options. This had two significant advantages: the method was tested in a large variety of real-life situations and found to be robust; and it evolved on the learning curve of successive generations in consultancy offices (Nathaniel Lichfield Associates (NLA) and Nathaniel Lichfield & Partners (NLP) 1962–92, Dalia & Nathaniel Lichfield Associates (DNLA) 1992–).

As the Appendix shows, just because they were commissions, the case studies

---

7. For independent reviews of the CIE method, see Alexander (1978) and Dobben (1990).

varied considerably in nature. They reflected different aspects of the planning process (discrete projects, contributions in a planning team, application as planners within the plan-making process). They related to different levels and scale of plan (national, regional, subregional, urban and project). They were applied to different kinds of topic in the planning process (roads, public transport, new towns, shopping, etc.). They were carried out with differing time and money budgets and therefore to differing degrees of intensity (from weeks to years). They were done by individuals on the "quick and dirty" approach or by teams using elaborate models. They were presented in different ways (as reports, in public consultation and participation, in planning inquiries).

By the same token, however, because of the variety, it is not practicable to demonstrate a *standardized* method for general adoption. Furthermore, that adoption has been made more difficult because of the tendency to publish in academic journals which required some rigour in presentation and analysis. This, and the fact that the published studies display differences in technique because of their varying context, has meant that other practitioners have made their own adaptation as they thought fit.[8] For that reason the method to be demonstrated here (Ch. 7) is not a universal but a *generic* process, for adaptation as necessary.

4.6.4 The decline in the 1970s and 1980s

From the preceding it is seen how evaluation in land-use planning in Britain took off from a very low point in the mid 1960s and then flourished. Its culmination was undoubtedly in the work of the Research Team of the Roskill Commission in 1968–70 (Roskill Commission 1971). But thereafter there was a decline to a low ebb in the following two decades. In retrospect can be seen four distinct but related reasons. First, since evaluation of the kind discussed here is identified with the rational model in planning, it declined with falling confidence in that model in the 1970s (Breheny & Hooper 1985). Secondly, this period also saw an attack on government planning generally, from economists who were creating very successful platforms for greater reliance on market forces at the expense of planning. This platform was well provided and supported by the New Right, which found consistent and well argued academic sinews (e.g. Seldon 1981, 1983, 1985). Of particular interest here was the attack on the use of cost–benefit analysis in the urban & regional field as being an inadequate substitute for the market (Peters 1971). This academic onslaught found very fertile political support in the 1980s from the Thatcher Government (Thornley 1990). Thirdly, it was in this atmosphere that the work of the Roskill Commission played a significant part. The Commission mounted, via its Research Team with the vigorous support of two of the market orientated Commissioners (Alan Walters and Alfred Goldstein), the boldest and best financed use of cost–benefit analysis in a planning issue than had been seen before (Roskill Commission 1971; for a critique, see Lichfield 1971). As such, the work of the Commission could have taken project evaluation in planning via cost–benefit analysis to a position of considerable strength. But unhappily, despite the undoubted technical competence and imagina-

---

8. See for example Alexander (1974), McMaster & Webb (1978).

tive innovation of the work, the study backfired against evaluation for a variety of reasons. This introduced in the 1970s a decade of distrust for cost–benefit analysis in planning. Fourthly, this tendency was reinforced by Peter Self's spirited and amusing polemic against the "econocrats" who use economics in policy making (Self 1975). But by the time that Self had modified his position (Self 1985), the damage was done.

The result was some twenty years in the wilderness in the UK (but not elsewhere; see McMaster & Webb 1978, for example) for plan evaluation via the cost–benefit family. However, there is evidence that the tide is turning, both for evaluation and the cost–benefit model (Hall 1988). The process of land-use planning saw a revival in the closing years of Thatcherism, which has been strengthened with its demise. The role of evaluation itself has been seen to be of importance in addressing the choice between options. And whereas the role of the market in deciding such choice has strengthened in areas which can be left to the market, it has weakened in those areas, such as land-use planning, where it is generally accepted that the market has limitations for the purpose.

4.7 Role of PBSA/CIA as planning analysis

In this chapter we have argued the case for PBSA/CIA being appropriate for planning analysis. We close Part I with a summary of the case made. From the preceding it is seen that:

- the cost–benefit approach is relevant to planning in that it enables choice to be made between options for the better advancement of welfare;
- the cost–benefit family has individual members who use the same approach for answering different kinds of questions (what kind, and whose cost and whose benefit, etc.);
- not all the members of the family can make an appropriate contribution to plan evaluation for choice, as visualized here.

The question then arises: given that the cost–benefit family has a useful role in aiding choice, why is it that planning balance-sheet analysis / community-impact analysis is seen as having the most helpful role in land-use planning of all the family members? This question would be the easier to answer if it were quite clear in planning theory and practice just what should be the criteria for choice between alternative plans, be they for whole communities or for projects within the communities. But whereas the literature on evaluation methods is very rich, the literature on the nature of choice to be made in planning, and the logic behind it, is not.

Part I of this book can therefore be seen as an attempt to make the case for CIE as a support system for choice and decision in planning. We pursue the attempt in Part II in relation to theory and principles, and in Part III in demonstrating their application. Before starting Part II we put forward some simple propositions on this topic, namely:

- The end objects of any land use planning are to advance the welfare of the community that is affected.

- The nature of the community is defined not administratively but functionally, by the delineation of the impacts to be felt by what is proposed.
- In functionally defining the impacts, it cannot be expected that a tidy boundary can be drawn for a community embracing all kinds of functions. The impacts on the natural environment could be quite local (noise, visual intrusion, etc.), whereas the impacts on the economy could be very widespread (introduction of economic activities which have import/export repercussions). Thus, any particular administrative boundaries cannot be self-contained for the analysis.
- The impact on the community is the impact on the people in that community. But it cannot be expected that all groups in the community will fare equally well or equally badly; indeed, any individual will experience a diversity of impacts, as indeed could any family.
- This will lead to conflicting attitudes on impacts. From this it follows that a choice of one option as opposed to others cannot satisfy all those impacted.
- From this it follows that any particular choice and decision must have regard to the array of sectors upon whom the impacts will fall and the nature of the benefits and costs to be experienced by them.
- The problem is: which choice to make between options? Should it be that:
  - Which benefits most people; if so, what about the rich and the poor?
  - Which shows the least adverse impacts; if so, are not beneficial impacts from other options being needlessly sacrificed?
  - Which benefits particular classes of people, the poor, landowner or motorist; if so, how take into account the adverse impacts on the others?
- The short answer could be that option which is in the "public" or "community" interest. That would be acceptable if it were only possible to define with general agreement just what that concept means. How are the interests between different groups to be traded off to achieve the overall "public interest"? Is the community favoured by benefiting those seeking to change as against those seeking to conserve?
- Given the difficulties of defining the "public" or "community" interest, and also the fact that the choice in the end will be made by representatives (elected or otherwise) on the basis of their own criteria, would it not help the decision-takers and stakeholders to have the following display of information about the options: their implications in terms of costs and benefits (disadvantage/advantage) to the various community sectors, with a delineation of differences as between the sectors. They would then be in a position to understand the implications of particular choices (the private and social opportunity cost) and also be warned that the choice which favoured certain community sectors/groups would arouse the opposition of others adversely affected.

If this be a useful manner of proceeding, then pragmatically PBSA/CIE can make its contribution. Indeed, that was the driving force behind the search for a method that, although based on the approach of cost–benefit analysis, nonetheless met the objections to the use of that analysis in its traditional form in the land-use planning process.

CHAPTER FIVE

Theory and principles of impact assessment

5.1 Origins and evolutions

The environmental effects from development and operation are many and of great variety. The consequences are to be seen in the evolution of the practice of impact assessment. For one thing, the scope of the impacts taken into account had been growing. For another, the scope of the planning process, within which they need to be considered, has for some time been extended from discrete projects to programmes, policies and plans, making the role of the assessment more ambitious (O'Riordan & Sewell 1981: 1; Breheny 1984: 20–24; Rickson et al. 1989; Lee & Walsh 1992). Because of this evolution, the practice of impact assessment has been lumpy and, as in all evolving disciplines, various lumps have originated from differing disciplines. This must accordingly be reflected in our account below of the origins from the various sources.

5.2 Some definitions

5.2.1 Effects and impacts

In everyday usage an *impact* is the *effect* of a specific *cause* (OED). A distinction can thus be made between *effects* which are ". . . the physical and natural changes resulting, directly or indirectly, from development ". . . and *impacts* which are ". . . the consequences or end products of those effects represented by attributes of the environment on which we can place an objective or subjective value" (Catlow & Thirlwall 1976: para. 2.16). This distinction has been brought out more specifically in relation to *physical impact*, which ". . . can be viewed as development – induced changes in the environment which affect natural resources and land and water and air quality" (Nelson 1988: 28). More broadly, the impact is "any alteration of environmental conditions or creation of a new set of environmental conditions, adverse or beneficial, caused or induced by the action or set of actions under consideration" (Rau & Wooten 1980: 1–26). Or, more particularly, it ". . . can be described as the change in an environmental parameter over a specified period and within a confined

61

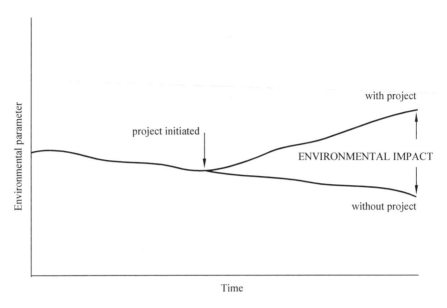

Figure 5.1   An impact (*source:* Wathern 1988: 8).

area, resulting from a particular activity compared with the situation which would
have occurred had the activity not been initiated" (Wathern 1988: 7). This comparison, known in the cost–benefit literature as "without and with a project", is illustrated
in Figure 5.1.

5.2.2 Categories of impact

The impacts just described comprise two major categories: direct (first order) which
fall directly on the environment as a result of the project input; or indirect (second
order, induced or secondary) which are generated by activities that result from the
project (Wathern 1988: 7). In terms of duration, such impacts can be cumulative;
short-, medium- or long-term; permanent or temporary; reversible or irreversible/
irretrievable. They can be positive (beneficial) or negative (harmful). In terms of their
timing they can arise before construction on a project has commenced (e.g. through
blight), during construction, following operation (commissioning) and at the end of
the operational life, as in the dismantling of a bridge or nuclear power station (decommissioning).

5.2.3 The environment in impact assessment

Whatever the diversity of impacts the approach to their assessment is broadly common: *ex ante* analysis to predict the scale, magnitude and significance of the impacts.
But the assessment of any particular impact can be made for a pre-selected section of
the environment. For some this is the physical surroundings in which they live, the
landscape urban and rural. For others it is simply the whole world in which they live,

the whole picture of human, animal and plant life and their interactions upon one another. It is thus a prior definition of the environment which conditions the score of the assessment.

5.3 The various impact assessments and their sources

The watershed of contemporary assessment is the US National Environment Protection Act 1969 (NEPA), which stimulated assessment on the environment around the world. However, this was preceded by assessment practices that we can now recognize as forerunners of formal impact assessments. Accordingly, our description of the various kinds of impact assessments and their source is made in three broad chronological divisions: pre-NEPA, NEPA and post-NEPA.

5.3.1 Pre-NEPA
The following assessment methods are of various kinds that share a common feature: they have not specifically addressed themselves to impact assessment as such, but in fact have implicitly done so.

Protection against pollution of the natural environment
The industrial revolution and growth of organization led to pollution of the air, water, in disposal of waste, and so on. The need for protection was recognized early, measures having been introduced in Britain, for example, in the 19th century (Royal Commission on Environmental Pollution 1971). These measures were administered by government regulation, which in contemporary language introduced controls against the environmental impacts.

Financial assessment
The principles, theory and practice of financial assessment grew up under the various titles of investment appraisal or financial analysis (Merrett & Sykes 1973). Financial assessment predicts impacts in terms of the financial resources required for the project and the income/capital flowing to the investor. In essence the method compares the financial costs that will be incurred by the investor with the financial return he would obtain, with all the costs and returns being those that would typically accrue to the investor, so ignoring other parties in the project and also externalities.

Fiscal assessment
Here the analysis is financial but the area of application is different. Typically, the term applies to municipal expenditure on the cost side and, on the return side, the municipal and governmental revenues that would result, typically geared to some form of tax base valuation (Mace et al. 1961). The impact is on the costs that need to be raised from the tax payers on the cost side, and the tax revenues that can be raised from the residents, businesses, and so on, for the new services.

Economic assessment
The principles, theory and practice of economic assessment grew up as cost–benefit analysis (Ch. 6), with the costs and benefits that are at the core of the assessment seen as the costs and benefits flowing from *economic impacts*. The assessment has widened over the years in scope, from narrowly economic to socioeconomic impacts (Schofield 1987).

The earliest essays in this field were introduced in 1902, for river and harbour projects in the USA, and extended under the New Deal by the Flood Control Act 1936 to water resource development in general (Eckstein 1958). From these beginnings, the practice grew to cover other kinds of public sector investment. From the USA spread to Britain, initially in the 1950s to transport, and by the mid-1960s to a wide array of public sector projects (urban development and renewal, transportation, recreation, health and social services, economic development, regional policy (Prest & Turvey 1965: pt III; Peters 1971; Schofield 1988) as well as urban & regional plans (Lichfield et al. 1975, Lichfield 1988a). And it has also been used extensively in the Third World for development projects of all kinds (Bridger & Winpenny 1983).

The cost–benefit family of project appraisal (Ch. 6) is singular for the huge amount of literature on theory and principles which supports the application in practice. As a result, compared with most other forms of impact analysis and assessment, it has decided theoretical rigour (Eckstein 1958, Krutilla & Eckstein 1958, McKean 1958, Pearce & Nash 1981, Mishan 1982, Pearce 1983b, Schofield 1987).

Technology assessment
Science and technology are the engines that drive the development that is the foundation for social and economic growth. As such their impacts on society are wide-ranging, for good or evil. It was in order to attempt to predict these impacts that the field of technology assessment arose, starting in the USA in the 1960s, as the systematic study of the effects on all sectors of society that may occur when a particular technology is introduced, extended, or modified, with special emphasis on any impacts that are unintended, indirect or delayed. For this assessment, a wide array of methods have grown up, which have been categorized as (Hetman 1977):
- partial: pre-selected secondary consequences
- futures: forecasting and planning
- problem orientated; focusing on a societal problem
- environmental impact: under the NEP Act
- wide scope: possible impacts of several categories relating to different disciplines.

5.3.2 *NEP Act 1969 and its influence*

The environmental dimension
Following a history of environmental concern that originated in the USA in the early part of the 20th century (O'Riordan 1976), contemporary practice relating to environmental impact statements stems from the US NEP Act of 1969. The Act required that

all Federal Agencies should prepare environmental impact statements (EIS) for ". . . legislation and other major Federal actions significantly affecting the quality of the human environment". The statements were to be prepared under the guidelines of the Council on Environmental Quality, which annually ". . . provides advice to the President on environmental management, prepares the President's Annual Environmental Quality Report and assists federal agencies in the implementation of NEPA" (Council on Environmental Quality).

This was the simple beginning of an international explosion in the field, which took different forms:

- in the USA, having started at the Federal level, the practice spread to States and municipalities who laid down their own code for the preparation of the statements. Unlike the Federal level, these applied also in certain states to the private sector (Clarke et al. 1980);
- the 1969 Act did not define the term "environment", but ". . . it is clear from Section 102 that the term is meant to be interpreted broadly and include the physical, social, cultural, economic dimension . . . which affect individuals and the community and ultimately determine their form, character, relationship to survival" (Rau & Wooten: 1–24). In the early years in most EIS, the environment ". . . was defined in a narrow manner and usually referred only to the natural environment" (Clarke et al. 1984: 192).
- US practice was echoed internationally following the first International Conference on the topic in Stockholm in 1972 (Ward & Dubois 1972). In consequence the US practice has (with modifications) spread to both the Anglo–Saxon world (e.g. Britain, Canada, Australia), the Far East (Japan), Europe (Holland, West Germany and more recently the whole of the EEC), the developing world and also the international agencies (O'Riordan & Sewell 1981; Clarke et al. 1980: sec. 4; Wathern 1988: pts IV, V).

In consequence environmental science (social and natural) has flourished (Lapides 1971). Although the environmental impact assessment procedures were initially in relation to projects, there has been a tendency for an extension from project appraisal to policies and plans (Therivel et al. 1992). Whereas environmental assessment was introduced in the USA alongside the planning machinery and procedures, in other countries (but not all) it has been integrated with the planning system and machinery. This is so in Britain, following the introduction of Regulations in response to the EEC Directive of 1985 (see Ch. 17). Here the environmental assessments are prepared for projects likely to have significant impacts, and the assessment must be considered as part of the grant of the planning permission (Lichfield 1989a).

5.3.3 Post-NEP Act
All the following kinds of assessment are made to some extent in the mould of environmental impact assessment. They reflect the recognition stimulated by the 1969 Act: it is the impacts introduced by change that are the link between the inputs and the outputs of projects.

The social dimension

Although the NEPA recognized that socioeconomic impacts were called for, these were not initially covered in the same way as those relating to the natural/biophysical environment. However, the scope was soon widened to do so, under pressure of the US governmental institutions (in particular the Council on Environmental Quality), the Courts (which had a large role in the environmental assessment procedures) and the public, when opposing the projects.

One contributory factor to the delay was the lag in application in the research and practitioner field, stemming in the main from sociology and anthropology. This was soon tackled in that *social* impact analysis (SIA) began to be developed, in principle and practice (Finsterbusch et al. 1983). From modest beginnings in 1984 the litera-ture has escalated, as has professional practice. SIA has been carving out a role which, although built on the foundation of the 1969 NEPA, has a distinctive flavour. Some features are (Carley & Bustelo 1984):

- In contrast to EIA, "the central question addressed in social impact assessment is: what difference is the proposed development making and likely to make in the lives of residents in the area targeted for development?" (Gold 1978: 105–116). Furthermore ". . . its orientation is to people who reside in the vicinity of the development site rather than to people elsewhere in the region or nation who might be the main consumers of the products of the developments".
- An extension of the impacts on people's lives is a prediction of their "social wellbeing" (Fitzsimmons et al. 1977). This isolates five different effects: on the individual, community, area, nature and society. The exploration is conducted through a series of subcriteria for each. Then there is an aggregation into three measures:
  - quality of life: physical and mental wellbeing which is seen as perception of the opportunities for further development of individual and family life (Campbell & Converse 1972)
  - relative social position: a function of equity in distribution
  - social wellbeing: which is the function of the community and its institutions.

The urban dimension

Following initial study by the Rand Corporation, in 1978 President Carter issued his National Urban Policy message which called for a new programme for the cities. Within this he asked for "a continuing mechanism . . . to make the analysis of urban and regional impacts of new programmes . . . an integral and permanent part of all policy development throughout our government" (Breheny 1984: 11–12). This mechanism was the "urban and community impact analysis" (UCIA), later shortened to "urban impact analysis" (UIA). In brief, it asked for a spatial prediction of the ". . . anticipated *unintended* consequences for urban areas of *non-urban* Federal initiatives to be assessed *before* the initiatives were finalized and implemented" (Breheny 1984: 12).

Again the research back-up for this completely new approach was not long in coming forward (Glickman 1980a,b,c, Lichfield 1983, Breheny 1984). In seeking to

identify the impacts of Federal policy four categories were detected (Glickman 1980a):
- general policy that had national orientations with indirect effects on urban life (e.g. defence and youth employment programmes)
- programmes targeted on localities (e.g. urban mass transit, administration programmes and community development grants)
- indirect, by changing relative prices, such as wage subsidies, air pollution regulations, gas price deregulation and minimum wage law
- direct influence on relative prices in localities (e.g. assisted housing of the US Department of Housing and Urban Development or urban mass transit administration operating subsidies).

These programmes produce an array of impacts (Glickman 1980a: 7):
- direct and indirect
- absolute and differential
- varying size
- qualitative as well as quantitative
- spatial (interregional, intra-regional, intra-metropolitan and intra-urban).

These impacts were by definition to be sought in "places". But it was appreciated that, even so, the *reasons* for the policies were to be found in *people* (unemployment rates, poverty rates, etc.), so that some way had to be found of exploring the "people impacts" (Edel 1980).

Unhappily, the methodology of urban impact analysis in the USA has not matured beyond this promising beginning, for the advent of the Reagan Government halted this particular programme (as others) and research and development in the field.

Development
By contrast with the diffused effects in urban impact analysis, development impact analysis relates to the impacts from carrying out *urban development itself*. Impact assessments of this kind were devised following the US Housing and Community Development Act of 1974. This made available the ". . . use of Federal funds for 'blocks' of development projects within a community rather than single and specific works. They were named Community Development Block Grant Projects, and required assessment of environmental impacts prior to releasing of funds" (Chatzimikes 1983). Various methodologies were used with varying degrees of success (Schaenaman 1976).

From the preceding it is seen that development impact analysis subsumes environmental Impact analysis for the development in question. A parallel approach considered not only the physical and activity impacts (that is, effects on people) but also the welfare impacts, which could be traced through planning balance-sheet analysis to aid the land-use planning decision (Lichfield & Marinov 1977).

Transportation
Although there are common impacts in transportation's many modes, each has its own distinguishing set: air, land (highways, including private and public transport;

rail; cycle and pedestrian); water (seas and rivers); underground (road and mass transit). Thus, transportation impacts can be of diffuse kinds. Here, by way of example, we consider only those from highways.

Whereas the assessment of narrow economic impacts for highways came early, their full-scale impact assessment is of more recent origin. In the USA they were brought into the net by the NEP Act 1969, and were concurrently catered for in specifically highway administration (Stopher & Meyburg 1976). The array of "social, economic and environmental impacts" were wide and could be categorized as:
- user: changes in travel times, speeds, congestion and accident rates
- non-user: economic, social and environmental impact, on the people, land uses and environment adjacent to the transportation.

Within this array a further categorization could be (Stopher & Meyburg 1976: 104):
- operational
- activity distribution
- monetary
- social
- environmental
- aesthetics
- institutional.

In Britain the concern with non-user impact was first seen in the reaction to the urban motorway programme in London around 1970 (Urban Motorways Committee 1972). But the concern did not lead to action until the acceptance by the Department of Transport of the recommendations of the Leitch Committee in 1977 (ACTRA 1977). This led to the adoption of the form of assessment known as framework appraisal (Ch. 4; SACTRA 1979, 1986), and also to environmental impacts from highway developments in terms of traffic noise, visual pollution, air pollution, community severance, effects on agriculture, heritage and conservation areas, ecology, disruption attributable to construction, pedestrians and cyclists, the view from the road, and driver stress (DOT 1993).

Risk/hazard
Risk assessment has its antecedents in the concern with mounting traffic accidents following the Second World War, and the need to incorporate accident assessment in the cost–benefit analysis that was then emerging. But as *hazards* grew (". . . the inherent property of a system that could cause injury or damage") from nuclear emissions, floods, chemical pollution, and so on, so did the need for *risk* assessment (". . . the chance that an event happens which demonstrates these harmful properties" (Ramsay 1984). And as with other techniques in the field, the sheer growth in the magnitude of the hazards, and the perceived potential risks, has escalated the technology.

Risk assessment is not seen as an overarching topic, as with economic or social impact, but rather as incidental to the assessments carried out for projects, such as a nuclear power station (Carley & Bustelo 1984: ch. 19). As such, the method and technique has its own characteristics (O'Riordan 1979): The following is an application in the planning system (Nelson 1988):

- risk identification, discovering and defining the risks associated with the project
- risk estimation, with its prediction of the likely consequences in time and space
- judgement on acceptable levels of risk
- balancing of this possible outcome against perceived and/or estimated gain
- risk control, setting of standards, monitoring discharges and ambient levels, enforcing regulations and codes of practice, and monitoring.

5.4 Methodology of impact assessment for projects

5.4.1 The impact assessment process overall

There is no uniform approach in the methodology of impact assessment. This is understandable because of the wide array in the different kinds of impacts that come into consideration in assessment; the contributions from a great variety of academics/ professionals, with each kind tending to work, as always, in some isolation from others; the spread of the method originating in the USA to other continents and countries, each of whom have made their own adaptation; and finally to the relatively short period and breakneck speed in which impact analysis and impact assessment have grown, not affording too much opportunity for consolidation. Nonetheless, the processes have similarities (Carley 1980: 54).

In order to convey the method, in this section we first attempt an overarching generalization built around the natural environment, and then more briefly the economic and the social.

5.4.2 Natural environment

A review
The environmental assessment process is part of the broad process summarized in Figure 5.2 (Wathern 1988: 17). "Although there may be variations in the detail procedures adopted within a particular country, most systems, in essence, conform to the pattern shown"

From this it is seen that the procedure starts with a definition of the proposal to carry out an environmental assessment of the project under consideration and ends with the monitoring and auditing following implementation of the project. Within this is the assessment proper, the method for which primarily concerns us here.

There are many methods in use (also called methodologies, technologies, approaches, manuals, guidelines; Clarke et al. 1980) which are not of equal/even merit (Atkins 1984). From these we now generalize a standard method.

A standard method[1]

Baseline studies   "The term usually refers to the collection of background informa-

---

1. For more recent treatment, see Glasson et al. (1994) and Morris & Therivel (1995).

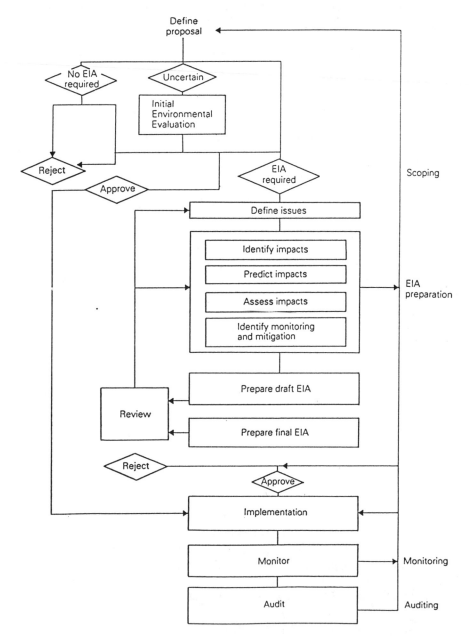

Figure 5.2   Flow diagram showing the main components of an EIA system (*source:* Wathern 1988: 18).

tion on the environmental and socioeconomic setting for a proposed development project and it is normally one of the first activities undertaken in an EIA.. . . In general, it is taken to refer to a description of some aspects of the physical, biological and social environments which could be affected by the development project under consideration.. . . The focus is often on information and data which are readily available rather than on what is needed." (Beanlands 1988, Bisset 1988).

Identification of impacts   Enough has been said above to show the great diversity of impacts: on the natural environment under its many subdivisions; on human beings in respect of their many different lives (social, economic, family, cultural, etc.); on the built environment, be it in general or on the cultural heritage. Their identification requires some systematic sieving process, for example by matrix (Leopold et al. 1971) or transparency overlay (McHarg 1969).

A variation is given in Figure 5.3. In this there is a somewhat random set of "environmental factors" (called characteristics of the existing situation) explored against the series of potential environmental effects that could arise in the construction and operational phases (called adaptors).

Identification of impacts to be assessed   Since the number, diversity and type of impacts are very wide, it is important not to attempt to deal with *all possible* impacts for a particular proposal, but rather to make some preliminary sieve in order to identify those impacts that should be assessed. This takes two forms:

- screening (Beanlands 1988): which projects are likely to give rise to impacts; and of the impacts so indicated, which are likely to be significant
- scoping (ibid.): of the array of impacts, which are likely to raise issues of importance to the local population.[2]

As regards screening for projects, there are five possible methods, sometimes used in combination (Tomlinson 1984):

- project thresholds: for example, project size, area of land influenced by the project, volume and height of the project, land take in comparison to the scale or character of surrounding environment, more than local importance, highly visible, project cost
- sensitivity of environment of project: for example, the carrying capacity in relation to degree of interference or disturbance, the quality of the individual components of the area
- positive and negative list, by drawing a boundary between types of projects which are subject to a mandatory environmental assessment or those that may be on the basis of stipulated characteristic.
- matrices, which consider possible effects in the rows against activities in various stages of the project development in the column with a view to identifying potential impacts
- initial environmental evaluation in order to establish possible relationships

---

2. More recently, "scoping" has widened and embraced both aspects (see DOE 1994c: ch.2).

Figure 5.3 An impact matrix.

Characteristics of the existing situation	Construction phase															
	Immigration	Severance	Transport of raw materials	Transport of employees	Site preparation	Dust and particulates	Employment	Local expenditure	Water demand	Vibration	Noise	Gaseous emissions	Odours	Aqueous discharges	Solid waste disposal	Hazard
Climate																
Land uses		X														
Water quality					X	X								X		
Ecological characteristics					X									X	X	
Population density																
Tourism																
Employment structure							X									
Unemployment							X									
Local economy								X								
Traffic			X	X												
Water supply									X							
Sewerage																
Finance	X															
Education	X										X					
Health service facilities	X															
Housing	X									X	X					
Emergency services	X															
Community structure	X															
Culture	X															

Characteristics of the existing situation	Operational phase																
	Immigration	Structures	Severance	Water demand	Local expenditure	Employment	Transport of raw materials	Transport of employees	Transport of products	Noise	Vibration	Gaseous emissions	Odours	Dust and particulates	Aqueous discharges	Solid waste disposal	Hazard
Climate																	
Land uses			X									X					X
Water quality															X	X	
Ecological characteristics		X													X	X	
Population density																X	
Tourism	X																
Employment structure																	
Unemployment						X											
Local economy						X											
Traffic					X								X				
Water supply							X	X	X								
Sewerage															X		
Finance	X																
Education	X																
Health service facilities	X									X							
Housing	X									X	X	X					
Emergency services	X																X
Community structure	X																
Culture	X																

Source: Clark et al. (1981: 14).

between the project and the environment, with a view to determining those elements or aspects which may be subject to important impacts.

For "scoping" the essential purpose is ". . . developing and selecting alternatives to a proposed action and identifying the issues to be considered in an EIA . . . and so establish the terms of reference for the EIA (Tomlinson 1984). There is no specific method ". . . but rather an assemblance of interactions and discussion between the public, various agencies and the project proponent" (ibid.) which are used in combination, such as:

- legislative framework: in some countries, for example, social and economic issues are excluded whereas in others issues, such as impacts on health, are mandatory
- consultation with involved bodies
- meetings with the public
- positive and negative checklists
- matrices.

Magnitude of impacts   Having selected the impacts for consideration, it is then necessary to make some estimate of the likely magnitude of their output. This is initiated by defining a unit that is appropriate to the predicted impact, to the greatest possible extent using the approaches of natural and social science (Rau & Wooten 1980). Examples are noise level in decibels, air pollution in parts per million, and agricultural land take-up (in hectares).

As with all prediction and measurement there must be considerable uncertainty as to the future, so that techniques are needed for handling the uncertainty (de Jongh 1988).

Significance   In the assessment literature the term *evaluation* ". . . is concerned with determining the significance of individual environmental impacts and/or the aggregate significance of all environmental impacts relative to the economic and social effects of a development" (Lee 1983: 20). The exact meaning of this elusive term was not defined in the NEP Act of 1969, but Section 102 conveyed that it is ". . . an action in which the overall cumulative primary and secondary consequences significantly alter the quality of the human environment, curtail the choices of beneficial uses of the human environment, or interfere with the attainment of long range human environmental goals". This required an analysis of such matters as magnitude (scale), exposure (both geographical extent or number of people affected), timing (one off, frequent repetition), enduring (short or long term), and irreversibility (whether or not the effects can be remedied).

Following this approach the judgement of significance is a two-part process. First, there is the significance of the magnitude of an individual impact. This can be explained in terms of four possible relationships between magnitude and significance (Fig. 5.4). Secondly, having assessed individual impacts, it is then necessary to determine their relative importance in the aggregate of impacts. For this some forms of scoring, scaling and weighting have been developed, such as, for example, that of

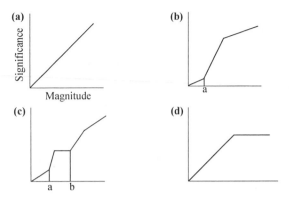

Figure 5.4   Relationship between magnitude and significance.
(a) A linear relationship, which is the simplest, e.g. in agricultural land take-up.
(b) A single threshold (x) beyond which significance rises dramatically, e.g. the effect of changing water levels to the point where a river breaks its banks.
(c) Several thresholds (x and y), e.g. where in water pollution each threshold might represent a concentration of pollutants at which another species begin to die off.
(d) Point of no return, where above a certain point, magnitude has no further effect on significance, e.g. the point where all life in an ecosystem has been destroyed.

Leopold (Leopold et al. 1971), environmental evaluation system (Dee et al. 1973), the Ram system (Solomon et al. 1977), and overlay on separate transparencies giving an aggregate significance (McHarg 1971).

An impact gauging scale has been devised for assessment of the aggregation of individual impacts. In order to indicate the significance of the proposed action on the human environment (Fig. 5.5). For this purpose each impact is taken in turn and ranked ordinally, with points scoring to bring out the degree to which it is beneficial or adverse or neutral, having regard to such factors as duration, population and geographical area affected. Each of the components is then judgmentally ranked in accordance with a points system and then the points aggregated to show the significance under three heads:
  • the overall net impact derived from a matrix
  • significant adverse impacts on natural systems and components
  • the overriding factors, which would lead to a veto of the project itself.
A wider role for evaluation in EIA is disputed from two extreme viewpoints (Lee 1983):
  • because of the value judgements involved the EIA report should be limited to a statement of the likely magnitude of the individual impacts
  • despite the value judgements involved, the analyst should attempt to derive a measure of environmental significance of the development as a whole.
A new dimension for evaluation from outside impact assessment was given by the Royal Commission on Environmental Pollution. It put forward a specific form of assessment in relation to pollution of air, water and land, which is not cost–benefit analysis but is based on a cost–benefit approach. Its First Report pointed out that

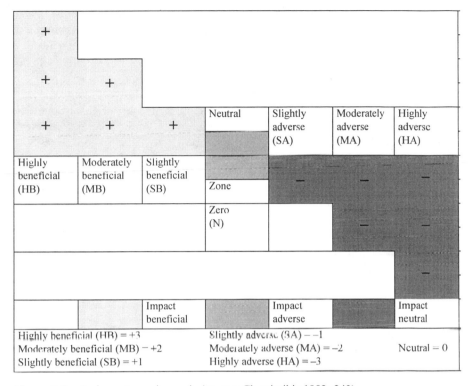

Figure 5.5   An impact gauging scale (source: Chatzimikis 1983: 243).

there is no completely scientific objective means of striking a balance between environmental and other conditions (Royal Commission on Environmental Pollution 1971). Accordingly, the reasons for a decision should be open and accountable, and the value judgements that necessarily underlie it should be clearly identified, and there should be the widest possible opportunity for others who may be affected to contribute to the decision. If the grounds for a particular decision are clearly understood, it can be more readily reviewed when assumptions or value judgements are changed. In the Fifth Report, the Commission amplified its approach and saw "best practicable environmental option" (BPEO) as an extension of the concept of "best practicable means" (BPM) which goes back to the Alkali Act of 1874, which has been the cornerstone of industrial air pollution control in England and Wales since that time (Royal Commission 1988: para. 1.3). The consideration of what constitutes BPM ". . . involves the balance of factors including the present state of development of the relevant control technology, the financial implications, current knowledge about effects of pollutants and local conditions and circumstances".

The BPEO procedure ". . . established, for a given set of objectives, the option that provides the most benefits or least damage to the environment as a whole at acceptable cost in the long term as well as in the short term" (ibid.: para. 2.1). The procedure has regard, amongst other things, to local conditions and circumstances, to the financial implications and to the current stage of technical knowledge. And it is accepted that ". . . the option chosen as best would depend upon the interpretation and evaluation of the predicted impacts by whoever takes the final decision. It is unlikely to be the best choice for all time" (ibid.: para. 2a).

This approach was adopted in the integrated pollution control (IPC) to be used by Her Majesty's Inspectorate of Pollution (HMIP), established in 1987 (Royal Commission 1988: para. 2.9). Under the EPA 1990, in order to prevent or minimize environmental damage, their authorization must be obtained for any industrial processes that release prescribed substances to the three environmental media of air, water and land. The objective is to secure the BPEO by use of BATNEEC, i.e. the best available techniques (BAT) not excluding excessive cost (NEEC). In this the cost of the BAT must be weighed against the environmental damage from the process. The concern is with what costs *in general* are excessive; the lack of profitability of a particular business should not affect the determination.

Mitigation   At this point judgements will be made as to whether or not the adverse impacts of significance can be mitigated or ameliorated, with a view to making them more acceptable.

Communication   Having reached the end of the assessment, it is then necessary to communicate the conclusions to those who are affected and concerned. This is conventionally done in an environmental impact statement, which is a popular version of the EIA.

Economic impacts
As indicated above (§5.3) economic impacts can be very widespread. But economic impact assessment proper is generally made more specific (DoE 1993a: ch.7). In this, the impacts relate to any exogenously produced effects on economic life, that is production (including distribution), consumption and exchange (including transportation), seen in their widest ramifications. The effects can arise in different ways: through the introduction of some project (input of resources to produce outputs of various kinds), or some change in activity without any significant works of buildings (such as increase in tourism following some new archaeological discovery). The impact can be either direct (primary) or indirect (secondary). It is the latter that is termed the "multiplier": the immediate effect on an endogenous variable of a change in an exogenous variable, as opposed to the total or long-run effects of such a change (Pearce 1983a).

The multiplier can be traced in various dimensions: incomes, outputs, or employment (Schofield 1987: §14.2). "The size of the local multiplier depends directly on the proportion of income spent locally or, inversely, on the proportion of income that each round of spending leaks out of the local spending stream into savings, taxation, reduced transfer payments, or important purchases. Its size, therefore, varies directly with the size of the local area in question, since important leakages decline as the size of the area increases".

There are various models for the multipliers, of which three types have played a prominent role in urban & regional impact analysis: economic base, Keynesian, and input–output (ibid.). The multipliers produce either "induced" or "generated" outputs. These can take either the pecuniary or technological form: the former relating to changes that are transfers, and the latter to real changes in economic life.

Social impacts
Social impact assessment evolved rapidly after the introduction of the narrower environmental assessment in the 1969 NEP Act (Gold 1978). But although emanating from the sociological disciplines, the processes were judged similar, involving the following steps (Carley 1980: 54):

1. the establishment of a database which describes the existing situation
2. development of the means of describing change related to the project
3. forecasting changes in the base situation with and without the given project, including qualitative and quantitative aspects.

The similarity is further seen in the form of SIA is set out in the Table 5.1.

Table 5.1   Social impact assessment: the main steps.

Assessment steps
1. *Scoping* How large a problem is it? How much is enough?
2. *Problem identification* What is the problem? What is causing it?
3. *Formulation of alternatives* What are the alternatives?
4. *Profiling* Who is being affected and how?
5. *Projection* What is it causing?
6. *Assessment* What difference does it make?
7. *Evaluation* How do you like it?
8. *Mitigation* What can you do about it if you don't like it?
9. *Monitoring* How good are your guesses?
10. *Management* Who's in charge here?
11. *Bottom line* Who benefits and who loses?

Source: summarized from Wolf 1983.

5.5 Impact assessment of plans, policies and programmes

5.5.1 Evolution from project impact assessment

In this Chapter so far we have described the evolution of impact assessment for projects since their introduction in the US NEPA 1969. Here we introduce proposals mentioned above (§3.2) for the evolution from environmental assessments of projects to that of plans, policies and programmes. Despite the fact that this need was seen already in NEPA 1969, the progress in this direction has been slow and the practice is far from well established.

NEPA 1969 visualized that EIA would be used not only for federal projects but also in their "land use plans, regional development programmes and economic programmes, including for those specific sectors" (Tomlinson 1986: 460). This concept was further articulated by the Council of Environmental Quality guidelines of 1978, which interpret the requirements of the NEPA, that EIA in addition to projects may be necessary for the following actions (Therivel et al. 1992).

- "Adoption of official policy, set of rules, regulations and interpretations".
- "Adoption of formal plans, such as official documents . . . which guide or prescribe alternative uses of federal resources".
- "Adoption of programmes, such as a group of concerted actions to implement a specific policy or plan".

Similarly, the EEC also visualized, from the outset of its environmental programme in its preliminary draft Directive of 1978, that the assessment system would apply also to plans as well as projects (Wood 1988). However, this proposal was not incorporated in the Directive of 1985, which set up the environmental assessment of projects. These intentions were clarified and emphasized in the Commission's Fourth Action Programme on the Environment in 1987 (Council of the European Communities 1992): "The Commission's concern will also be extended, as rapidly as possible, to cover policies and policy statements, plans and their implementation, procedures, programmes . . . as well as individual projects . . .". This was endorsed in the Fifth Action Programme on the Environment, agreed in 1992, as a necessary adjunct of sustainable development, namely (Commission of the European Communities 1992). "Given the goal of achieving sustainable development it seems only logical, if not essential, to apply an assessment of the environmental implications of all relevant policies, plans and programmes . . .".

Pressures in this direction have also come from other international agencies such as the World Bank, the Asian Bank and OECD (Lee & Walsh 1992). The approach has also been officially adopted by the Department of the Environment in Britain, despite British resistance to a Directive by the EEC on this topic, on the grounds that it could be incorporated readily in the British planning system. Initially the approach was set out in the White Paper on the Environment of 1990 (DoE et al. 1990), which committed the Government to publishing guidance on the topic:

There is scope for a more systematic approach within Government for the

appraisal of the environmental costs and benefits before decisions are taken.
The Government has therefore set work in hand to produce guidelines for the
policy appraisal where there are significant implications for the environment
. . .

The resulting guidelines were presented by the DoE in 1991 (DoE 1991a) and then
again more specifically in relation to land-use planning in 1993 (DoE 1993b), again
linked to sustainable development:

The planning system, and the preparation of development plans in particular,
can contribute to the objectives of ensuring that development and growth are
sustainable. The sum total of decisions in the planning field, as elsewhere,
should not deny future generations the best of today's environment. This
should be expressed through the policies adopted in development planning.

Under these authoritative pressures, the practice of what has come to be called
"strategic environmental assessment" (SEA) has been slowly evolving in its theory,
principles and practice (Lee & Walsh 1992) and applied around the world to a variety
of strategic situations.

5.5.2 Methodology of SEA

That the extension of the environmental assessment beyond projects is desirable there
can be no doubt. For one thing it would avoid the criticism of "projectitis", as
opposed to more comprehensive planning. For another, it fits naturally as a sectoral
stream of "environmental planning" into the development planning process, in which
are employed varied sectoral streams, such as planning for traffic, open space, and
education. Thus employed, it would ease the preparation of project assessments and
their approval, since these could be carried out within the framework of a general
environmental assessment, just as development projects are in relation to a plan.

However, there is still no commonly agreed method for carrying out an assessment. This is perhaps not surprising, given the widely varying circumstances within
with the SEA needs to be applied, as brought in Figure 5.7 below. As the authors indicate, the categorization of action and type of assessment "is a simplified representation of what, in reality, could be a more complex set of relationships". From this, it
is understandable why the SEA method cannot be standardized but must be capable of
adaptation to both the particular purpose for which the SEA is being carried out and
the level of government that is seeking the assessment.

Considering the evolution of EIA and SEA from their beginnings, it seems apparent
that EIA can be regarded as a generic method of environmental assessment, which
then needs to be adapted for the SEA in the particular situation that determines the category of actions and type of assessment. Given this, it is to be expected that there
would be similarities and differences between the EIA and SEA processes.

Clearly, it is the *differences* that give rise to the need for variations in the SEA from
the generic model of project assessment. This situation has been explored (Lee &

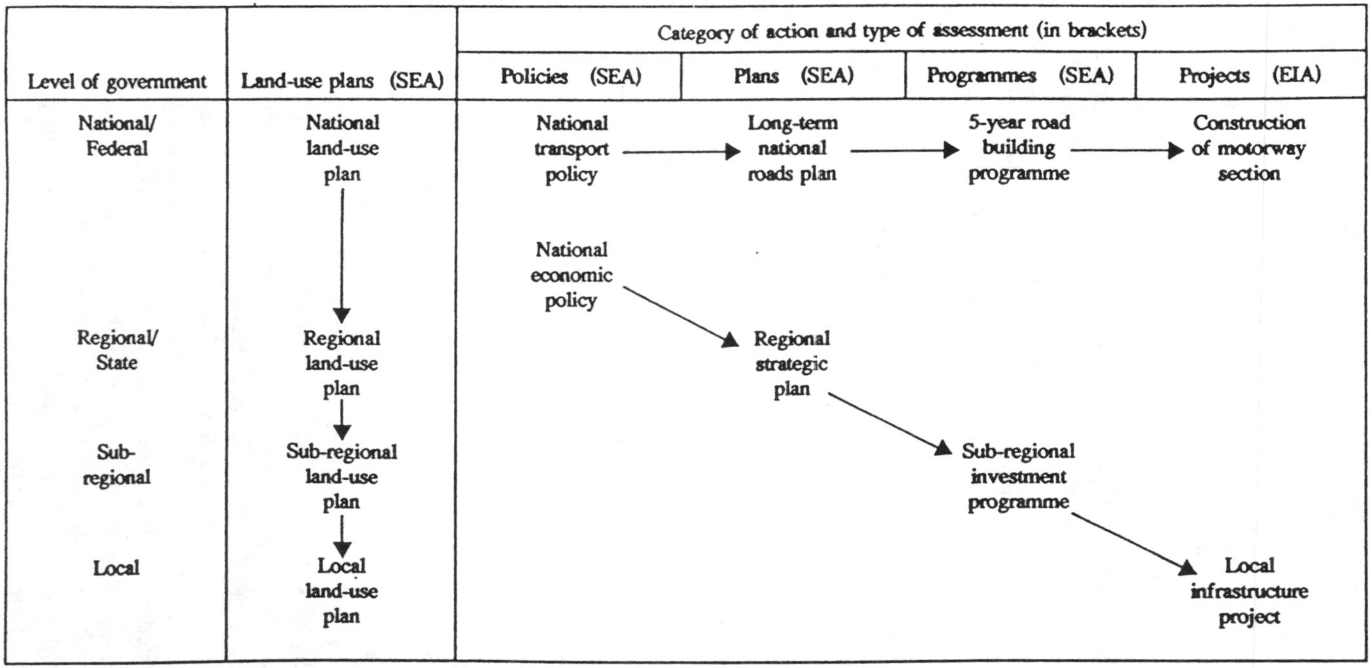

Category of action and type of assessment (in brackets)
Level of government
Land-use plans (SEA)
Policies (SEA)
Plans (SEA)
Programmes (SEA)
Projects (EIA)
National/ Federal
National land-use plan
National transport policy
Long-term national roads plan
5-year road building programme
Construction of motorway section
National economic policy
Regional/ State
Regional land-use plan
Regional strategic plan
Sub-regional
Sub-regional land-use plan
Sub-regional investment programme
Local
Local land-use plan
Local infrastructure project

N.B. **This is a simplified representation of what, in reality, could be a more complex set of relationships. In general, those actions at the highest tier level (e.g. national policies) are likely to require the broadest and least detailed form of strategic environmental assessment.**

Figure 5.6 Sequence of actions and assessments within a tiered planning and assessment system. (*Source:* Lee & Wood 1978a).

Wood 1992; Therivel et al. 1992: appendix C). Figure 5.7 gives three particular models of SEA methodology, with contrasts between them. This situation is comparable to that arising in evaluation analysis, of the kind being explored in this book, since the methods for particular circumstances can be seen as a necessary adaptation of the standard model. Such adaptation in evaluation to the different circumstances arising in the planning process (plans, policies, proposals, sectoral and programmatic programmes, and strategies) is explored below in relation to evaluation (Ch. 12). The approach there links the appropriate method of evaluation to its role in the comprehensive planning process as a basis for identifying the changes needed from the generic model itself.

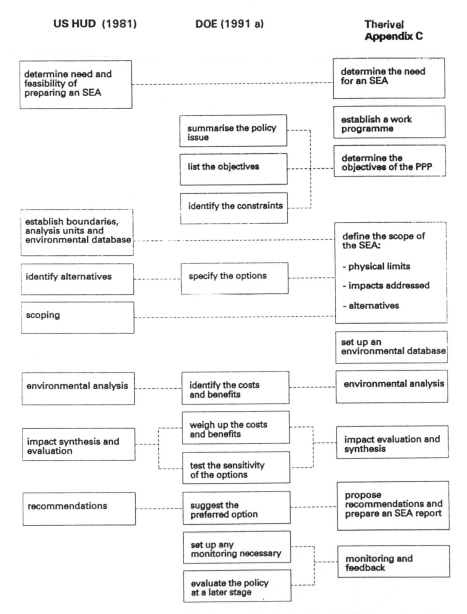

US HUD (1981) **DOE (1991 a)** **Therivel Appendix C**

determine need and feasibility of preparing an SEA ---- determine the need for an SEA

summarise the policy issue ---- establish a work programme

list the objectives ---- determine the objectives of the PPP

identify the constraints

establish boundaries, analysis units and environmental database ---- define the scope of the SEA:

identify alternatives ---- specify the options ---- - physical limits

- impacts addressed

scoping ---- - alternatives

set up an environmental database

environmental analysis ---- identify the costs and benefits ---- environmental analysis

impact synthesis and evaluation ---- weigh up the costs and benefits ---- impact evaluation and synthesis

test the sensitivity of the options

recommendations ---- suggest the preferred option ---- propose recommendations and prepare an SEA report

set up any monitoring necessary ---- monitoring and feedback

evaluate the policy at a later stage

Figure 5.7 Steps in EIA methodology (*source:* Therivel et al. (1992: appendix C)).

CHAPTER SIX
Theory and principles of cost–benefit analysis

6.1 Origins and applications

Cost–benefit analysis is primarily used for the *economic assessment appraisal or evaluation* of the *economic* impacts of *projects, programmes and policies* which, being in the public sector, cannot rely upon market indicators of profitability for the purpose. The term is used interchangeably with "investment planning", "project appraisal" or "project assessment", the distinction being terminological rather than actual.

The analysis is designed as an aid to choice between options of the decision-taker. It is ". . . a process of investigation and reasoning designed to assist a decision-maker to reach an informed and rational choice" (Sugden & Williams 1978: 3) or ". . . a practical way of assessing the desirability of projects, where it is important to take a long view (in the sense of looking at the repercussions in the further, as well as the nearer, future); and a wide view (in the sense of allowing for side effects of many kinds on many persons, industries, regions, etc.)" (Prest & Turvey 1966: 683).

The theory and practice had simple beginnings in the USA at the opening of the present century (Prest & Turvey 1966: 1–3), although its theoretical foundation derived from the middle of the 19th century in France (Dupuit 1844). The US River and Harbour Act 1902 required engineers to report on the desirability of the Army's River and Harbour Projects, taking into account the amount of commerce benefited and the cost. This led to the Flood Control Act of 1936 which authorized Federal participation in flood control schemes "if the benefits to whomsoever they accrue are in excess of the estimated costs". Ensuing practice was codified in the "Green Book" (Inter-Agency River Basin Committee 1950). By that time the analysis had spread to other kinds of projects, notably highways. Since then the fields for application of the analysis have exploded.[1]

This wide-ranging application creates difficulties in the way of presenting a clear statement of the theory and principles of cost–benefit analysis in evaluation. The literature is enormous. There are variations according to the fields in which the analysis is exercised. The practice is being continually developed and new fields opened up, as in public transport and the environment. There are contributions from different academic and professional disciplines outside economics (systems analysis, operations research, financial analysts and law).

6.2 Procedures in cost–benefit analysis

From all this it follows that, while the various writers and practitioners show internal consistency in their methods of analysis, it is difficult to trace a standard set of procedures that would apply in all circumstances. Accordingly, it is not surprising that: ". . . there can be no uniquely 'proper' way to do cost–benefit analysis; it is to be expected (and in our view welcome) that the studies will be done in many different ways" (Nash et al. 1975a,b).[2] And, furthermore, "Each technique must reflect one or more value judgements. We describe the ethical basis for the technique. The relevant value judgements pertaining to any one technique we call an underlying value set. These value sets may in turn be thought of as deriving logically from some set of underlying moral notion" (ibid.).

This in itself is open to controversy. Some argue that cost–benefit analysis should be carried out by one consistent method, whose value set is agreed on academic/professional grounds by economists (Mishan 1982). Here the notion is that all decision-takers concerned with public sector projects, and different kinds of such projects, should have the same underlying value set. As against this it is argued that the various schools of economic thought (neoclassical, Marxist, the new economics) would each have their own values that would affect method (Cole et al. 1983).

In this book we take the latter view, that values must influence and enter into the analysis. On this basis, our primary purpose in this chapter is to show how cost–benefit analysis has been employed as a second foundation for community impact evaluation, the first foundation being impact assessment (Ch. 5).

For this purpose we summarize the essentials of cost–benefit analysis around a generalized sequence of procedures that could be followed in making the analysis. These steps are designed to echo the approach used below for the sequential steps in community impact evaluation (Fig. 7.2). Figure 6.1 accordingly shows four sequential steps (A–D) and in 12 boxes what should be pursued in order to reach a conclusion from the evaluation.

In the description that follows, the aim is not the impossible one of comprehensiveness but rather a guide to the essentials for use as appropriate *within* the community impact analysis itself. As in that analysis, we first present the steps and boxes that provide the setting for the analysis (A–B and boxes 1–6), then the analysis (C and boxes 7–11) and finally the evaluation itself (D and box 12). We now proceed to

---

1. For example, Prest & Turvey (1968), present applications in irrigations, flood control, hydroelectric power, roads, rail, inland waterways, urban renewal, recreation, health, education, research and development, defence; Peters (1971) adds urban projects; Lichfield et al. (1975), explored urban and regional plans; Pearce (1980) concentrates on the environment; Bridger & Winpenny (1983), adds ports, airports, housing, wastewater, industry, livestock, tourism; Schofield (1987) covers residential renewal, transportation, recreation, land-use planning, health and social services, regional policy, local economic development; Pearce et al. (1989) enters the wide ranging instances of environmental pollution; Lichfield (1992) reviews transport in its different modes.
2. See for example McKean (1958), Eckstein (1958), Layard (1970), Abelson (1979), Mishan (1982), Pearce (1983b).

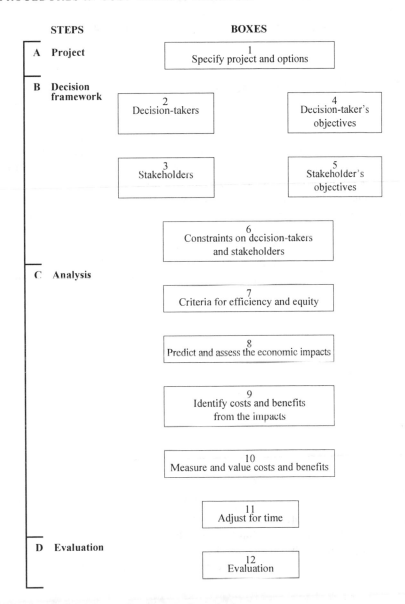

Figure 6.1   Sequential steps in cost–benefit analysis.

describe each of the steps and boxes in turn. In the description of boxes 7–11 the
abundance of footnotes conveys the wealth of background literature to cost–benefit
analysis for the past half century. In order to place them in an historic time sequence,
many of these references are necessarily quite old. Accordingly they may not reflect
the major breakthroughs on particular aspects in the past 20 years or so.[3] However,
this has not undermined the basic intention, which is to give guidelines for the use of

the analysis as a foundation for the economic evaluation inherent in community impact evaluation.

6.3 The analysis and evaluation

A The project

Box 1: Specify project and options

Since the analysis relates strictly to a particular project, it is necessary from the outset for this to be specified (the "with project" situation), together with any alternative formulations for the project (the options). One of the options is always the possibility of "do nothing" or "do minimum" (that is allowing for unavoidable maintenance and repair). For this, it is necessary to specify the situation that would exist in future if no project were undertaken at all (i.e. the "without situation"). This will not necessarily be the same as that which exists at the time of the appraisal, for the benefits and costs need to be related to the projected or predicted situation, pending which certain conditions are likely to change, for only that way are the true future differences estimated (see Fig. 5.1).

B The decision framework

Box 2: The decision-takers

Since the analysis aims at facilitating the choice amongst options by decision-takers, it is necessary at the outset to specify who they will be. As indicated, in the main they will fall into the public sectors, such as the various departments of central or local government, quasi-public *ad hoc* bodies set up by them, or any other body guided by considerations outside the market.

Box 3: The stakeholders

Although a decision-taker clearly has the ultimate responsibility for a project, his choice is affected by what have been termed "stakeholders". These are responsible for taking decisions on particular *aspects* of the project, so having the possibility of affecting the choice of the prime decision-taker, on whom they thus act as a constraint, or perhaps offer an opportunity. For example, unless those responsible for the finance make a favourable decision, the decision-taker's choice could be sterile, unless, for example, a charitable fund which is sympathetic to the project could fill the gap.

Boxes 4 and 5: The decision-taker's and stakeholder's objectives

Since we are concerned with economic analysis, the underlying presumption is that the generalized objective of the decision-taker is achieving "value for money" (Ch. 4). This is interpreted differently in the private and the public sectors.

In the former, the objective is to secure the maximum or a satisfactory surplus of

---

3. I am indebted to David Pearce for this warning.

money value of the output over the money costs of the input, to the benefit of those who own or operate the undertaking (individuals, families or shareholders). But the objective is not always simply financial. Promoters can also have other objectives: "psychic" benefits, such as securing a good image in being a patron of the arts; having no discriminations against women or ethnic groups; providing sponsorship in the inner cities, etc.; or "political" benefits, such as securing loyalty and performance in the management and workforce.

In the latter, the public sector, the prime objective is also "value for money". Since the decision-taker's objectives are presumed to relate to the economy as a whole, it estimates the costs and benefits in "economic" rather than financial terms (Sugden & Wiliams 1978). But while the public agency seeks "value for money" in pursuit of financial viability it will also have supplementary aims of "social value for money". In addition to the psychic and financial benefits of the private sector could be added advancement of the public interest, social welfare, non-economic governmental policies, attracting votes.[4] In this there are many variants. The state agency supplying utilities may be directed by its parent department to act as though it were in the private sector (financial value for money), but subject to the directives of the minister, which would regulate the activities in the "public interest". Or, where a privately owned monopoly utility service (gas, water, electricity) has the objective of maximizing revenues for shareholders, it could be constrained by the direction of a regulator as to price bands. Or, where a government department is responsible for services that cannot be expressed in money value, it could specify intermediate objectives that are presumed to advance the ultimate objectives, such as the number of school places or bed places in relation to education and health.

Box 6: Constraints on the decision-takers or stakeholders
Since the decision relates to a specified project and its options, it can have physical constraints that flow into the costs and benefits that will be generated: examples are the site topography, availability of access and utilities, design, legal impediments to implementation. They can be translated into *economic* constraints.

There could also be *choice* constraints, deriving not so much from the project itself but from limitations imposed upon or by the decision-taker. The former could be insufficient finance to cover the project in question, or the balance of a programme of projects in contemplation (for example, links in a road system). The latter could be pre-emptive strikes by the decision-taker against certain possible options. Such constraints can affect the choice of priorities in projects, and the precise nature of the decision criteria chosen, and thereby the decision itself.

---

4. Downs (1957): politicians use an economics type calculus to maximize their potential votes.

C: Analysis

Box 7: The criteria for efficiency and equity

General criteria   An initial stand needs to be taken that relates to the considerations raised above (Ch. 3) as to whether the analysis will relate to both costs and benefits as variables (CBA), or to the cost input being fixed (cost effectiveness or CE), or
the output being fixed (cost minimization or CM). In each case, bearing in mind the
objectives and constraints, decision-takers needs to adopt criteria for selection of the
option that will achieve their objectives.

In this a distinction is made between those relating to efficiency (value for money,
ignoring distribution effects) and equity (having regard to such effects). Then arises
the question of trading off these two in any particular evaluation for decision. There
is another distinction (Sugden & Wiliams 1978: 91–2). To some ". . . cost–benefit
analysis is seen as a process of appraising decision problems in the light of the objective chosen by the decision-taker, which stem from his value content". Others would
follow the precept, primarily from Mishan (1981a), which ". . . starts from a distinct
position about what the objectives of social decision-makers ought to be . . . distilled
from a consensus of the value judgements of the individuals who make up society,
. . . This interpretation of cost–benefit analysis sets it apart from the process from
which actual decisions are taken in practice from any point in space or time". Our
approach here is certainly the former.

Efficiency   The fundamental efficiency criterion of cost–benefit analysis is to identify projects where the value of the output is greater than the cost of the input; and,
among options for a particular project, where the surplus is the greatest (the optimum). By definition, such projects will increase the net wealth of society and so provide the net growth that would facilitate an increase in welfare. It is this concept that
roots cost–benefit analysis in the theory of welfare economics (the generic name for
"normative economics") which provides prescriptions or statements about what
"should be" rather than "what is" (Mishan 1981b).

Pareto is the generally accepted father of the "new welfare economics" (Schofield
1987: §2.5.2). He initially formulated the rigorous rule whereby a judgement could
be made (on the explicit concept that society is made up of the sum of individuals)
that society would be better off through some ". . . reallocation of resources which
makes at least one person better off while making no one worse off . . ." This constitutes a Pareto improvement. When no such improvement is possible, then a Pareto
optimum is said to exist. Where there are both gainers and losers, the strict Pareto
criterion is impracticable since it involves comparing people's subjective valuation
of benefits, that is, interpersonal comparisons. Much of the new welfare economics
has been concerned over recent decades with analyzing conditions under which the
Pareto optimum and Pareto improvements can be said to come about. A more practicable approach is the Hicks–Kaldor compensation principle of potential Pareto
improvements: that gainers would be in a position to compensate losers after imple-

mentation and still remain better off, even if they did not in fact pay the compensation (Kaldor 1939).

In order to judge whether a particular project will in fact increase net wealth, a variety of efficiency criteria have been developed (Merrett & Sykes 1973). Each is addressed to the same issue, the net increase of wealth (income or capital), discounted to take account of time over the years: yields, net present value (NPV), net terminal value (NTV), annuity, and payback period. Among these, net present value is generally used, taking the "present" to be the time of decision. NPV is then the difference between the present value of benefits (PVB) and the present value of cost (PVC), using the appropriate discount rate.

Net present value is aimed at "optimizing", that is achieving the *best* situation or state of affairs, with the "best" being defined by the decision-taker according to his objectives. Where the optimal *cannot* be achieved, then the best *attainable* situation is the "second best", for example, achieving only *satisfactory* aspiration levels, which is then "satisficing" (Simon 1957). One form of "satisficing" criteria is the achievement of the appropriate criterion in the year following completion of the project (also called "optimal timing"). In that case there is a gamble with uncertainty, in that the future years may in fact turn out to show a worse benefit–cost relationship.

Such "optima" are most difficult to achieve in practice because of the very rigorous conditions that need to be met (Schofield 1987: ch. 2). Therefore, there is the tendency to relax as many of the conditions as practicable, to achieve a "constrained optimum". However, it has been argued, the best *attainable* situation (the second best) may not enhance social welfare if any of the rigorous optimality conditions remain unfulfilled elsewhere in the economy, as they typically do.

Equity or distribution   At any moment in time there is in any society a particular distribution of economic welfare (income, wealth, etc.) which inevitably leads to inequality in conditions, opportunities, and so on. These inequities can be illustrated by reference to residents in a town.

First comes the "geographic" or "horizontal". Because the activities of the town are spread over an area, it is not possible for all residents to be evenly accommodated in the level of service offered. Some residents will have long walks to bus stops, and infrequent services, whereas others will not. Thus, in choosing where to live, a household will weigh up the various attributes of different locations against each other, of which accessibility by transport is one (others being proximity to countryside, availability of schools, shops, and local environmental amenities); and, having chosen, the total attributes of the package will then be traded off against price.

Second is the "income" or "vertical". The town's residents do not compete evenly, because of varying income and wealth levels, and consequential access to information and professional help. Accordingly, those with low income are disadvantaged in the competition.

Third is the "social". The young, elderly or infirm are disadvantaged in that they cannot afford to run a car, or are incapable of driving. This further reduces their mobility and accessibility compared with, say, a car-borne next-door neighbour.

It is into this contemporary maldistribution of welfare that the allocation of resources to any particular project, programme of projects, policies or plans, will distribute the output in a predetermined way. If this distribution is guided by the rules of allocation, which are implicit in the efficiency criteria for cost–benefit analysis, it will be found that the contemporary distribution is made more regressive, since the values of the costs and the benefits built into the analysis reflect the current distribution of income and wealth and thereby favour the better off.

Faced with this situation, four issues arise, which we take in turn:
- Should economics include considerations of equity?
- If so, what rules to adopt?
- Should compensation be made for losses?
- How can efficiency and equity be combined?

SHOULD ECONOMICS INCLUDE CONSIDERATIONS OF EQUITY?   From the above it follows that, within the methods and techniques of cost–benefit analysis, the economists' rules for distribution/equity/social justice are not as clear-cut as those for efficiency. A brief account follows. Whereas the potential Pareto improvement criterion has made it practicable to select projects that will advance net aggregate welfare, welfare theory has not been of equal help in offering normative rules for distribution of the net increase. One obstacle, to which much of welfare economics theory has been devoted, is finding a way by which economics could show how the necessary "interpersonal", "intergroup" and "intertemporal" comparisons could be made (Kaldor 1939). Furthermore, the Pareto optimum/improvements criteria have the defect for our purpose of being based on the aggregation of individual consumer demand as expressed in the market, so that additions to net wealth do not take account of the externalities (such as environmental pollution) which are a feature of the project. In practice, there are many differing views as to how the issue should be tackled. Some are here presented (in ascending order of application to the problems involved).
- Ignore the issue, not simply on the fact that economics can offer limited guidelines, but on the basis that the market knows best. The fact that, in the market, the economic vote of the poor must be more feeble than that of the rich is not considered to be an objection, nor that the market cannot, by definition, fully register externalities.
- Those who are not so enchanted with reliance on the market, and also not with what economics can offer as guidelines, seek the answer from various sources: analysis from positive economics to issues and questions raised by decision-takers; leave the responsibility to the bureaucracy, subject to review by higher political authority, or to the analyst and expert.
- The nettle is grasped by asking the decision-taker to postulate constraints on the efficiency outcome. These could specify earmarked benefits to particular regions or groups; or specify the distributional criteria that will be subject to an aggregate efficiency constraint (e.g. that the proposal at least breaks even in efficiency terms).

- The nettle is grasped even more firmly by asking the decision-taker to adopt specific rules, by which distribution effects are to be judged, as giving the social optimum (the social welfare function, SFW; Bergson 1938). For example, there would be differential "social pricing", or weighting of net benefits accruing to the different parties affected by a project or programme, the weighting possibly relating to varying considerations such as income, future as against the present, favouring environment as against growth, or vice versa. On the same approach, but coming from philosophy, is that of Rawlsian justice, based upon a clearly articulated set of principles (e.g. the least well-off person in society be made as well-off as possible, not simply in terms of utility but of primary goods, which can be defined as rights, freedoms, opportunities, income and wealth (Pearce 1983a: 374).

IF SO, WHAT RULES TO ADOPT?   Accordingly, unless the consequences of efficiency analysis alone are to be followed, it is necessary to make a specific decision in relation to each project as to whether it should continue, intensify or reverse the distribution of economic welfare that exists without the project. Given that there is the intention to do so, there must be a search for the ethical principles on which it is based. These are older and richer than that implicit in the Pareto criterion and its variants. The arguments started with the Greeks (Aristotle), have been continued by the philosophers (Hume), the churchmen (Aquinas) and the political economists (Marx) (Sabine 1963). It continues today, since there is no general agreement on the basic guideline as to what would constitute "equity" (Miller 1976). In a particular country there will be variations in the views of the political parties, and between local government administrations (even with a common political label), and over time (compare changes within this century in Britain). And there will be variations as between countries assembled around the tables of international agencies, as seen in the discussions on the EEC Social Chapter.

But even if there were such agreement in the past two millennia of debate, it would not necessarily apply to any contemporary society; there is the contemporary disagreement between the First, Second and Third Worlds, and in any country between governments in power and those in opposition, who in their approach to the principles of social justice follow their own ideologies and values.

In the contemporary search for the ethical principles, a variety of basic questions arise.

- Should the distribution as between sectors of the community follow the status quo (the customary) or should there be some shift in distribution towards a different order? An example is the current practice of weighting benefits and costs by the marginal utility of income of those experiencing them.
- Given the latter (which would be generally accepted), in what direction should the redistribution be made? Should it be progressive or regressive?
- And given agreement here, what are the criteria for subdivision of the product among particular groups or members of the group. This could, for example, be based on (ibid.):

- rights (on legal, institutional and moral rules);
- deserts (on actions and personal qualities, moral virtues, productive efforts, capacities);
- need (from levels of deficiency, e.g. income, size of family, etc.).
- Another approach, which is consistent with positive economics, is to avoid formulating equity / social justice rules, but to offer the decision-takers/-makers an *ex post* analysis of the past decisions they have taken in order to bring out their actual distributional philosophy and ideology in practice with a view to asking whether they wish to follow or modify them.[5]

SHOULD COMPENSATION BE MADE TO THE LOSERS?   The Hicks–Kaldor compensation principle does not in practice make compensation available to losers, and thus does not in practice ameliorate the inequity. But such compensation is provided for by governments, who are thus implicitly making judgements on equity and taking action on them. Examples in the planning/environmental field are money compensation for taking property; the imposition of penalties on the beneficiaries from the project, in favour of those disadvantaged (e.g. amelioration of the disbenefit through conditions or agreements on planning permissions) (see Ch. 13); or economic penalties to "make the polluter pay" for damage to the environment.

COMBINING EFFICIENCY AND EQUITY   Since it is rare today that decisions are made simply on efficiency criteria alone, it is necessary to integrate the two kinds of criteria in the decision rules. Although this makes for complexity, since the equity criteria are vaguer than those for efficiency and there is no consistent approach to their formulation, there are many approaches to the integration. Examples are constraints, weighting of distribution, specifics of compensation and redistribution (Squire & Van der Tak 1975: pt II). Against this general background we now examine how economists contribute to the rationale for influencing the distribution from projects (Dasgupta & Pearce 1972: 62).

Starting with the common assumption that, whatever view is taken of this distribution, the project that is subject to the CBA will not alter that distribution in a significant fashion:

- Even if the redistribution criterion be accepted, it should be ignored, since it requires a value judgement which is outside economic analysis.
- Make an explicit attempt to allow for distribution of consequences, either by trying to observe social preferences concerning this distribution or asking the decision-taker to apply his own weights to the gains and losses of the various sections of the community.
- Impose an explicit value judgement in the social utility function, for example by scaling down higher incomes and scaling up lower incomes to equalize their influence on the cost–benefit distribution, perhaps by weighting the ratio of the average national personal income to the individual's income.

---

5. Essays of this kind have been made in relation to water resource projects (see McGuire & Garn 1969).

Box 8: Predict and assess the economic impacts
An assessment of the project's economic impact would be a preliminary to the cost–benefit analysis. As such, the topic is covered above (§6.1), but an amplification is needed on prediction. As an aid to decisions on projects, in contrast to an audit of what had happened, cost–benefit analysis requires prediction of the future in the many ways brought out in the above analysis, such as the profile of inputs and outputs, their measurement and valuation, their impacts (initial, second round and third round), and price levels. In all this prediction the art and science of forecasting is involved, by the use of a large array of techniques. Since prediction for cost–benefit analysis has grown up around the requirements of the problems that have been tackled by engineers, economists, natural scientists and statisticians, in an innumerable array of situations (market research, urban and regional planning, regional and national development planning, demography, etc.) prediction and forecasting for cost–benefit analysis has a wide base in theory and practice to draw on (Encell et al. 1975, Miles 1975, Whiston 1979).

Box 9: Identify the costs and benefits from the impacts
So far we have introduced the terms costs and benefits in their common-sense usage (input/output or disadvantage/advantage). Since they can be of a varied kind, misleading results will flow from the analysis unless there is precision in definition. That is the focus of this section. It has been placed after the earlier one on criteria (§6.3.7), for the simple reason that the nature and scope of the costs and benefits under review will be influenced by the criteria to be adopted in the specific case.

Direct costs Any project requires an *input* of resources (costs, disutilities) that are consumed/given up in order that the *outputs* (benefits, utilities) are to be derived. Such costs can be financial (in relation to purchase or exchange) or economic (in relation to production, operation and consumption).

Financial costs are those money payments that must be made to the owners of resources to ensure that they will be employed in the project (e.g. payments for all factors: land, savings for investment, materials, etc.). But these are not necessarily *all* costs from the viewpoint of society as a whole, for some are *transfers*, in that while income or wealth may pass between individuals or organizations or nations, they do not result in net increase in wealth to society as a whole. Examples are payment for land or natural resources, interest on loans, taxes, subsidies, depreciation of assets.

Direct benefits The benefits from the project are the direct value of the output that is produced for consumption – or indirect, such as raw materials that will be initially used for production for later consumption. Whereas costs are incurred once and for all (be they capital or operating costs at intervals), benefits are related to flows over time, be they instantaneous (consuming an ice-cream) or enduring (although fluctuating), as in services from a building.

Whereas the consumption of resources as inputs can have repercussions on society (the building of major commercially attractive projects can divert resources from less profitable needs), such repercussions are much more clearly identifiable in the out-puts. They are the effects/impacts from the project (see Ch. 5). As such they can be primary or first round (the building of dwellings), secondary or second round (the provision of work to the labour supply), and tertiary or third round (environmental pollution which results from the factories). Such impacts can be either *pecuniary*, in that they relate to *transfers* of wealth or income (associated financial), or technolog-ical (associated real), in that they ". . . alter the physical production possibilities of other producers or the satisfaction the consumers can get from given resources" (Prest & Turvey 1968, quoting McKean 1958: 68).

The transfers would not be counted in the efficiency criteria, since the CBA is con-cerned with real increases in wealth and income, which would leave society in the aggregate better off or worse off. But financial transfers do have a role in assessing the financial viability of the project to the public authority, and in influencing their decision in terms of equity (e.g. taxing the rich to pay the poor, or vice versa).

Externalities   A decision-taker in the private sector would on the whole confine his objectives and criteria to his *private* costs and benefits: the costs he *must* bear and the benefits for which he *can* charge, since these would fall upon the organization (fam-ily, company, etc.) for which he is responsible (Pigou 1932, Kapp 1950, 1963). He will tend to ignore those costs and benefits that would fall upon others. These are the "externalities" (with its synonyms of external effects, external economies or disecon-omies, spillovers, neighbourhood effects). Examples are, the benefits in consumption to passers by of a pleasantly landscaped garden; or the disbenefits in production caused by its generation of traffic that would give rise to congestion costs amongst other traffic on the roads. Where externalities exist a *divergence* arises between the private and social costs and benefits, with the externality being the total cost/benefit minus the private ones.

Although decision-takers in the public sector, in seeking value for money of *their* resources, have concerns similar to those in the private sector, they are more likely, because of their perception of their public responsibilities, to take the divergence into account in their objectives and criteria for decision. An authority building roads, for example, might not confine itself strictly to the output of the road investment but also take into account the repercussions on other modes of transportation serving the same traffic demands, such as buses or underground railways (Beesley & Foster 1965).

This basic concept of externalities outlined by Pigou in the early part of the cen-tury has since been extensively amplified in the literature (Kapp 1950, 1963, Turvey 1963, Little & Mirlees 1969, Mishan 1982). Their implications for public policy stem from the Pareto rule that, for *maximum social efficiency* in the allocation of resources, each activity in society should be adjusted to the amount at which the marginal social benefit equals the marginal social cost. Thus, where private and social costs/benefits diverge, decision-taking in the private sector, which tended to ignore externalities, would lead to a misallocation of resources. This instance of *market failure* gives rise to the need for public intervention, for which economists have suggested various models (Cheung 1978: 23–39).

- Pigou's input tax: restraining the owner's behaviour by such devices as taxation or compulsory compensation to the neighbours.
- Viner's output tax: the marginal balancing of output price and social output cost.
- Coase's transaction costs: weighing the gain for one party against the loss of another, in the recognition that social cost effects are mutual (restraining the producer by Pigouvian taxation or compensation inflicted harm on others), and policy should aim at avoiding the more serious harm.
- Cheung's aggregated balance: offset the cost of the action incurred by the performer against the sum of the values of its generated effects, whether these be included in the calculus of the promoter (contracted) or not (impacted).

Intervention along these lines is practised in everyday planning and environmental control, with a view to internalizing the externalities and making the polluter pay. Instances are ensuring that the promoter of a shopping centre provides car parking within his site and so avoids imposing costs directly on the public authority, and imposing conditions on the operations of a factory so that its effluent is treated before discharge into the air or water.

Thus, while the fairly simple Pigouvian concept has been considerably analyzed, criticized and amplified, it has not been rejected in principle, even by its contemporary trenchant critics, who have developed a critical onslaught on the concept (Coase 1960, Rowley et al. 1967, Cheung 1978, Littlechild 1986). To understand this, it is useful to digress on the nature of an externality and its treatment.

In functional terms, the externality is the consequence of any socioeconomic activity: in production (atmosphere pollution), distribution (noise from lorries and motorcars), exchange (deposit of litter from packages), or consumption (generation of radio or television noise). If those generating the impact are in the same unit (company, household, etc.) as those receiving it, then there will be internal adjustment (members of the family negotiating over noise levels within the one dwelling). But if the externality is on another proprietary unit (noise on the neighbour next door), there needs to be an identification of and discussion on mutual property rights.

In this last instance, if there is no recourse to higher authority (environmental controls, planning controls, taxation, the courts), then the property owners will negotiate and bargain to their mutual satisfaction, just as they would over the wish to allocate the costs and benefits in building a garden fence. Thus, from any starting point of the incidence of costs and benefits of different property owners, it would be practicable for the market through normal negotiation to resolve the externality issue. A major difficulty in practice is the high level of transaction costs which would impede the workings of the market. This can arise from various sources (Cheung 1978):

- acquiring information (the uncertainty of the cause and effects of discharge to the atmosphere)
- negotiating the price to be paid (where there are large numbers of people involved, as with football crowds or holiday-makers on the beach)
- means of charging for the use of the resources involved
- excluding "free riders" from consuming the resources they have neither provided nor paid for.

Thus, if the transaction costs can be reduced in scale, the operations of the market are facilitated, so that there is the possibility of relying on the market to resolve externalities; this, it is argued, would be simpler than leaving the task to government or the courts. Taken to its limit, where resources are "free" in the sense that they are held in common with common access, and cannot be charged for (e.g. under the seas), this approach requires the introduction of property ownership and the creation of a market as a prerequisite (Denman 1984).

This argument is reinforced by asserting not only the superiority of "market planning" in the allocation and distribution of resources, but by emphasizing the defects of government failure in dealing with market failure, which have been so dramatically exposed by the public-choice school (Niskanen 1973, Buchanan 1978). For one thing, although the tools of cost–benefit analysis needed to replace those of the market have been well developed, they have their trenchant critics and are still far from generally agreed (Self 1975). For another, the bureaucracy involved would be influenced by their own goals, which are independent or separate from those of their political masters and the electorate, leading them to decisions that are not necessarily in the public interest (Niskanen 1973). For another, politicians will not necessarily follow the advice of the analysts and bureaucrats, but will reach a "political decision" that could suffer from "short-termism" in having too much regard to anticipated voter behaviour (Downs 1957); and, even if they earnestly wish to make a decision on behalf of the community, in contemporary democracy they would find it difficult to formulate appropriate rules for so doing (Arrow 1967).

All this might be thought to lead to the conclusion that the private property / public choice attack has, to the satisfaction of their trenchant proponents, removed externalities from all reckoning. However, this is not so. One critic, Littechild (1986: 104), accepts that externalities exist but maintains that, since they arise from "loopholes in the system of private ownership", it is better to seek the private property remedy and not regulation. And another (Cheung 1978: 67) states "my criticism of the analysis in the Pigouvian tradition should not be construed as an argument against government control".

What seems to remain from the controversy is this. Externalities do exist, almost by definition. But what constitutes an externality is a moving feast in particular cases, depending on property rights and the result of government intervention. Opposing views on their identification fall back on the balance between the market and government intervention. But it remains true that both the analysis of externalities and of the prescription for their resolution require very careful treatment in practice (Turvey 1963).

Box 10: Measurement and valuation of costs and benefits
The measurements of the inputs and outputs in a project relate to *physical* resources, being once for all (capital) or continuing (operation, maintenance, etc.). Together, these are *costs in use* (Stone 1983). Given such measurements for each of the options, the differences between them can be identified and compared, using as a common datum the consequences of the "do nothing" or, more realistically, the "do minimum"

option. This produces different conclusions from a "before and after comparison", since both input or output could change during the life of the do nothing option.

For the comparison, an initial step is the measurement of physical resources involved. Depending on the data available, one or more of the following recognized measurement scales can be used (Kaplan 1964):

- *Nominal.* Numbers are assigned only as labels or names, which distinguish the individual from the generality. Examples are the numbers on credit cards, bank accounts, home addresses, names.
- *Partially ordered.* Within the generality some relatively homogeneous subset is selected, to which the scale assigns a number that distinguishes the subset from the others. The ordering is achieved by *binary* means (yes/no, better, larger, more attractive). Examples are the ordering of hotels in terms of numbers of stars, or people in terms of sex, age and family size.
- *Ordinal scale.* The subsets are further distinguished by placing them in some order on selected criteria, with any one number for that order fixing an upper or lower bound for all the other numbers to be assigned. In this the *difference* between the categories is decided by reference to some standard, but not the magnitude of the differences. An example is the hardness of minerals, measured by the capacity to produce a scratch in another mineral; or the heights of people (tall by reference to others).
- *Ordered metric.* The ordinal scale is taken further by measuring the *intervals* between the differing groups, as in the categorization of chess players, from beginners to Grand Masters, on the presumption that a player of a given rank can usually defeat a player of any lower rank.
- *Interval scale.* The differences between the categories are subject to quantitative as well as qualitative measurement, each being of equal dimension from some *arbitrary origin.* A familiar example here is the thermometer, in which the intervals in temperature are equal on the arbitrary assignment of freezing (0°C) to boiling (100°C).
- *Ratio scale.* Here there are also equal intervals, but the zero point is no longer chosen arbitrarily but in relation to some fixed point. This provides *absolute measurement* and *absolute comparisons,* as in the height of land above sea level.

These then are the array of scales available for a measurement. They range from the qualitative, giving an opportunity for ordering (*partially ordered* or *ordinal*) to the quantitative, giving order plus arithmetic dimension (*ordered metric, interval scale* of *ratio scale*). Clearly, the cardinal scales give more precise information on attributes than the ordinal. Accordingly, it is useful to be able to translate the ordinal to the cardinal. This is achieved by a variety of techniques associated with multi-criteria analysis, which can be applied within a wide range of disciplines. For example where costs and value statements are used to derive quantitative inferences via cross tabulations (Nikamp 1975, 1982, Nikamp et al. 1991).

For economic analysis it is necessary to go beyond physical measurement to the *utility* or *disutility* to *people* of the resources. For this purpose, in order to avoid comparing apples and pears, it is necessary to express the varying utilities and disutilities by the "measuring rod of money", i.e. their costs and value in money terms, which we call *valuation*.

In this, a starting point is market data. Inputs (costs) can be valued in terms of the resources required in market prices. Outputs (benefits) can be valued in terms of the willingness of people to pay for them (WTP) or to accept payment for relinquishing them (WTA). These in practice are not identical and WTA tends to exceed WTP.

Such valuations traditionally relate to the perceived benefits of *use* or *participation*, leading to preferences. As such, the perceptions can rely on only a limited range of experience and prediction, in particular where the valuation has been to extend WTP/WTA to the following (Pearce et al. 1989):

- option value: the probability that the non-user, direct or indirect, will wish to make use of the utility at a later date
- bequest value: to preserve the utility for the benefit of descendants
- existence value: to ensure that the value will continue to exist, even if no actual intended future use is in mind.

However, such market indicators are only a beginning. The *true* cost of using the resources for the construction and operation of the project is its *opportunity cost*, namely the value that the resources would have produced in an alternative use, for these are the benefits that are foregone (Schofield 1987: 44). Likewise benefits are valued at the opportunity cost of the payments that are made for them, since that is a measure of what is foregone by the consumer.

> Thus, the costs of a project that are relevant for project appraisal are those that would be incurred if the project were undertaken but would not be incurred if it were not . . . similarly the revenue of a project is the extra amount which would be earned if the project were undertaken rather than not undertaken. (Sugden & Williams 1978: 13).

But even this may not be the full measurement of cost and value. Willingness to pay as expressed in the market does not reveal "consumers surplus", namely the amount that consumers are willing to pay for a product over the price they are required to pay (Schofield 1987: 38, 44–52). Alternatively, consumers surplus may be seen as compensation required above product price in the event that the consumer is denied the product. There are parallel considerations on the cost side. The producer's willingness to supply is not necessarily at market price; where he would be willing to supply at a lower price he can be said to enjoy "producers surplus" when he sells at market price.

As indicated above, the approach to valuation of costs and benefits starts with prices in the market. But there are many reasons why this starting point is not reliable. First, in economic theory the market expression of value generally assumes that the market is "perfect", which rarely exists in the private sector and certainly not in the public (Mishan 1982: 22–3); or the project may be of such a size that the very quantity of its output will affect market prices on both the supply and demand side; or there are the prices that, even as registered in the market, need to be adjusted (since

they reflect subsidies or taxation which represent financial transfers and not economic value. Secondly, there may be no market for the goods and services in question, since these are being offered by a public authority at zero price (e.g. an urban park); or they are public goods which, if supplied to one person, does not reduce their availability to anyone else (e.g. a public road; Pearce 1983a: 360); or there will be the externalities which are just not exchanged in the market at all. In this second category, the goods and services may be described as (Sinden & Worrell 1979: 10–12):
- *non-market*, where the goods and services are not exchanged in organized markets at regular market prices, or
- *extra-market*, where the goods and services are exchanged, but not in a free market for monetary prices, as in museum charges, or
- *indirect*, where the goods and services are "exchanged", but not through the market (externalities, spillovers, etc.).

In all these instances and others, given the desirability of providing some kind of money price, to facilitate the comparison of costs and benefits in the common measuring rod of money it is necessary to devise some surrogate way of assessing WTP or WTA. It is here that the techniques of "shadow, accounting or social" prices have grown up in cost–benefit analysis (Sugden & Williams 1978: ch. 8; Schofield 1987 ch. 5). The hypothetical values are explored through surrogates, derived for example by travel costs approaches, observed behaviour, hedonic prices, contingent or stated preference valuation (Pearce et al. 1989).

All the above conceptual and practical difficulties in the measurement and valuation of benefits and costs have provided obstacles in the application of the analysis to practical problems, in the situation where the built-in drive in economics is to value in money terms in order to produce sound positive or normative conclusions. The difficulties have been attacked in theory and application, with results that have been impressive and also invaluable in facilitating the use and application of the analysis (Sinden & Worrell 1979, Pearce et al. 1983, 1993). Of necessity, these essays in "measuring the unmeasurable" have ventured into fields traditionally thought of as outside economics, such as the value of life (Nijkamp et al. 1986). As such, they have attracted spirited criticism for attempting just that (Self 1970), and in spirited counter-attack (Williams 1972).

In the valuations referred to, the theory of welfare economics, on which cost–benefit analysis is grounded, has the assumption that valuation in cost–benefit analysis aims at an aggregation of the valuations of those individuals who are affected. But the decision-taker, in reaching his conclusion for choice can, and often does, depart from this assumption. He could take the view that the aggregation of the individual valuations does not reflect the collective interests (as judged by himself, his party or advisors) of the community on whose behalf he is taking the decision, or of future generations whose welfare he must take into account. This we call "social/collective" valuation as opposed to the aggregation of individual valuations.

This distinction is provided for in cost–benefit analysis through the principles of "weighting" or "social pricing", which is an adjustment of the aggregate individual valuations to take account of such sociopolitical issues. Apart from the equity issues introduced above, which can also be handled by "weighting", examples are the merits of different groups (children as against adults), the conservation of natural resources or the cultural built heritage against erosion (sustainable development), strengthening national defence as against consumption (guns before butter), and savings as against consumption (developing countries).

Box 11: Taking account of time

The context   So far the analysis has been discussed as though the costs and benefits were concurrent and instantaneous. Whereas this could be true of benefits in purchase for immediate consumption (an ice-cream from a street vendor), it is not so for any development project, with its long project cycle for preparation and implementation, and long life over which its services are made available. This brings time into the typical development project cycle (Baum & Tolbert 1985).

The cycle begins with A, the identification of the need for the project. At B the project has been matured, by the typical team of developmental skills, including the economist. The project is then implemented through the development process and completed at C. It is during B–C that the outlay on capital costs is completed, following study costs between A and B. Following completion of C, the operating costs continue over the life of a project and the benefits are obtained from the services, until the end of the life-cycle at D, when the use is terminated (involving decommissioning costs and perhaps scrap values).

The application   From the development project life-cycle, it follows that time enters into the analysis in at least six ways:
- *"Without" and "with" comparison:* For the former it is necessary to establish the datum in the "do nothing" or "do minimum" situation, for the valuation of the costs and benefits that would arise over the period of the project life-cycle. In this without situation, there could be benefits in the absence of a project (continuation of the open rural scene or of life in a village which is to be immersed under a reservoir) and there could be costs (the maintenance of an ageing, obsolescent structure would be avoided if it were replaced). Furthermore, such costs and benefits would not remain constant but could vary in real terms; for example the rising value for an unimpeded view because of its increasing scarcity in a rapidly urbanizing area, or increasing maintenance costs for an ageing structure.
- *Double counting:* Whereas initial costs are deployed in the "construction" phase, the benefits emerge thereafter as a stream, as first, second and third round, and as either pecuniary or technological. It is these benefits that can lead in practice to double counting, that is counting a single element of benefit more than once in the analysis, so inflating the benefit stream.

    The rules to avoid double counting are not readily formulated; thus, what is required is awareness by the analyst of the pitfalls and how to avoid them. For example, an improved transportation link will reduce the costs of transportation of travellers (a benefit). From the savings, the travellers could afford increased rents or purchase prices for property, whose accessibility has been improved, for which they will in due course be called upon by the owners of property. This results in a transfer of the travel savings to the owners of the property. The resulting benefits can be measured by either the cost savings, or increased land and property prices. It would be incorrect to employ both, unless, for example, there are economies of scale.

- *Price level:* One of the few certainties in predicting costs and benefits is that the prices will not remain constant over the considerable period necessarily involved. The expected changes are of two kinds (Sugden & Williams 1978):
  - general level: where *all* prices are expected to change in much the same ratios, as in inflation. Here, the future price changes can be ignored, since the relationship between costs and benefits can be assumed to be constant. The usual practice is to take the price level at the point of appraisal for decision, and assume "constant prices";
  - relative level: where changes can be expected between the *relative* level of major elements in the project, either between costs and benefits as a whole or between different costs or benefits. For example, significant improvements in the technology of construction could bring down the *real* costs of the resource input; or expected scarcity of certain of the outputs (for example, demolition of dwellings in a town, where replacement would be difficult owing to severe land limitation) will cause a *real* increase in house prices. Forecasts of these differentials are difficult but must be attempted in some way.
- *Discounting:* Since the analysis is of necessity prepared in advance of the flow of costs and benefits, either in an initial appraisal at A in the project cycle, or more firmly at the point of decision at B, it is necessary to treat with consistency the flow of costs and benefits in the future. These differ in timing (the costs being incurred first for achieving the benefits), in character (the costs between B and C would tend to be capital and between C and D operating), and in uniformity (the benefits between C and D are unlikely to be constant in amount). Thus, it is necessary to consider these costs and benefits in the future beyond a particular point in time, be it at A or B (on appraisal or decision), or at D (on termination of the project) (Pearce 1983b: ch. 4). The general presumption is that, even if price levels can be expected to be constant, any individual would prefer to spend a particular sum of money tomorrow rather than today, and have its benefits today rather than tomorrow. This being so, there needs to be some adjustment for the asymmetry in time in flow of costs and benefits. The question then arises: how heavy (i.e. at what annual rate) should the discounting be?

For individuals, the discount rate is generally taken to be the observable real rates of interest that prevail in the market for the deferment of consumption, that is investment of savings. The discount rate is the amount that should be invested today to receive £1 in the future, which is the reciprocal of the accumulation of £1 from today which is invested at that particular interest rate (Pearce 1983: 39). A discount rate established in this way gets around the effect of inflation on such expectations, since the market is presumed to fix the rates against a background of expected inflation: interest at 10% in a year of 5% inflation produces a 5% real discount rate.

For society, the rationale for the rate is not so clear, and therefore for cost–benefit analysis in general. Here there is much more controversy (Pearce 1983b: ch. 4; Sugden & Williams 1987: ch. 15; Lichfield 1988c: ch. 14). Should the rate be equivalent to that in the private sector, in order not to distort the comparison between returns on the use of private and public sector money? Should it be the rate at which government borrows money, in order not to distort the relation between different forms of government investment? Should it recognize the distinction that public sector projects are orientated towards investments with limited commercial attractiveness, and therefore should be discounted at a lower rate than in the market? Should it be ignored on the proposition that discounting differentiates between present and future generations, and that these should be treated equally? Or should the reverse stand be taken, that it is the current generation which is bearing the costs, whereas the future generation will receive the bulk of the benefits or disbenefits?

- *Risk and uncertainty:* In predicting future costs and benefits there will necessarily be assumptions about the future. These tend to be based upon another assumption: that the future can be known with *certainty* (for example, July weather in a Mediterranean country) or with *uncertainty* (July weather in Britain). Such assumptions, and accordingly the predictions, can be readily falsified by events. Accordingly, at the point of appraisal for decision, the assumptions may not be so robust as to be left untested by risk analysis. The test requires asking a series of "What if?" possibilities and building into the appraisal an allowance for uncertainty. This is done by offering a range of outcomes in costs and benefits on varying assumptions about the future state (sensitivity analysis), so that the decision-taker can choose according to the state he visualizes or will gamble on; at least he will then know whether in fact the outcome of the analysis is sensitive to possible future variations in the states and, if so, the consequences of choosing what proves to be the wrong state (Sugden & Williams 1978: ch. 15). In order to supply more than guesses to varying states, and to be able to offer his "best guess estimate", the analyst has two approaches (Pearce 1983b: ch. 6):
  - *Risk*: where the *probability* can be predicted that the benefit or cost will take on particular values. The prediction could be *objective*, that is based upon data of past events (e.g. river flooding); or, where this is not possible, *subjective*, that is an assessment by the analyst, some other expert or by a group on the Delphi method. Whatever the origin of the probability estimate, it is practicable to present the expected values of the costs and benefits in some form of *probability distribution*. And from these it is possible to infer an *expected* value criterion, which is the weighted average of the possible outcomes, the weight representing the probability.
  - *Uncertainty*: where the probabilities cannot be predicted, and accordingly no probability distribution or expected value criteria can be derived. Here there must be reliance upon different assumptions (with no back-up of predictions

about the future state), which are presented in a *pay-off matrix* comparing uncertainties (for example, benefits from a road project against changes in GNP and therefore vehicle ownership). It is to this array of *possibilities* that a decision-maker can apply his criteria for choice, which will reflect his own attitude to future uncertainties. Some examples are:

- maximax: conveying considerable optimism (the largest or the maximum pay-off)
- maximin: conveying pessimism (the largest of the minimum pay-offs)
- index of pessimism: probability related to the worst outcomes
- Laplace: affording equal probability to the outcomes
- minimax regret: minimizing the regret from the wrong choice.

- *Irreversibility or not?* Project decisions have differing degrees of reversibility/ irreversibility, and thereby differing costs of remedying past decisions, should the need arise. The decision to build a surface carpark on open land can readily be reversed (at clearance costs), perhaps with considerable benefit in the new higher land value achieved. At the other extreme, the use of open land for capital-intensive development (urban expansion) is reversible only at heavy costs of replacement; this is particularly so in the decommissioning of nuclear power stations.

In such project decisions, the replacement cost of the land use in question comes into the appraisal. The loss in the value of open land for agriculture can be replaced by growing extra food elsewhere (Wibberley 1959). But when the open land has particularly high amenity or recreation value (wilderness with wildlife, etc.), its use cannot be replaced, so that the potential loss from irreversibility is much greater. In such a situation, the "opportunity cost" approach is used to compare the benefits of using the wilderness in perpetuity against the benefits forgone from not having the proposed project itself (Krutilla & Eckstein 1958). In this, options for the development of the site in question will need to be evaluated. If one option is locating the development instead on a site inferior for the purpose, and thereby of less occupation and investment value, it is the diminution of the project value which is the opportunity cost.

Box 12: Evaluation

Here the conclusions from the preceding analysis (Boxes 8–11) are brought together in the *evaluation*, which is the culmination. In essence, the comparison is made between the aggregate preferences for each of the options, in terms of the measured and valued benefits minus costs, to which weightings may have been applied to reflect equity, in order to provide the basis for advice and recommendations to the decision-takers.

CHAPTER SEVEN

The generic method of community impact evaluation applied to projects

7.1 What is a project?

In its everyday use, a project is a plan, design or scheme for doing something in the future. Within this array, the term is more tightly defined for specific purposes. To the Major Projects Association (Morris & Hough 1987: 3) "a project is an undertaking to achieve a specified objective, defined usually in terms of technical performance, budget and schedule. There are basically two kinds of project: those that are complete in themselves, like an oil platform or a tunnel, those that represent a series, or programme, of products or projects, like an aircraft or an aid programme". To the World Bank, in their financing programmes since around 1950 (Baum & Tolbert 1985: 8) ". . . a project is taken to be a discrete package of investments, policy measures, and institutional and other actions, designed to achieve a specific development objective (or set of objectives) within the designated period". Within this concept they also see a range of project types, such as (op. cit.: 8): "capital investment in civil works, equipment, or both (the so called bricks and mortar of the project); provision of services for design and engineering, supervision of construction, and improvement of operations and maintenance; strengthening of local institutions concerned with implementing and operating the project, including the training of local managers and staff; improvements in policies – such as those on pricing, subsidies, and cost recovery – that affect project performance and the relationship of the project both to the sector in which it falls and to broader national development objectives; a plan for implementing the above activities to achieve a project's objectives within a given time".

In this book, a *project* is any planned activity that uses resources to produce output, with the intention of favourably satisfying human needs, wants or desires.[1] A project could also relate to the abandonment of resources in disinvestment or decommissioning (e.g. closing down a coal mine, steel works, department store). In the planning field, a project may comprise one or more of the following: *physical development* (house, factory, school, road) in which the developer/entrepreneur/promoter

---

1. See Lichfield (1988c: 16) for a discussion on the complexities behind this simple formulation.

finances the use of the factors of development production (land, the building industry, investment resources) in order to create the finished product, as a contribution to
production, distribution or consumption; *activity,* where the starting point is established physical development (building, road, etc.) in which an operator changes (in
quantity or kind) the *activities,* with no or only modest investment of resources
(an instance is a change in frequency of bus services on an existing road system);
linkage, where the *activities* on one or more sites are changed as a result of a new traffic or infrastructure link (new drainage, telecommunications, etc.); *decommissioning,*
where *any* of the three kinds just mentioned are abandoned. Here, the investment
resources required could be minimal (closing a railway line), or heavy (closing down
and clearing away a coal mine, steel works or nuclear power station).

By definition, any project and its associated activity must be carried out on a particular project site, usually defined within actual or potential ownership boundaries,
i.e. "on site". Here there will be the use of resources (at varying levels) with which
are associated the human activities/uses. Clearly, no project can fail to have repercussions "off site", be they in terms of the linkages with the outside world or with more
diffuse effects and impacts. This enables us to define the totality of the project elements. The onsite/offsite distinction is relevant to the concept of externalities (see Ch.
6). These may arise *within* the project site (traffic noise from a shopping centre carpark to a school) or *outside* (traffic generation onto a major road). Thus, they could
arise functionally *within* one ownership, even though the costs and benefits would all
be *internal.*

7.2 Project planning in the project life-cycle

To use the project input in order to produce the intended output, any project must be
planned, in the sense of applying forethought. This planning endures from inception
of the project, through its implementation via construction, its operation and management following completion, when it produces the outputs that were envisaged in setting it up. In appropriate projects it will extend also to the decommissioning (e.g. a
power station; Lichfield & Darin-Drabkin 1980). To the World Bank this is the
"project cycle": identification, preparation, appraisal, implementation and evaluation
(op. cit.: 332). To the Major Projects Association it is the "project life-cycle" where
"every project, no matter of what kind or for what duration, essentially follows the
activity sequence of pre-feasibility/feasibility, design and contract negotiation,
implementation, hand-over and in service support" (Morris & Hough: 3)

The nature of this cycle/process will vary enormously according to the scale, complexity and purpose of the project. It requires very little time and professional help to
plan the use of leisure time over the weekend. It takes much time, and payment to a
large array of professions, to plan for the building of a road, or a motor car or aeroplane,
or for the closing down of a mine or nuclear power station. By the same token, the
development process/project cycle might involve as decision-taker only the individual concerned (as in a family leisure time project) or a complex array of governmental
institutions, interest groups and public at large, as in the building of a motorway.

7.3 The nature of project planning

Project planning falls within the general plan-making process described above (§1.6.2). It is aimed at the production of a specific output for detailed design and construction, within the project development process/cycle (§7.2 above). In this it differs from area planning, which is visualized as a framework for the continuing process of planning and implementation of development projects, and indeed individual buildings. There is no standard process for project design; that will vary from project to project. But, as with plan-making generally, it is useful to devise a model for heuristic purposes, which will not necessarily be followed sequentially in practice (Fig. 7.1). The model has four stages that relate to the general plan-making model in Figure 1.4, but are more site specific. In the four stages, there are eight studies and fourteen tasks, in each of which there could be consultation. We now describe the tasks in turn, which would follow the client brief.

Figure 7.1 Simplified project planning model.

Stages	Studies	Tasks
A. Analytical	1. Site context study	(1) (2)
	2. Site study	(3) (4) (5) (6)
	3. Demand study	(7)
B. Planning and design	4. Planning policies and brief	(8)
	5. Design	(9) (10)
C. Feasibility and evaluation	6. Feasibility	(11)
	7. Evaluation	(12) (13)
D. Conclusion	8. Development brief	(14)

Note: the tasks in the third column are described in the text.

Analytical

Site context study
1. Study the area surrounding the project area, and carry out consultations for its bearing on the project design, for example in relation to utilities (drainage and water supply), access and communications, function (journey to work or shopping catchment areas), visual aspects (views from or to the site), and landscaping (surrounding and into the site).
2. Study the likely changes in the study area, not only those that are known but also those that are likely to happen. For this there could be guidance from the relevant authorities in respect of proposals for new roads or utility infrastructure, or in urban and regional plans. These, together with the planning authority's views of the development of the site itself, are typically contained in a *planning brief* (Ch. 1). But in the absence of specific guidance in such plans and briefs, the potential for planned change in the surrounding area should not be ignored. It becomes necessary at least to speculate on such possible future changes, and to register assumptions on them as a basis for the project design. It could be necessary for

such speculation to devise a tentative plan for the surroundings as a basis for discussion with the planning authorities, in so far as they have not yet formulated
views of their own.

Site study
3. Identify the specific site for the project, within the context of the surrounding area.
4. Identify the current situation on the site, in terms of built-up parcels and open land,
   and for each identify the current usage and stages of obsolescence.[2]
5. From the preceding, identify what is to be retained (hard) and the land/buildings
   available for change (soft).
6. Estimate the capacity of this land, in site area and building volume at assumed
   densities.

Demand study
7. Forecast the demand for the various public and private uses. The techniques will
   vary for the different uses according to the state of the art. And there will be some
   contribution by use of statistics, direct enquiry of potential uses, observation in the
   market.

Planning and design
Planning policies and briefs
8. Prepare assessment of official plans, policies, planning briefs, views of consultees
   in order to provide planning problems, opportunities and constraints. Within this
   would come the negotiations with the authorities on planning gain/obligations
   (Ch. 15).

Design
9. Analyze site characteristics, including in relation to surrounding land, as a basis
   for the design of options.
10. Generate the options, having regard to constraints and opportunities from the
    preceding, in land use, layout, access, built form views, landscaping. Within this
    would be undertaken the environmental assessment, be it required under the Reg-
    ulations or not (see Ch. 17), in order to predict the effects from the proposals on
    the environment, which need to be ameliorated or enhanced. Although the project
    plan/design must result in proposals within the site boundaries (on site), it recog-
    nizes that the future of the site must be visualized within its context (off site). The
    latter will have constituents as follows:
    • linkages: to utilities, transport, telecommunications and urban services
    • effects/impacts: economic, social, cultural, linkage, environmental, risk/
      hazard.

---

2. For an analysis of obsolescence see Lichfield (1988c: 21–5).

Feasibility and evaluation

Feasibility

11.Carry out feasibility studies of the options in terms, for example, of physical site conditions, financial cost and return, public acceptability, political acceptability, planning permissions.

Evaluation

12.Prepare a matrix for the options having a reasonable degree of feasibility in the above terms.

13.Evaluate for choice and decision.

Conclusion

Development brief

14.Prepare the development brief for the chosen option, from the above, comprising programme and illustrative drawings for the development content on the site.

7.4 What is project appraisal/assessment/evaluation?

This comprehensive description of a project and its repercussions can now be used to illuminate the role of the project assessment, appraisal or evaluation (§3.3.7). This, in its simplest terms, is helping to judge whether a project is a worthwhile investment. The criteria for this judgement are varied. To some it would be profitability; to others it would be sensible to build a monument to national heroes; to others to attract prestige and influence voting; to others to carry out research for no clear immediate return beyond the advancement of basic knowledge. Here we wish to concentrate on the criterion most commonly associated with the cost–benefit family of methods in project appraisal: the comparison of the predicted outputs with the needed inputs, in terms of benefits and cost, in order to answer the three standard questions (§4.1 above):

Such questions can be asked, it is seen from §4.3 above, by an array of decision-makers (stakeholders), who would use the member of the cost–benefit family of methods most suited to their own decision rules. Some may use more than one.

In the remainder of this chapter we concentrate on one member of the cost–benefit family, namely, *community impact analysis/evaluation*. We do so by devising a *generic* method which can be adapted as appropriate to specific projects.

7.5 The generic method of project evaluation by CIA

7.5.1 Overview

The generic method is shown in Table 7.1 and an algorithm in Figure 5.2 of:

(a) 10 discrete steps A–J comprising project description, analysis & decision, and

(b) 16 boxes, and their sub-boxes, giving tasks identified with a specific step, which represent intermediate stages for the data in the evaluation process.

On (a), the essential purpose of the sequence of steps is to provide a method of reaching conclusions for the evaluation report, following consultation (I), by way of answers to the questions raised in the decision framework (E). In order to achieve this purpose, the steps are taken in sequence. First comes the project description (A–E) which starts with the planning process involved in the project (A), then a description of the projects in the system (B) together with their options (C). These will introduce the visualized change in the system (D), leading to the decision framework (E) in respect of which the analysis is made and conclusions drawn (F–H). It is this analysis and conclusions that are presented for consultation leading to the evaluation report and its recommendations (I). It is this that is presented to the decision-takers for choice and decision and then decision communication to all the affected parties (J). But not all the steps are needed in all situations. Some can be omitted in order to simplify the process, to provide a simplified "short cut" or "quick and dirty" approach that still follows the logical sequence of the algorithm (see Ch. 8).

On (b), within each of the boxes and sub-boxes is the title that indicates the content of the data required in relation to the step to which it is allocated, data that would be in the form of textual description, tables, plans and graphs. The step and the box can be linked by numbering (e.g. C7), with divisions being described, for example, as C7.1, C7.2, and subdivisions as C7.1.1.

In the process described in (a), the content will need to be presented in full. For some of the boxes this will require certain input data in advance. This would be provided in "background tables" or figures (bt). Some background tables, which are distinguished as "demonstration tables", are actual examples from relevant case studies (dt). Both background and demonstration tables will be introduced in the following presentation of the method. For ease of reference, a list is presented in Table 7.1.

Table 7.1 The generic method showing relation of steps, boxes, and tables and figures.

	Steps	Box	Title	Background (bt)	Demonstration (dt)
Project	A	1	Planning process in the project		
	B	2	Project description		
		3	Project options description	F7.2	T7.3, T7.4
		4	Pre-project system		
		5	Post-project system		
	C	6	Plan variable	T7.5	
		7	Specification of project option	T7.7	T7.6
	D	8	System change	T7.8	
	E	9	Framework for decision		
Analysis	F	10	Effect assessment	T7.3	T7.9
	G	11	Impact evaluation	T7.4	T7.10, T7.11, T7.12
Decision	H	12	Decision analysis		T7.13, T7.14
	I	13–14	Evaluation report		
	J	15–16	Decision communication		

Figure 7.2   Algorithm for generic method of CIA applied to projects

1 Plan-making implementation and operation process for project

2 Project description

2.1 Physical	2.2 Activity	2.3 Linkage	2.4 Decommission

3 Project options description

3.1 X (datum)	3.2 Y	3.3 Z	3.4 N

6 Plan variables

6.1 Use				6.2 Implementation		6.3 Operation	
6.1.1 A	6.1.2 B	6.1.3 C	6.1.4 D	6.2.1 E	6.2.2 F	6.2.3 G	6.2.4 H

7 Specification of project option by plan variable

7.1 Options				7.2 Specification			
7.1.1 X (datum)	7.1.2 Y	7.1.3 Z	7.1.4 N	7.2.1 X (datum)	7.2.2 Y	7.2.3 Z	7.2.4 N

8 Changes in the urban and regional system through project option

8.1 System element	8.2 Datum		8.3 Change through project				
	8.2.1 Current	8.2.2 Without project	8.3.1 Blight	8.3.2 Displaced	8.3.3 Retained	8.3.4 Construction	8.3.5 Completed

9 Questions to be answered in the evaluation

10 Effect assessment of options by plan variable

10.1 Plan variable			10.2 Community sector			10.3 Summary of effect change on affected community sector by option	10.4 Size of effect		10.5 Characteristics of effect						10.6 Community affected
10.1.1 Use	10.1.2 Imple-mentation	10.1.3 Mgt	10.2.1 Item	10.2.2 Descrip-tion	10.2.3 No.		10.4.1 Unit	10.4.2 Amount	10.5.1 Kind	10.5.2 Type	10.5.3 Timing	10.5.4 Probable	10.5.5 Rev.	10.5.6 Signifi-cant	

11 Impact evaluation of options by plan variable

11.1 Community sector participating			11.2 Plan variables			11.3 Summary of impact change on impacted sector	11.4 Difference in impacts between options				11.5 Sectoral objectives for impacts	11.6 Difference in net benefit of options				11.7 Prefer by sector on options			11.8 Ranking of options by sector
11.1.1 Item	11.1.2 Sector	11.1.3 No.	11.2.1 Use	11.2.2 Impact	11.2.3 Op.		11.4.1	11.4.2	11.4.3	11.4.4		11.6.1 X (datum)	11.6.2 Y	11.6.3 Z	11.6.4 N	11.7.1 Line	11.7.2 Subsec.	11.7.3 Section	

12 Summary of sectoral preferences and ranking

12.1 Community sectors			12.2 Preference by sector				12.3 Ranking of options			
12.1.1 Item	12.1.2 Description	12.1.3 No.	12.2.1 X (datum)	12.2.2 Y	12.2.3 Z	12.2.4 N	12.3.1 X (datum)	12.3.2 Y	12.3.3 Z	12.3.4 N

13 Consultation bargain negotiation

14 Evaluation report

15 Choice decision

16 Decision communication

Notes: Project description: Steps A–E, Boxes 1–9.
Analysis: Steps F–G, Boxes 10–11.
Decision: Steps H–I, Boxes 12–16.

From this description it will be seen that the process builds up in a hierarchical manner, with information provided in particular boxes being used in later ones, so providing a logical sequence. To maintain this sequence it is necessary for the content of each box to be described in map, plan and written form, so that the completed evaluation is placed on record. The presentation and description will be confined in this chapter to the technique within the overall method, without attempting to give the rationale in theory and principle. This is left for Chapter 10.

7.5.2 The generic method in Figure 7.2
Steps A–E: Project description

A Planning process in project   BOX 1: As seen above, project evaluation is a step in the plan-making process to aid choice and decision on options, as a preliminary to action (implementation). Accordingly, it is necessary from the outset to be clear as to the particular nature of the planning, implementation, operation (which includes management) and decommissioning process to be applied from the variety that are available to the particular project (§7.3).

B Project description   BOX 2: Because of the great variety of projects, there can be no standardized description. Here an indication is given relating to the four different kinds of project introduced above (§7.1), which can be used in combination.
1. *Physical development (e.g. new school)*
    • project site: boundary; existing infrastructure; buildings and onsite activity; location in relation to the remainder of the urban system;
    • tenure: the current owners and occupiers, in their various relationships;
    • project related site: boundary of land on which related works will necessarily be involved (e.g. road connection for generated traffic);
    • proposed development: new buildings and works, with area and new activities;
    • development process: a description of the way in which the change on the site would be brought about, and its timing and stages.
2. *Activity (e.g. introduction of night shift)*
    • project site: as in (1);
    • project related site: as in (1), relating not to works but activities (e.g. use of offsite carpark at night);
    • proposed activity: change in activity related to "project site" above;
    • operational process: change in manner of operation (e.g. introduction of new machinery);
    • institutional: change in operating agency (e.g. a large organization taking over a small one).
3. *Linkage (e.g. introduction of new road)*
    • project site: the streets in question and their adjoining/nearby buildings;
    • project related site: the other streets affected;
    • proposed linkage: the new traffic system;

- operational process: how the change would be brought about, its timing and stages;
- institutional: any change in agency.

4. *Decommissioning*
   - The project description would vary according to which of the three kinds of preceding projects is to be decommissioned. In illustration the following would apply to type 1, physical development:
     - project site: as in (1);
     - project related site: as in (1)
     - development visualized at the end of the project life: describe as far as practicable, distinguishing between those elements that will still be retained and those which will be abandoned, demolished, etc.;
     - process of decommissioning: the way in which the development would be taken out of the system, including removal from site of contaminated material, etc.

BOX 3: Data relating to the project options will be available from the project planning process (§7.3). They need to be presented in a systematic form as a basis for a precise comparison of the options. Since the differences between options are critical for the evaluation, they need to be clearly grasped. In illustration, Box 3 shows four options (X, Y, Z, N), which typically require the analysis illustrated in Table 7.2 by the system elements, on site and off site, shown in column 1. Column 2 shows the current situation on and off site, in relation to each of the project elements that are pertinent to the option. Columns 3–6 show the absolute amounts or the difference in the content of the pertinent elements for each of the project options, with one as datum (X), in column 3. For this purpose it is useful to determine the elements in the option that are common and those that are variations. In this regard only significant variations should be noted, significance being defined as of sufficient size to make possible a meaningful comparison, after allowing for the inevitable margins of error in the particular evaluation process being pursued.

In all cases data will be needed for the comparison. This will require the preparation of background. Tables 7.3 and 7.4 are examples, relating respectively to physical and financial data. For the latter not all the data were measured. Therefore, symbols were used to indicate the relative change and its direction from the datum: capital letters for capital sums and lower case for annual/recurring (see §10.4.1).

BOXES 4–5: A project is seen as a specific intervention into the current urban & regional system, which will in itself create changes to the future (post-project) system. The elements in that system are shown in Table 7.2 above. Fortunately for our purpose there is no need to describe the pre-project and post-project systems in their complexity (that is the "baseline" and "scenario" respectively), for it is only the *changes* to that system (Box 8), flowing from the project (Box 3), that are to be evaluated. The post-project change will emerge from Box 10.3 below.

Table 7.2   Project options description (based on Table 1.3).

System elements	Project options				
	Current	X (datum)	Y	Z	N
1	2	3	4	5	6
1. ON SITE					
(a) Population					
Living					
Working					
Visiting					
(b) Physical fabric					
Land					
Natural resources					
Utilities					
Transportation					
Telecommunications					
Buildings and sites					
Open space					
(c) Activities/uses					
Residential					
Production					
Distribution					
Consumption					
2. OFF SITE					
(a) Linkages					
Utilities					
Transport					
Telecommunications					
Urban services					
(b) Effects/impacts					
Economic					
Social					
Cultural					
Linkage					
Environmental					
Risk/hazard					

C Options specification by plan variables

Box 6: Since we need to define clearly the change to be introduced into the system, and since in any project the change will come about from particular system elements in the process (called plan variables) but not others, we need to articulate what these are, individually or in groups. The approach is illustrated in Box 6 and Table 7.5.

The variables can be of four kinds: implementation (comprising development and construction), use on completion, operation of those uses, and perhaps decommissioning. *Use variables* relate to the project elements, as in Table 7.2 on site. However, it is not the *plan* itself that will in fact introduce the change but the *implementation* (development, construction) of what it proposes. These could be uniform as between the options. But in certain projects there will be *implementation* options, for example

Table 7.3 Comparison of available measured material for project options: Ipswich.

Proposal	1 (W)	2 (Bel/M)	4 (Br/M)	7 (Bel/Br/M)	8 (E)^2
POPULATION					
1 *Existing population affected*					
2 Belstead	705	705	–	610	310
3 Bramford	3330	–	3300	3330	–
4 Martlesham	–	8775	8775	8695	9435
5 Total	4035	9480	12105	12635	9745
6 *New population (design purposes)*	120000	120000	120000	120000	120000
7 *LAND*					
Estimated area required for expansion					
8 Belstead	5100	5100	–	3750	1800
9 Bramford	4600	–	4600	4600	–
10 Martlesham	–	6600	6600	5250	9350
11 Total	9700	11700	11200	13600	11150
Agricultural land taken					
12 *Good quality*					
13 Belstead	3400	3400	–	}4400	1100*
14 Bramford	3400	–	3400	–	–
15 Martlesham	–	–	–	–	–
16 Total	6800	3400	3400	4400	1100*
17 *Medium quality*					
18 Belstead	–	–	–	–	–
19 Bramford	–	–	–	–	–
20 Martlesham	–	2200	2200	1200	3800*
21 Total	–	2200	2200	1200	3800*
22 *Designation area, total*	6800	5600	5600	5600	4900*
23 *COMMUNICATIONS* (roads: annual vehicle miles in thousands)					
24 Motorway	46402	79677	76825	45360	46401/57971*
25 Dual carriageway	75277	109080	100845	93657	75277/119844*
26 Single carriageway	27595	16405	30152	41604	27595/25140*
27 Total	149274	205162	207823	180621	149275/202956*
28 *Peak hour cross-town movements in PCUs*	1300	2570	2430	2600	1300/2550*

Notes:
1. Does not include figures of money cost; see Table 7.4.
2. Where figures have not been prepared for proposal 8, we have made estimates based on the other proposals; in particular 1 and 6, which are the nearest comparisons to 8 in terms of concentration of the new development and siting, respectively. Figures where this has been done are shown with an asterisk (*).
Items:
1–11: As estimated by the County Planning Officer and County Surveyor; Report on the Supplementary Report on the Ipswich Expansion, 1967, Table VI. Figures relate to possible Designation Areas for the Proposal in question.
12–22: As estimated by Shankland, Cox and Associates in their Supplementary Report, Table 4.1 and Appendix B. Estimates relate to possible Designation areas.
23–28: Shankland Cox; Supplementary Report, tables 3.1 and 3.2, supported by tables A1–A10 excluding external traffic
Source: Lichfield & Chapman (1970: 180)..

Table 7.4 Comparison of capital costs and annual costs & revenues for project options: Ipswich.

		1 (W)	2 (Bel/M)	4 (Br/M)	7 (Bel/Br/M)	8 (E)[3]
	Proposal					
	Date of commencement of major development	Year 1	Year 1	Year 1	Year 1	Year 5
	Capital cost[2]	£000s	£000s	£000s	£000s	£000s
	LAND					
	Acquisition of agricultural land					
1	Belstead	1020	1020	–	660	330*
2	Bramford	1020	–	1020	650	–
3	Martlesham	–	614	614	364	1140*
4	Land for regional recreation	M_1+	M_1	M_1	M_1	M_1
5	Subtotal	2040	1634	1634	1684	1470
6		M_1+	M_1	M_1	M_1	M_1
	Acquisition of developed land in Ipswich					
7	For roads	M_2	M_2+	M_2+	M_2++	M_2+
8	Within town centre	M_3	M_3+	M_3+	M_3+	M_3+
	CONSTRUCTION AND SITE ENGINEERING					
	Roads					
9	for the expanded Ipswich	12508	22133	20012	16008	12508/18776
10	future new or extended by-pass	M_4+	M_4+	M_4+	M_4+	M_4+
	Main services					
11	sewage works and sewerage system	6310	9225	9225	8800	9000*
12	gas/electricity supply	M_5	M_5++	M_5++	M_5	M_5+++
13	water supply	M_6	M_6+	M_6+	M_6	M_6++
14	Building construction	M_7++	M_7+	M_7+	M_7++	M_7
	TOTAL CAPITAL COST	20858 plus M_1+ M_2 M_3 M_4+ M_5 M_6 M_7++	32992 plus M_1 M_2+ M_3+ M_4+ M_5++ M_6+ M_7+	30871 plus M_1 M_2+ M_3+ M_4 M_5++ M_6+ M_7+	26492 plus M_1 M_2++ M_3+ M_4+ M_5+ M_6+ M_7+	22978+/27246+ plus M_1 M_2+ M_3+ M_4+ M_5+++ M_6+ M_7+
	Annual cost	£000s	£000s	£000s	£000s	£000s
16	VEHICULAR TRAVEL COST	4354	5984	6060	5268	4354/5919
17	ROAD MAINTENANCE	48	66	64	65	48/66
18	ACCIDENT COST	234	289	309	297	289/297
	DRAINAGE					
19	Sewage works	55	70	70	710	73
20	Sewerage system	100	165	165	150	155
21	GAS/WATER/ELECTRICITY SUPPLY	m_5	m_5++	m_5++	m_5+	m_5+++
22	Total annual cost	4591 m_5	6574 m_5++	6387 m_5++	5544 m_5+	4999+/6510+ m_5+++
23	Annual gross value of agricultural production lost	408	281	281	306	200
	Annual revenue	£000s	£000s	£000s	£000s	£000s
24	COMMERCIAL DEVELOPMENT Shops, offices, industry, warehousing, etc.	Assumed to be the same for all proposals				
25	RESIDENTIAL DEVELOPMENT					

Source: Lichfield & Chapman (1970: 182).

Table 7.5   Plan variables.

Variables			
6.1 Implementation	6.2 Use	6.3 Operational	6.4 Decommissioning
6.1.1 E	6.2.1 A	6.3.1 G	6.4.1 J
6.1.2 F	6.2.2 B	6.3.2 H	6.4.2 K
	6.2.3 C		
	6.2.4 N		

local authority acquisition compared with owners/developers acting on their own parcels within an overall plan. There could be options in the *operation* after implementation, for example, factories or carparks could be open during the daytime or over 24 hours, including *management* options, where the residents of a freehold housing area set up a management trust for the dwellings as opposed to leaving the management to the individual owners. And there could be *decommission* options, as where a temporary access system is to be ploughed up following the operation, or continued. Box 7: The articulation of the use, implementation and decommissioning variables in the way described (Box 6) now enables us to specify more clearly the project options (Box 7), by bringing out the *differences* between the options in terms of the use variables. Table 7.6 is a simple form showing directional differences in ten bus service options on a given route, which result from changes in three 1985/86 operation variables, with the datum being 1984 (Lichfield 1987b). Table 7.7 is more complex, specifying where there are differences from the datum (0 at X) in the project options defined in plan variables, with a zero indicating no difference.

Table 7.6   Specification of project options: directional.

Options	Level of service	Fare	Subsidy
1	0	↓	↑
2	0	0	↑
3A	0	↑	↓
3B	0	↑	↓
4A	0	↑	↓
4B	0	↑	↓
5A	↑	↑	↓
5B	↓	↑	↓
6A	↑	↑	↓
6B	↓	↑	↓

System change   Box 8: The nature of the urban & regional system was presented above (Table 1.3) distinguishing between the supply (physical stock) and the demand (flow of activities by uses and consumers). As just indicated in Box B4–B5, we are not concerned to describe the system as a whole, but only the *change* to be brought about by the project, through the development process. Table 7.8 shows how this is identified, for each option or, in complex cases, by a background table for each plan variable (Box 6).

Table 7.7   Specification of project options by plan variable.

Plan variables	X (datum)	Project options Y	Z	N
Use				
(a) Population	0	0	A	B
(b) Physical fabric	0	A	0	C
(c) Activities/uses	0	A	0	D
Implementation				
(a) Local authority development	0	0	0	0
(b) Local authority / private developer	0	A	B	C
(c) Private sector alone	0	0	0	0
Management/operation				
(a) Overall management	0	0	0	0
(b) Trust	0	A	B	C
(c) Management scheme	0	0	0	0

Notes: 0 is the datum on option X. 0 is options X, Y, Z indicates the same as datum. A, B, C indicates degrees of change from datum; these are not related vertically.

The system elements reflect those of Table 1.3 and also Table 7.2. The *offsite* elements may be adjoining or well away from the site; for instance, generation of traffic from an onsite residential complex to urban facilities in the nearby town centre or elsewhere, which we call *linkages* (b). As such they are different from those *effects/impacts* (c) that can penetrate throughout the community without physical linkage, or indeed across frontiers (for example, the diversion of labour demand from economic enterprises, or the pollution of the natural environment in the seas or ozone layer).

From the typical development process it is apparent that the change which the project will bring into the current urban & regional system is not felt immediately but by phases over time. This is brought out in Table 7.8. The changes can be predicted for any of the system elements, comparing the situation in any of the rows with the current (Column 1) or, where the development process will last some time, during which the current situation will change, by reference to the projected situation without the project (column 2). The various change situations are:

- Column 3 Blight: of the current situation in the planning stage prior to implementation
- Column 4 Displaced: removal off site to accommodate the project
- Column 5 Retained: system elements as part of the new project
- Column 6 Construction: carrying out of constructional work
- Column 7 Completion: occupation and operation/management of the development and activities for which the project was designed (commissioning)
- in certain instances (a power station or dated steel works) there could be: Column 8 Decommissioning: where the change can be visualized, even though not the actual date.

Table 7.8   Change in the urban and regional system through project options (effect matrix).

Datum			Change through project					
System elements	Current 1	Without project 2	Blight 3	Displaced 4	Retained 5	Construction 6	Completion 7	Operational 8
1. ON SITE								
(a) Population Living Working Visiting								
(b) Physical fabric Land Natural resources Utilities Transportation Telecommuni- cations Buildings and sites Open space								
(c) Activities/uses Residential Production Distribution Consumption								
2. OFF SITE								
(a) Linkages Utilities Transport Telecommuni- cations Urban services								
(b) Effects/impacts Economic Social Cultural Linkage Environmental Risk/hazard								

Source: adapted from Lichfield (1988c: table IV.1).

E The decision framework/space

BOX 9: We saw above (Ch. 2) that any choice/decision needs to be tackled within a particular context. This therefore needs to be articulated for any specific evaluation analysis in order to produce results that are meaningful to the decision-taker. In addition to the generic questions of any evaluation (§7.4), the evaluation will need to address specific questions that will be *ad hoc* to the project (§8.3). They could be posed from various sources, such as the following: (a) the decision-taker, as an aid to his reaching conclusions; (b) the planning authority, local or central, as an aid to planning assessment, including indication of how the adverse impacts could be ameliorated; (c) the local authority for the area in which the project is found, as a basis for their view as government; (d) consultees on the project, official or otherwise; (e) groups of the public, with whom consultation has raised concerns; (f) the analyst, who could see pertinent issues that arise outside the above.

Box F–J Analysis, evaluation and decision

F Effect assessment   Box 10 generates the *effects* from the project which will be evaluated in Box G. As the title conveys, the distinction is followed between *effects and impacts,* which was introduced above (§5.4), with the effects being ". . . the physical and natural changes resulting, directly or indirectly, from development". It is these effects, stemming from one or other of the plan variables (Box 10.1) which are predicted. As seen above (Ch. 5) such predictions are made via what is conventionally called an "impact matrix". But, as stated there, no standard method has emerged, so that we have here adopted an individual approach in what we term an "effect matrix", using the distinction just made. This approach is to use Table 7.8 for the purpose, by linking the *changes* in the system with the *effects* generated by the system elements, following from the project.

BOX 10.1: PLAN VARIABLES   The plan variables are brought down as necessary from Box C6, where the project is complex in content. The generation of the effects can be simplified by predicting them from one variable at a time, as shown in Table 7.9 (column 1). If so, a separate Box 10 would be needed for each. Because of the range of variables possible, it is useful to prepare a background table of those in mind, as in Table 7.5. It could be that the Table would be much simpler, as in 7.6.

BOX 10.2: There are clearly innumerable ways of subdividing people in communities: income groups, ethnic origins, functions, and so on. In practice, experience in CIA has shown the last approach to be of practical value: the division of the relevant community into owners/producers/operators on the one hand and consumers on the other. Not only does this have some relationship to socio-economics, which is the foundation on which the evaluation rests, but it is also of pragmatic value in the analysis. Visualizing the sectors who will be involved on the one hand in providing the inputs to the production and operation of the project (including the owners who input their resources), and the consumption of its outputs on the other, is a practical way of ensuring that all the appropriate sectors can be readily visualized and listed, and none forgotten. In this we are not concerned simply with those producing and consuming on site but also others who are involved off site. For example, in the Sha'arey Tsedek project in Table 7.9, the municipality would be involved in the provision of roads and services, both on and off site; just as consumers (users of the site) will have offsite repercussions in traffic, multiplier effect, and so on. Thus, the production/consump-

Table 7.9   Sha'arey Tsedek, Jerusalem: summary from three project variable of effects by sectors, options A (rehabilitation) and B (redevelopment).

1	2		3	4
	Sector		Summary description of effect	Change
	No.	Name	Option A	Optn B
1 New roads	5(1)	Municipality on site	More roads	3=
	9(1)	Municipality off site	More accessibility	3=
	11(1)	Adjoining landowners	More accessibility	3=
	13(2)	Urban services	More accessibility, more users	3=
	6(1)	Users of site	Traffic noise	3=
	10(1) & (2)	Traffic	More accessibility	3–
	12(1)	Adjoining occupiers	Increase in accessibility	3=
	14(4)	Users of urban services	Increase in accessibility	3=
2 Current	5(2)	Municipality on site	New museum	–
buildings	5(3)	Municipality on site	New Grove	–
	5(4)	Municipality on site	New open space	3
	7	Government on site	Retain CBH	–
	13(1)	Jerusalem employers/firms	More business	–
	15	Government	Increase tax income	3+
	6(2)	Visitors to NHM	Enjoy new NHM	–
	6(3)	Visitors to Grove	Enjoy Grove	–
	6(4)	Visitors to open space	Enjoy open space	3
	8	Tourists/visitors to NHM	Enjoy national heritage	3
	14(1)	Jerusalem workforce	Jobs in NHS and space	–
	14(2)	Nearby residents	Enjoy fresh air & visual impact	3–
3 New	1/2	Landowner and occupier	Increase site value	
buildings	3	Developer	Increase profit	3+
	5	Municipality on site	Betterment tax	
	11(1)	Landowners adjoining	Increase land value	3–
	11(2)	Landowners elsewhere	Increase land value	3+
	12(2)	Occupiers elsewhere	Maintain value of flats	3–
	13(2)	Jerusalem urban services	Increase business locally	3+
	4	New residents	New access in central location	3+
	14(1)	Jerusalem workforce	Increase employment in flats and servicing cars	3+
	14(2)	Nearby residents	Increase environmental attraction	3–
	14(3)	Downtown users	Increase enjoyment through activity at night	3–
	14/16	Government/tax-payer	Less taxes	3+

Notes. Col. 4, change of effect in option B compared with Option A: same kind: 3= less: 3– more. 3+ none: –
Source: Lichfield (1988c: table 15.8).

tion link is not simply that of direct exchange in the market. It takes in outputs that are not bought and sold (the externalities). CIA experience also shows that there is a reasonable degree of uniformity of sectors in the various evaluation studies. Because of this, it is possible to prepare a preliminary checklist of community sectors for Box 10.3. Those in the Sha'arey Tsedek project are fairly typical.

Although the purpose of the analysis is advanced by the segregation into producers, and so on, and consumers, the two groups are not identified in the same way. In the former, the sector tends to be an institution or organization (landowner, development company, local authority, nation as a whole, etc.), and the latter groups of individuals who are likely to take advantage of (consume) the outputs from the project. Alternatively, instead of identifying the production/operation function, which is translated into outputs that are available for consumption, the consumption from the outputs (that is the effects) could be identified first and then traced back to those who would be concerned with the production or operation. Within this general structure, refinements are possible. There may be the need to disaggregate particular sectors, as for example by income level or relative deprivation or disadvantage (e.g. accessibility to jobs, or the countryside, for those not having motor-cars).

BOX 10.3: Here are summarized the conclusions from the *effect assessment* (Fig. 7.3, line 9), which is conventionally made under the name *impact assessment* (see §5.4 above). As shown there, different methods can be used for the prediction from the individual system elements. For our purpose there is no need to depart from the methods of conventional *impact* assessment (Ch. 5). But it is important to have a uniformly presented summary of the various *effects* and to ensure that their scope embraces all that are relevant to urban & regional planning as opposed to just effect assessment proper, as in the Prospect Park case study (see Ch. 14.) The assessment needs to be made separately in the critical phases of the project cycle: construction, finished development and operation post-completion, and in some cases also in pre-construction blight. The process that was chosen for doing this is shown in Figure 7.3.

As part of this summary it is necessary to identify from the predictions the characteristics of the effects that are inserted in the remainder of Box 10, as follows:

BOX 10.4: The size of the effect is clearly of importance. Accordingly, the Box provides for:

- unit of measure: physical, money value, intangible, etc. (see Ch. 10);
- size: described on the nominal, ordinal, interval or ratio scale (Ch. 6.2.10).

BOX 10.5: While the information shown in Box 10.3 gives a certain impression of the effect, it is necessary for decision-taking to know more of its characteristics. Based on the literature and practice of impact assessment (Ch. 5) these are:

- *kind*: financial, fiscal, economic, sociological, cultural, natural environment, hazard;
- *type*: direct (on site), indirect (off site), associated real (technological), or associated financial (pecuniary);
- *timing*: the date of commencement and duration from the project life-cycle;
- *probability*: the degree of certainty about the effects;
- *reversibility*: whether or not the effect is for ever, or reversible;
- *significance*: by reference to some standard, or to number of people affected or other considerations underlying significance (Fig. 5.4).

BOX 10.6: The preceding identification of the relevant community sectors, in particular of consumers which are impacted, provides a way of identifying the *functional*

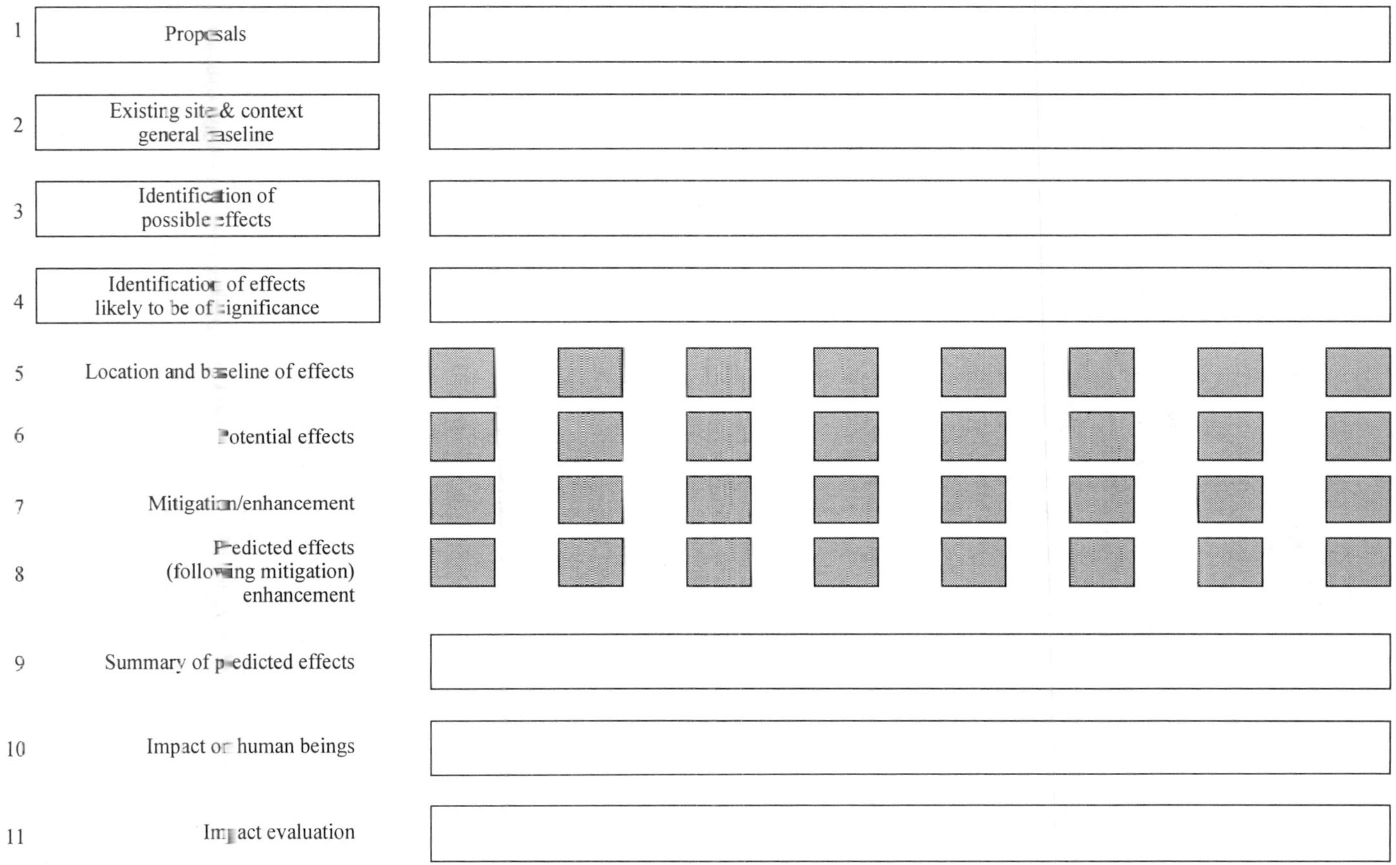

1 Proposals

2 Existing site & context
   general baseline

3 Identification of
   possible effects

4 Identification of effects
   likely to be of significance

5 Location and baseline of effects

6 Potential effects

7 Mitigation/enhancement

8 Predicted effects
   (following mitigation)
   enhancement

9 Summary of predicted effects

10 Impact on human beings

11 Impact evaluation

Figure 7.3 Environmental and planning assessment of effects: framework for evaluation (*Source:* Lichfield & Lichfield 1992).

community to which the project relates. For example, its shopping public and its labour market for a shopping centre; or the nation or the world for the restoration of a historic monument. This is described in Box 10.6.

Because of the varied "spheres of influence" of the various impacts on particular community sectors, it will not be practicable to define *the* geographical community tidily, by showing one boundary on a map. Several boundaries would be needed, the rare exception being the isolated self-contained town, which has no attraction for tourists, in the remote desert. But however untidy the *geographical* community, it is useful to compare it with the *administrative* community, namely that which is the administrative decision-making unit, based on legal boundaries and, in projects requiring interrelated decisions at more than one level, the overlapping *administrative* communities: local authority, county, region and central government, perhaps international (e.g. UNESCO for world historic sites).

To these will be added in all cases the *decision participating* community, viz. those who would be involved in consultation and comment. Their geographical spread should be closely related to the *functional* community. But in practice it is often confined to the potential voters in the *administrative* community for which the decision-taking authority is responsible, without reference to those who are impacted outside. This raises implications for the concept of democracy in decision-making in planning (Ch. 11).

G Impact evaluation   Box 11: Having in Box 10 identified, predicted and described the *effects* from the project, it is the purpose of this Box to translate the *effects* into *impacts* on sectors in order to evaluate the options with a view to showing preferences by sectors. By *impact* here is meant not *just* the ". . . consequences or end products of those effects . . . on which we can place an objective or subjective value" (§5.2), but the consequences of the effects on people which will lead to a change in their way of life, on which the sector's valuation can be based.

An example will illustrate. Construction works, or traffic on a motorway or airport, will raise the level of noise; these affect the way of life of people living nearby, in that they would keep their windows shut in the summer and so experience higher temperatures and lose the benefit of fresh air, or open the windows and be irritated or worse by the noise.

Thus, it is necessary to distinguish between different sectors of the community to distinguish the reception of the effects. Accordingly, Box 11.3 aims to produce a summary of the impacts on the various community sectors, and Box 11.4 a summary of the differences between them, as a preliminary to their valuation, by those sectors in Box 11.5–11.8. The translation from *effect* to *impact* is by use of a concept specifically devised for the generic model, namely the impact chain, which is a bridge between Boxes 10.3 and 11.3 (Lichfield 1985). This is presented diagrammatically in Figure 7.4. The chain starts with the introduction of the project as a whole, which will cause the effect and impact, or in the more complex cases by the plan variable (be it in use, implementation, operation/management or decommissioning). These give rise to certain *effects* which will fall on one or more of the community sectors.

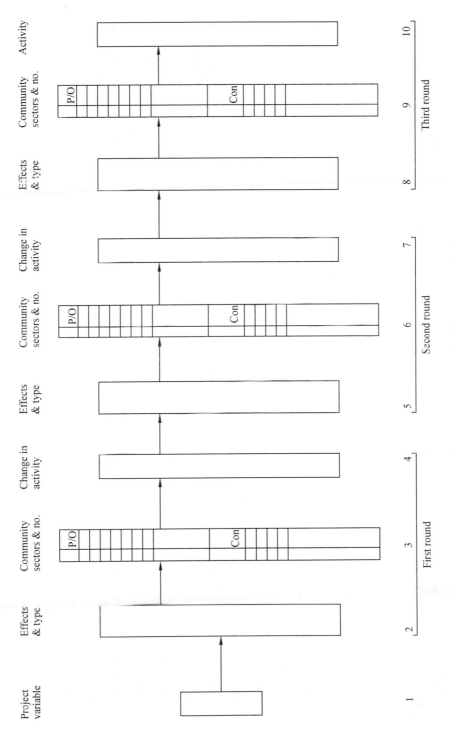

Figure 7.4 Impact chain: generalized diagram for all effects.

Table 7.10 Paestum: impact chain – excavation/construction phase.

System elements 1	Effects 2	Community sector 3	Impact 4	Effect 5	Community sector 6	Impact 7
ON SITE	Round 1				Round 2	
(a) Population						
living						
working						
visiting						
(b) Physical fabric						
land	Occupy land for works	Landowner	Give up possession	More activity on site	Surrounding occupiers	Interference with continuing activities
natural resources	Destroy resources	Natural heritage	Diminish quality	Visual appearance	Passers by	Less amenity
utilities	–	–	–	–	–	–
transportation	–	–	–	–	–	–
telecommunications	–	–	–	–	–	–
buildings and sites	Defer deterioration	Cultural heritage	Enlarge cultural heritage	Attract archaeological works	Archaeological works	Archaeological education advanced
open space	–	–	–	–	–	–
(c) Activity/uses						
residential	–	–	–	–	–	–
production	Excavation and construction works	Firms, employees, archaeologists, educ.	Business, jobs, education	Greater economic activity	Local and regional enterprises	Increased incomes, jobs and taxes
distribution	–	–	–	–	–	–
consumption	–	–	–	–	–	–

Table 7.10 Paestum: impact chain – excavation/construction phase.

System elements 1	Effects 2	Community sector 3	Impact 4	Effect 5	Community sector 6	Impact 7
OFF SITE (a) Physical linkages		Round 1			Round 2	
utilities	Greater use	Utilities operator	Increased profits	Enlarge services	Utility consumers	Better living conditions
transportation	Improved services	Transport operators	Increased profits	Enlarge services	Travelling public	Greater mobility and accessibility
telecommunications	Improved services	Telecomms operators	Increased profits	Enlarge services	Public using telecommunications	Greater freedom in communication
urban services	Improved services	Service operators	Increased profits	Enlarge services	Consuming public	Higher standard of living
(b) Effects/impacts						
economic	Increased – tourism – demand for museum owners – tourism services	Visitors, tourists Ministry operation	More visitors, tourists More . . . activity	Secondary economic effects on jobs, income, spending	Local and regional economy	Higher standard of living
social	Increased activity on site affecting local and regional community	Local and regional community	Social/economic change good/bad	Immigration, out-migration, social cohesion	Establishing community Newcomers	Social disturbance, social organization
cultural	Increased supply and use of cultural built heritage	Visitors and tourists	Greater cultural activity and awareness Enrich cultural tourism	Improve tourist-related business	Local and regional business operators, travel companies	Increased local economic impact
linkage	–	–	–	–	–	–
natural environment	Deteriorated	Surrounding occupiers	Reduced amenity	–	–	–
risk/hazard	Enhanced traffic	Greater danger	–	–	–	–

Source: Lichfield et al. (1993a,b).

Table 7.11 Sha'arey Tsedek, Jerusalem: evaluation of two options (on completion).

No.	Description	No.	Project variables	Impact type	Sectoral objective	Units	B–A	Subsector	Sector
1	2	3	4	5	6	7	8	9	10
	Producers/operators			D					
1	Current owner of site		1,2,3	D	Increased land value*	£	–	A	A
3	Developer/financier		3	D	Increase development profits	£	+	B	B
5	Municipality on site			D,					
	1. Roads/utilities		1	AF	More municipal services		0	=	
	2. NHM		2	D	Ditto		–	A	
	3. Grove		2	D	Ditto		–	A	
	4. Open space		2	AF	Ditto		+	B	
	5. New flats		3	D	More betterment tax	£	+	A	
					Preference for Sector 5				A
7	Government on site			AR					
	National heritage		2	AF	Conserve heritage		–	A	A
9	Municipality off site		1	AR	Reduce traffic congestion		0	=	=
	Other landowners			AR					
	1. Adjoining		3	AR	Increase land value	£	0	=	
	2. Elsewhere		3	AR	Increase land value	£	–	A	
					Preference for Sector 9				A
13	Jerusalem economy				More business		–	A	
	1. Employers/firms		2,3	AF	More accessibility		0	=	
	2. Urban services		1	D	More business		–	A	
			2		More business		+	B	
			3		*Preference for Sector 15*				N/C
15	Government budget		3		Greater financial contrib. to	£			
					Sha'arey Tsedek Hospital from				
					landowner		–	A	A
	Consumers								
2	Current site occupiers		1,2,3	D	Minimize disturbance		0	=	=
4	Residents in flats		3	AF	Secure flats in good location		+	B	B
6	Users of site								
	1. Traffic on site		1	D	Minimize traffic nuisance		0	=	
	2. Visitors to NHM		2	AR	Enjoy NHM		–	A	
	3. ——— Grove		2	D	Enjoy Grove		–	A	
	4. ——— open space		3	D	Enjoy new open space		+	B	
	5. Passers by		3	AF	Enjoy new view over town		+	B	
					Preference for Sector 6				A
8	Tourists and visitors		2	AR	Enjoy cultural built heritage		0	A	A
10	Traffic								
	1. To site			AR	Reduce congestion		0	=	
	2. General			AR	Increase accessibility		0	=	
					Preference for Sector 10				=
12	Other occupiers								
	1. Adjoining		2,3	AR	Increase occupation value		0	=	
	2. Elsewhere		2,3	AF	Maintain occupation value		–	A	
					Preference for Sector 12				A
14	Jerusalem economy								
	1. Workforce		1,2,3	AR	Greater number of jobs		?	N/C	
	2. Nearby residents – air/visual		2,3	AF	Greater environmental attraction		–	A	
	3. Downtown users		2,3	AR	Greater interest		–	A	
	4. Users of urban services		2,3	AR	Greater accessibility		+	B	
					Preference for Sector 14				N/C
16	Taxpayers		2,3	AF		£	–	A	A

Notes: Columns 4–7: the gap shows measurement other than in £; 8: +,B is better than A, and –,B is worse than A, and 0,B equals A; 9: ?, non-certain; 9 and 10: A and B indicate which preferred, and N/C indicates preference not certain. *: net of betterment tax.
Source: Lichfield (1988c: table 15.9).

These in turn will lead to changes in the activities of the sector, which is the impact. This completes the first round. The changes in the activities themselves will generate other *effects* of the secondary or indirect kind, which again are transmitted to one or more community sectors (the same as or different from the sector experiencing the initial effect) resulting in a related change in activity, or *secondary effects*. This completes the second round. The same process can then be visualized entering a third round, producing third-order *effects* on community sectors, resulting in a change to the activities and therefore tertiary impacts. And so on, subject to ability to predict in the light of uncertainty.

A demonstration is given in Table 7.10, taken from a case study of the archaeological restoration of a Greek town at Paestum, near Naples. The impact chain relates to the phase of excavation and construction and shows how system elements generate effects and impacts.

From the preceding, it will be seen that Box 11 is similar in intent to Table 7.11, which in the simpler form of table A has consistently appeared in all the PBSA case studies. But there are two significant differences with the maturing of the method, namely the specific rather than implied reference to effects/impacts, and the distinction between them via the impact chain (Table 7.10).

BOX 11.2: Instead of the checklist of community sectors in Box 10.2, this box shows the participating sectors, that is, those affected from Box 10.3

BOX 11.2: Plan variables are brought down from Box 10.1, which in turn were brought down from Box 6. Their insertion here makes it possible in complex cases to apply the ensuing analysis by each plan variable at a time. This will clearly result in several tables relating to Box 11, which must then be aggregated.

BOX 11.3: Certain steps are necessary to translate the *effects* from Box 10.2 to the *impacts* in Box 11.3. Taking each effect in turn, identify the community sector which will be *affected* from the tentative checklist (Box 11.1). This can be done formally via the impact chain, as in Table 7.10; or more intuitively without the formal chain as in Figure 7.4 which goes from the plan variables in turn to the community sector and impact. In this table the difference in impacts between the options is also shown.

BOX 11.4: Whereas the totality of impacts on sectors is of interest for particular planning purposes, in CIE it is the *difference* in the impacts between the options (Y, Z and N), which are traced by reference to the common datum, option X; provided there is consistency in this rule then any of the options can be used as the datum. Table 7.11 presents a simple example which shows how the impacts under Option B (redevelopment) compare with those under Option A (conservation). It will be noted that the Box 11 does not provide for tracing through the effect characteristics from Boxes 10.4 and 10.5. This is simply to avoid adding further detail to Box 11, since this information is available, and can be read from Box 10 by noting any difference when considering preferences (Box 11.7 below).

BOX 11.5: The description of the *effects* enumerated in Box 10.3 and their characteristics in Boxes 10.4 and 10.5 are *objective,* in the sense of being capable of specification and measurement. To make the judgement as to how they will be perceived by the people who are impacted, it is necessary to identify for each of the sectors their *sectoral objectives* (in the earlier case studies termed *instrumental objectives*) by which, it is thought, they would implicitly or explicitly perceive and judge each impact. These objectives are not formulated in abstract or general terms, but rather as a reaction based on the sector's objectives (values) to the *particular impact* which the sector will perceive, or will experience. More specifically, given the impact as predicted under Box 11.3, with characteristics under Boxes 10.4 and 10.5, would they wish to enjoy more (maximize) or less (minimize) the impact.

BOX 11.6: The application of the sectoral objective to the impact will show whether the impact will bring an advance in welfare (utility/negative cost) or a retardation (dis-utility/positive cost). The former would be desirable and thereby produce benefits; and the latter undesirable and thereby produce costs. From this it is possible to strike a *net benefit* for each impact; and by the same token a net benefit between sub-sectors. This is done on the same rules as tracing difference in impacts in Box 11.4, that is always by reference to a common datum X. Table 7.11 illustrates how the comparison is made. As in conventional cost–benefit analysis, it is necessary to take a "with or without" view of the future (Ch. 6 above). This, however, is very difficult in urban & regional planning, since the prediction of the latter is impossible at worst, or very time-consuming at best; it would require complex modelling, compounded by the complexity of deciding in the "without" scenario upon the effect of likely social intervention in the absence of planning. Accordingly, the approach is adopted of comparing options not in terms of absolutes in each end state and the "without situation", but each end state with the *other* end states, by means of selecting one as a datum. The essential purpose is to get at the *differences.*

Once the need to obtain absolute measures is abandoned, another simplification becomes possible in certain cases, namely where the trends from the current situation to the end state are consistently positive or negative. In this situation it is possible to abandon the "without" situation as datum and substitute the "current" or "before", the "after" still relating to the "with". The differences would be consistent with each other, even though the size of difference would be changed. These adaptations have one obvious disadvantage. Even if all the costs and benefits were put in money terms, which is rare, it is not possible to generate any rate of return, internal rate of return, cost–benefit ratio, or net present value. This disadvantage is not so critical if the decision to introduce a plan has been taken (as opposed to no plan) and it is not necessary to show that its implementation would exceed a stipulated rate of return in order to be "worthwhile" (§5.1 above). However, it is still necessary to pose the other two standard questions ("this way" or "now"), which leads to judgement as between options.

BOX 11.7: In Box 11.4 we have regard to the *size* of the impacts from the effects brought out in Box 10.4, and their *characteristics* in Box 10.5. In striking a *preference*, account must also be taken of the *characteristics* of the impacts insofar as they differ from those of the effects in Box 10.4. In relation to these, the preference between the options for the sectors would be influenced by the difference given to weight in the impacts, for example:

- *Size:* the greater the impact or the number of people impacted, the greater the weight.
- *Kind:* financial and economic impacts might or might not weigh more in the

judgement of the sectors than say, for example, environmental or cultural impact. Impacts valued for direct *use* by the sector may or may not weigh more than the *option*, *bequest* or *existence* values (see Ch. 6).

- *Type:* direct impacts are likely to weigh more than indirect.
- *Timing:* impacts of short duration would be less significant than long, as would impacts further away in time (e.g. second round as opposed to first).
- *Probability:* The more probable the impact the more significant it will appear to the sector.
- *Reversibility:* An irreversible impact clearly would weigh more significantly on the sector than one which can be removed later on, if need be.
- *Significance:* Clarification of this elusive concept (Fig. 5.4, above) will be helped by considering many of the preceding characteristics in order to judge how important the impact would be, and therefore its perceived change in the perceptions of the sector.

The analysis of differences in net benefits for sectors or subsectors in Box 11.6 will produce *preferences* (i.e. greater *utility*) for each impact, as shown in columns 9 and 10 of the Sha'arey Tsedek case study in Table 7.11 or the Ipswich case study in Table 7.12. From these it is seen that the analysis for preference proceeds impact by impact on the community sectors. In the comparison, each sector or subsector is treated in isolation from the others, whether or not the individuals so considered are members of other sectors (as they generally will be). For each the question is posed: Given the impacts fully described in Box 11.3 and 4, how would the relevant sector or subsector judge it by reference to its sectoral objective in 11.5? Will its welfare be perceived as advanced or retarded, the advance being an advantage (a benefit or minus cost) and the retardation a disadvantage (cost or minus benefit). This leads to the preference in Box 11.7, and the ranking in Box 11.8. In each option the comparison is against a datum, that is the situation with which the other options are respectively compared. Since the purpose of the comparison is to predict differences between options, the actual datum is chosen for convenience. It could be the current situation (which gives a "before and after" comparison), or the predicted situation (a "without and with" comparison), or one of the options (giving simply differences in future situations). Clearly, the datum chosen will affect the interpretation of the results of the comparison. The comparison is made easier the greater the degree of measurement in Box 10.4, and the greater the specification of the effect characteristics in Box 10.5. But even if there be no measurement, a *direction* can be inferred from the sectoral objective in relation to the impact and the *difference* conveyed.

As Table 7.11 shows, there can be a considerable amount of variation within sectors. It is therefore desirable to aggregate into sectors, or at least subsectors, called "reduction" in the case studies. This is done by inspection. In certain instances all the preferences will be in one direction, so that the conclusion will be clear. But in others there will be countervailing preferences. Where all the items are in money values they can be summed algebraically for conclusion. Where not, uncertainty arises. In such cases the aggregation by sector cannot proceed further. As Table 7.11 shows, the end result will therefore be preference by the sector or subsector for particular

	Possible unit of measurement		Numbers					Instrumental objectives	Order of preference — Greatest benefit or least net cost				
Sector	Cost	Benefit	1(W)	2(Bel/M)	4(Br/M)	7(Bel/Br/M)	3(E)		1(W)	2(Bel/M)	4(Br/M)	7(Bel/Br/M)	8(E)
Producers/operators													
1.0 Development agency	M		£19+ M_1–M_7 Year 1	£31+ M_1–M_7 Year 2	£29+ M_1–M_7 Year 3	£25+ M_1–M_7 Year 4	£25+ M_1–M_7 Year 5	Low net fin. cost	0	-2	-1	-1	-1
								Early devt of the land	0	0	-1	-2	-3
								Further growth	0	-2	-2	-2	-3
1.0 Reduction													
3.0 Current landowners													
3.1 Displaced													
3.1.1 In urban and village areas	m	M	n	n+	n+	n++	n+	Nil net cost	0	0	0	0	0
3.1.2 Agricultural Landowners								Minimum annual loss agricultural output	0	+3	+1	+2	+4
Farmers	m	M	n+++++	n+	n+++	n++	n						
3.1 Reduction									0	+3	+1	+2	+4
3.3 Not displaced													
3.3.1 Pecuniary charges	M	i	n++	n+	n+	n+	n		0	0	0	0	0
Potential devt value	M		n	n	n	n	n		0	0	0	0	0
Real changes: within town adjoining town	M								Elsewhere: item 2.0				
In urban and village areas	m	i	n+	n+	n+	n+	n	Choice	0	-1	-1	-1	-2
			n	n	n	n	n	Environment	0	1	1	1	0
3.3.1 Reduction	m		n+++	n+	n+	n++	n		0	-1	-1	-1	-2
3.3.2 Farmers	m	i	n+++	n+	n+	n++	n	Agric. output	0	+2	+2	+2	+3
								Accessibility	Elsewhere: item 2.8				
3.3.2 Reduction									0	+2	+2	+1	+3
3.3 Reduction									0	+1	+1	0	+1
5.0 Local authorities and ratepayers	m		203+	301+++	299+++	285+++	276/294 +++	Municipal cost	0	-4	-2	-1	-3

Order of preference

		Possible unit of measurement		Numbers					Instrumental objectives	Greatest benefit or least net cost				
	Sector	Cost	Benefit	1(W)	2(Bel/M)	4(Br/M)	7(Bel/Br/M)	8(E)		1(W)	2(Bel/M)	4(Br/M)	7(Bel/Br/M)	8(E)
2.0	Producers/operators Overall reduction									0	-4	-4	-4	-2
2.1	Consumers The public in the expanded town In the town and district centres Commercial occupiers and users of public/ private buildings	n	i	n	n	n	n	n	Choice Location (a) (b) Environment	0 0 0 0	-2 -2 -1 -1	0 -2 -1 -1	0 -3 -2 -2	0 -1 -3 -3
2.2	Reduction									0	-4	-4	-7	-7
2.4	In principal resid. areas Remaining or new residents	n	i	n++ n+	n+ n	n+ n+	n n	n+ n+	Severance Aircraft noise Environment	0 0 0	-1 -1 -2	-1 -1 -2	-2 -1 -1	-1 -1 -3
2.4	Reduction									0	-4	-4	-4	-5
2.6	In principal indust. areas Industrialists and workers	n	i	n	n	n	n	n	Occupational qual. Expansion space	0	0	0	0	0
2.6	Reduction									0	0	0	0	0
2.8	In principal open space and recreational areas Users of open space and countryside	n	i	n+	n	n	n	n	Choice Environment	0 0	-1 0	-1 0	-1 0	-1 0
2.8	Reduction									0	-1	-1	-1	-1

Table 7.12 Ipswich planning balance sheet (table A). (*Source*: Lichfield & Chapman 1970: 186–7)

options, with non-certainty in others. The obstacle to reaching conclusions from non-certainty can be overcome in particular cases by making judgements, as for example that the non-certain items would not be sufficiently critical to prevent conclusions being drawn by reference to others which are more certain. It is in Box 11.7 that the issue of *weighting* becomes relevant and important. Either all entries are treated on an equal basis (equal weight) or some are regarded as more important than others (unequal weights). This can be either "horizontal", for example in comparing unmeasured impacts between options, or "vertical", as for example comparing different subsectors within a sector. Some of these weightings are clearly the province of the decision-taker (for example as between sectors). Here, the choice for the analyst is either to avoid any weighting, and leave it to the decision-takers to provide their own to the results of the analysis; or to introduce his own weighting, specifying the assumptions and reasons, and leaving it to the decision-takers to accept or otherwise. BOX 11.8 : The judgements on preferences by sectoral objectives just brought out are a preliminary to the critical next step: of ordinally ranking the options in terms of the preference by the community sector or subsector in question. The ranking is made judgementally and subjectively. The decision in ranking is influenced by the kind of measurement adopted to identify the differences between impacts, costs and benefits and preferences (be it on nominal, interval, or ratio scales; the size of the difference in money or other measure; and the characteristics of the effects (Box 10.5). Box 11 provides for comparison between any number of options, by always comparing an option with the datum. Where these are many, simplification can be sought by proceeding group-wise or pair-wise to compare the preferences from each sector.

BOX 12: Since Box 11 analyzes a complex situation, and since the conclusions on the various sectors and subsectors are not likely to be clear, the outcome cannot be readily grasped. Some simplification of the analysis is therefore necessary to aid a conclusion. Using the analogy of company accounts, it would hardly be appropriate to present the Board with the whole package of financial accounts and ask them for decisions on the critical issues facing the company. Some simplification is required. This is shown in Box 12, which is the Table B of the case studies.

Box 12.1 reproduces the community sectors from Box 11.1, the sectoral preferences in Box 12.2 from Box 11.7 and the ranking in Box 12.3 from Box 11.8. Other examples are Tables 7.13 and 7.14 giving sectoral preferences. It is on this Box that conclusions are drawn in relation to the generic or standard questions raised in Box 9, and perhaps also some of the specific questions. From experience of the case studies, the degree of clarity in which the conclusions emerge is very much an accident of the case itself. In some, conclusions can be drawn with confidence in complex situations and the evaluation can lead to recommendations. In others there will be more uncertainty. We return to this critical issue below (Ch. 8).

BOX 13: During the evaluation process, as with the plan-making process, there could be many occasions for consultation or full-scale public participation (conveying information, eliciting reactions, obtaining views on options). It is here that a simplified form of Boxes 11.3 and 11.4 would be of use, in being able to convey the change in the way of life that would emerge from the options.

Table 7.13 Summary of sectoral preferences with option ranking (table B).

H/12: Summary of sectoral preferences and ranking										
12.1 Community sectors			12.2 Preference by sector between options				12.3 Ranking of options by sector			
Item	Descrip-tion	No.	X (datum)	Y	Z	N	X (datum)	Y	Z	N
Producers/operators/owners										
Consumers										

Table 7.14 Sha'arey Tsedek, Jerusalem. Conclusion on options: table B (completion).

Producers/operators		
1 Landowners	3	
3 Developer/financier		3
5 Municipality on site	3	
7 Government: national heritage	3	
9 Municipality off site	=	=
11 Other landowners	3	
13 Jerusalem economy		?
15 Government budget	3	
Consumers		
2 Current occupiers on site	=	=
4 Residents in flats		3
6 Users of site	3	
8 Tourists and visitors	3	
10 Traffic in general	=	=
12 Other occupiers	3	
14 Jerusalem economy		?
16 Taxpayer	3	

Notes: 3 = preference from 9; = A and B equal; ? preference not certain.
Source: Lichfield (1988c: table 15.10)

During the evaluation process there could also be negotiation and bargaining. As brought out above the planning process can be seen as a process of complex bargaining between a large variety of actors. Such bargaining proceeds whatever the level of communication. In essence it often amounts to sectoral views being expressed against (more rarely for) any proposal for change, be it by the local residents against intrusion into a residential area or by statutory consultees in respect of water, drainage, roads, etc. Whatever the stage in the planning process in which the bargaining is made (comment on planning application, presentation of planning application, public inquiry, public participation generally), the bargaining will achieve greater efficiency and clarity if it takes place around a central statement which lends itself to the purpose. Such a statement could be that in Box 11.6 and/or Box 12, which can be interpreted to portray to all the predicted change in the quality of life to different sectors which will emerge, for it is this that is of relevance to people. This would be an improvement on what typically emerges in a mass of conflicting statements, points of view, values, etc. through which it is difficult to reach a satisfactory conclusion.

The use of this vehicle for communication will be of value whatever the stage setting for the bargaining process itself. For example, it would be relevant in the resolution of impacts through their mutual trading-off (Susskind et al. 1978); in the "in camera" discussion on analysis between representatives of the decision-makers and -takers (Phillips 1987); at a public inquiry where there is arbitration about the basic options of permission or refusal.

BOX 14: The evaluation is framed in answer to the questions in Box E9, which will have been amended as necessary through Steps F to H. For this purpose there will be all the information generated in those and earlier stages, and also the background in the description of the project, supported as it would be by supplementary studies, by the planning team or others, in particular in relation to impact assessments. For the answers to these questions to be accepted as robust and justifiable, it will be necessary to present as appendices the analysis itself, which has been made by following through steps A–H in the generic model, with support from supplementary studies.

BOX 15: The scene is now set for the decision-takers to make their choice from the evaluation and their decision (see Ch. 2). For this purpose they will have available all the material referred to in Box 13, suitably simplified to facilitate the discussion and the decision. In practice it is found best to present the content of Box 12, the summary of sectoral preferences and ranking, as a basis for conclusions. This would be accompanied by the summary report which would contain the following essentials from the generic model:

- Box 2: project description
- Box 3: project options
- Box 8: effects in the system
- Box 9: questions to be answered in the evaluation
- Box 12: summary of preferences and ranking with assumptions, ordering and weighting.

It is here that the paradox of community impact evaluation arises. On the one hand, in order to achieve its purposes it must have a considerable array of information, some quite specific and some quite intangible. On the other hand, without the display, and also an open indication of the values that have entered into the analysis, there can be no confidence in the recommendation. To this we return below (Ch. 8).

BOX 16: The consequences of the process of choice and decision needs to be formulated for communication with those who are affected. The mode of communication will clearly be adapted to urban & regional planning. Given the complexities of decision-taking/-making in this field, it is apparent that there is compelling need for the communications to be made in the appropriate form and language, in order that the friction of misunderstandings can be minimized as efficiently and effectively as possible as the process unfolds. This in itself will contribute towards decisions not expending themselves into non-decisions (see Ch. 2). Of relevance here is Ashby's "law of requisite variety", which states that "R's capacity as a regulator cannot exceed R's capacity as a channel of communication" (Ashby 1956). This could be of various kinds:

- *Instructions:* Written and verbal instructions from the planning authorities to the staff, consultants, etc. who are to prepare the plans.
- *Team discussion:* Communication between specialist skills of considerable variety within the professional planning process (ecologists, geographers, economists, general planners, etc.), each of which is endowed with its own academic/professional language and means of communication: reports, maps, plans, charts, diagrams, computer programmes, etc.
- *Reports:* The results of work customarily contained in reports to the decision-taker and other interested parties (who may or may not have been consulted in the process). The nature of the reports and communication aids will vary with the recipient: from the technical report fully supported by technical appendices which are of interest to the professional, a summary for executives, the proverbial one page for the Minister, a popular summary for press release, or popularization for the public at large.
- *Consultation:* communication and consultation during the plan-making process with a large number of people and bodies who are affected, in order for example to obtain views, information, collaboration, and so on. This could involve communication by the applicant to the Authority in the form of plans, illustrations, technical documentation, etc.; the dissemination of the information for consultation by the planning authority, formal or informal; the technical communication between the various parties who have to be notified; summarization of the issues in the Planning Officer's report to his Committee; the discussion in Committee and the communication of conclusions by means of minutes, resolutions, etc. within the Authority; the issue of the decision notice with its technical statement of conditions on the grant of permission or statement of reasons for refusal. For each such situation there will be variation in the means of communication.
- *Appeal:* Should the applicant choose to appeal against a refusal, or unacceptable conditions, the complexities of an appeal hearing are set in train, with the formal communication of grounds of appeal and the setting up of the procedures for the appeal. Here the CIE could be tabled in full as an independent statement, in which certain inputs have been agreed by the parties, as a basis for the disputation, to replace the conventional mutual rubbishing on all sides. This would influence the production, in consultation between Counsel and the experts, of technical reports in the form of proofs of evidence; their presentation at the inquiry by Counsel and the experts, together with cross-examination; the verbal exchange of the site visit; the eventual inspector's report and, where relevant, Secretary of State decision letter; the appeal if any to the Courts for judicial review, which in itself requires quite different forms of communication appropriate to legal proceedings.
- *Implementation:* When the decision is made, there is need for a line communication between those responsible for implementation and those for execution,

for example, development control by the planning authority in relation to a would be landowner/developer. Where the implementation is to be by public authority through any of its possible measures, each could require a distinct mode of communication. For general purposes there would be planning briefs, or written, audio or visual statements, in the attempt to guide landowners, developers and the public in the direction of the aims sought by the plan. At the other extreme, the building of a trunk road with the use of public funds by the Department of Transport would involve another complex administrative process, with deposit of plans and other information, objections, representations, public inquiries, and so on, which could last for years.

CHAPTER EIGHT

Comprehending the conclusions of a CIE

8.1 Context

In the generic model described in Chapter 7 (at Box 9) we introduced the decision framework as the context for evaluation analysis. There would be enumerated the questions which the particular evaluation would be addressing in its conclusions. These questions we saw would be of two kinds: standard questions typically addressed in the cost–benefit family in evaluation; and questions specific to the case studied, which could be raised from various sources.

The ensuing description of the generic model at Box 11, with the preceding analysis and discussion, presents the raw material from which the answers to the questions can be drawn. The standard questions, the sectoral preferences in the community between the options, and the ranking of the options by this criterion, would be answered from Box 12. All the other questions need to be answered from the material presented in Box 11, with its textual support. It is the totality of answers to the questions which present the basis for comprehending the conclusions from the evaluation.

8.2 The approach

Any community impact evaluation produces a large amount of complex information. This, however, is not so much inherent in the method of analysis as in the substance being analyzed. Any urban area constitutes a complex system of physical arrangements and human activity within its region. Any plan for such an urban area presumes change in this future system which, being in the future, must be uncertain. The changes relate not just to physical facilities, such as a new road or airport, but to all elements in the system which are interdependent in geographical space, and which include human activity of various kinds over time.

Added to this, the demands on the evaluation method becomes complex because of the way in which society insists on dealing with its future in the contemporary world. There was a time, not long ago, when the expert urban & regional planners would present for approval one single option in the tentative plans they had prepared, backed up by technical data but not justified by an evaluation of that option or others. Today, that is completely inadequate. There is an insistence on the presentation of alternatives and on some formal evaluation of the options as a basis for discussion and recommendation. Furthermore, the presentation is not simply to the decision-takers but to the stakeholders and public at large, in public participation. It is this situation that moulds the nature of the evaluation process, its complexity and its openness.

In such a situation, the planners and evaluation analysts are far removed from, say, the civil engineer who puts forward a solution for a bridge over a river. No one would expect him to present the huge amount of background information (drawings, calculations, costings), which he has assessed. Only the summary would be of importance. This is certainly not so in urban & regional planning. The planners and analysts must be ready to make available the full story in open discussion for justification of choice. The result is a large display of information, which is an important feature of CIE, earning for it the description of "display method". Consequently it is necessary to have an approach for drawing out conclusions from the display.

8.3 Display or index

Boxes 11 and 12 of the generic model contain a wealth of information which, to those who can read them, with their supporting text, summarize all that is relevant for the CIE of the planning options. But to others the Tables are too complex and have attracted the following criticism (Peters 1971, quoted in Lichfield 1990a: 81).

> In Professor Lichfield's studies, as they now stand, one is bombarded with information and invited to judge the relative merits of alternative schemes. Monetary measures, quantitative measures and qualitative judgements are superimposed one upon the other in a somewhat confusing array.

Peters contrasts this display with other evaluation methods that produce one grand index number to convey conclusions. For example (Lichfield 1990a: 82):

> In the Buchanan Report, on the other hand, the points scoring system aims to produce an overall index of benefits, although based essentially on the value judgements of the analyst to which anyone might justifiably object. The end results are simple to interpret, but it is a pseudo simplicity which masks the basic components of the analysis.[1]

This "pseudo simplicity" arises in the "grand index" figure which is typically presented in three leading methods of evaluations: multi-criteria (Nijkamp & Spronk 1981), goal achievement matrix (Hill 1968), and conventional benefit–cost analysis

---

1. The contrast is amusing, since the reference is to Appendix 2 of the Buchanan Report by Crompton & Lichfield (1963).

(Ch. 6). These convey a simple enough message, but their full understanding can be grasped by the decision-taker, stakeholders or the public only if the various assumptions, calculations, and so on behind the index are fully explained. A simple analogy will illustrate. The shareholders of a company can readily grasp the "grand index" of percentage growth in profits, profit-earnings ratio or dividend declared. But to understand the background behind these simple figures it would be necessary to go to the full financial accounts of the company and the tabulations behind them. These they would find difficult to grasp, without adequate training, for the same reason as they would the complexities of a CIE display. It is for that reason that the chairman of the company in his annual report will clarify the background, but even then at a simple level. The corresponding report for CIE are the facts behind Box 14, which addresses the critical questions raised for the evaluation, in Box 8.

The proponents of the three leading evaluation methods just indicated are conscious of the limitations of the grand index (Lichfield 1990a). For cost–benefit analysis we have noted the critique in value context above (Nash et al. 1975). For multicriteria evaluation (Voogd 1983: 353): "The quality of a choice/possibility can never be solely represented by one number, since this would deny differences of opinion and the multi/dimensional nature of the choice/possibilities". For goals achievement matrix (Hill 1985: 174):

Although in some of my earlier writings I have suggested a weighted overall-performance score as a possible option when assessing alternative plans in terms of multiple objectives (Hill 1968, 1973), I have come to have serious reservations about this. A weighted overall-performance score may result in the loss of important information which might be relevant for arriving at a decision. In addition, it provides a numerical value which is extremely difficult to interpret.

In brief, wherever the grand index is employed it has a recognized defect in terms of communication to the decision-takers and public, and thereby of democratic planning. To clarify its meaning it is therefore necessary to spell out what lies behind the index. This could be as complex as the CIE display.

Another reason for favouring the display as opposed to the index is the importance of distinguishing in any evaluation between *fact* and *value*. Each evaluation analysis has its own unique statement of the values which are incorporated, just as each decision (whether based on an evaluation or not) is a unique statement on the values which are implied. This calls for the openness in value implication and to the suggestion (Nash et al. 1975a):

The failure of cost–benefit analysis to give a unique answer to the question of whether a scheme is desirable is in no way a criticism of the technique itself. On the contrary, whenever there is a dispute as to the moral notions to be used in evaluating a scheme, it is likely that the results of a cost–benefit study will vary according to which of the opposing value systems is adopted. . . .

Although we have questioned the logic of some of the arguments used to jus-

tify particular procedures, ultimately each procedure can be justified with reference to some set of moral notions, and thus the choice between methods is in
the end and ethical one. . . .

The implication of this is that cost–benefit analysis will be most useful if
results are presented in a disaggregated format, such as that of Lichfield's Planning Balance Sheet (Lichfield 1968) with incidence of effects and methods of
valuation clearly stated. Then it is likely that the reader will be able to adjust
the results to conform with his own moral notions, and thus to reach an informed
opinion on the desirability of the scheme. Indeed, in all value judgements, such
as those relating to equity, known to be controversial, it will be helpful if the
evaluator himself conducts, and presents the results of, a sensitivity analysis on
the effect on his results of varying the value judgements used. On the other hand,
quoting a single net present value or rate of return figure will be of use only to
those who are aware of and agree with the value judgements used in the study.

This argument in favour of the display is that it facilitates the comparison of
options in terms of not only efficiency but also equity. It allows an authority to consider a trade-off between the more efficient options and those less efficient options
that would provide a more equitable distribution of costs and benefits. It is for this
latter reason that Nash reiterates some 20 years after the statement of 1975 his support
for the community impact analysis, this time more specifically in relation to transportation (Nash 1992):

What we would advocate is that a thorough attempt should always be made to
identify all relevant effects of projects and to trace these back to appropriate
incidence groups. For this purpose we strongly commend the planning balance
sheet or community impact evaluation approach. At this stage where reliable
monetary valuations of externalities can be shown they provide useful information on the preferences of those affected regarding how much is worth spending
to secure those particular benefits or to remove those particular costs.

This paradox in relation to the display or grand index is seen (McAllister 1980:
68–9) as part of the "evaluation dilemma" of which there is no general resolution:

To understand the implications of proposed action it is very useful to divide the
impacts into many component parts, but to arrive at a judgement about its desirability it is necessary in some way to reassemble or synthesise the parts into an
understandable whole.

The former McAllister sees as "objective" and the latter as "subjective". But the
dilemma would appear to be deeper than this. Although it is desirable to make an
objective assessment of each of the impacts, it may not be practicable to do so for the
reasons given above (Ch. 8) when presenting the circumstances where a simplified
method of evaluation is needed. In such circumstances, the objective analysis could

be of limited extent, so that subjective judgements must enter. By the same token, when it is necessary to re-assemble or synthesise the parts of the analysis in order to reach conclusions, the area and role of subjective judgements is reduced. Then, the greater becomes the degree of objectivity on the analysis.

Indeed, in the whole spectrum of the evaluation process, however well resourced, there is the continuing need to exercise judgement, both as part of the "in design" and "post design" process (McAllister 1980: 5). Judgements are needed in the most elaborate of impact assessments, as well as in the re-assembling and synthesizing, the conclusions to be recommended to the decision-takers, and the decisions taken by them on the basis of the evaluation. There is thus the need for a continuing balance of the role of objective analysis and subjective synthesis throughout.

8.4 Drawing a conclusion from a CIE display

8.4.1 The case studies

In order to advance the ability to comprehend the conclusions from a CIE we present the following guidelines, derived from studies of CIE in actual cases in practice since 1962. In all these studies the approach has in general been consistent and standardized around what is called above (Ch. 7) the generic model. But the model itself has been adapted over this long experience, the major change being the introduction of formal impact assessment for the reasons described above. Within this evolution it has been necessary to adapt the generic method to the specific study, because of variations in cases such as the following which are brought out in the Appendix:

- the need to treat projects differently from plans
- the scale of plans (from the regional to the local)
- the kind of plans (strategy or policy to plan)
- the content of the plan (a new town, renewal, road building)
- different aspects of a particular kind of content (private motor car, public transport or pedestrian crossing)
- time and money resources available for the study (in depth to rapid assessment).

Despite the diversity of substantive areas that have been analyzed, one generalization can be made. In *all* cases it has been possible to draw conclusions from the display material, however voluminous and complex it has been. Furthermore, in *all* cases the conclusions have thrown light on the choice that is indicated between the options, so that the choice itself is likely to have been more closely aligned to the merits and the de-merits of the proposal than would have occurred in the absence of this analysis. And since, by definition, the bulk of the case studies have offered complexity in making the choice (or otherwise they not have been commissioned from consultants) it is apparent that the ability to draw conclusions from the display has been thoroughly tested.

This is not to say that the conclusions are uniform in their certainty or in providing the basis for clear-cut recommendations. The degree to which this occurs is very much the luck of the draw in the particular case. The following examples will illustrate.[2]

Worthing: alternative road proposals (1966)
In this case two alternative road proposals for central Worthing were offered by the
County Planning Officer and Borough Planning Officer respectively. Discussion on
them had been protracted without resolution. The PBSA showed that the dispute over
the road alignments was not the critical issue, since the options had been designed
with different priorities for traffic in mind. Naturally, the County planning option had
through-traffic in mind, whereas the Borough option favoured local traffic. Thus, the
basis for the dispute needed to be on the relative weights to be given to these two categories of traffic.

Edgware: road proposal for a shopping centre (1968)
Although the issue was comparatively confined, being road proposals for a suburban
shopping centre, the consultants established that there were many options, amounting
to a possible 28. Thus, the choice on standard engineering criteria (relief of through-
traffic, effective pedestranization of the centre, etc.) was difficult. But the application
of the PBSA led to a reasonably clear-cut conclusion in favour of a particular group of
options, and then one in particular, which led to a decision as to the best way forward.

Stevenage: private or public transport (1971)
In this new town, initially designed in the 1940s, it was found that the vehicular traffic on the roads had been underestimated. Accordingly, to accommodate the new
conditions it would be necessary to build extensive elevated roads over existing
roads. But this led to the question: If instead of motor cars there would be an increase
in buses, would the roads then be capable of taking the growth in motor cars without
double-decking them? In order to evaluate these options, considerable studies were
needed from corporation staff and consultants on the public transport options, and the
implications these would have on the volumes of motor car traffic. The study was carried out in consultation with the PBSA analysts, spreading over some three years. The
display material was accordingly complex. But a clear conclusion emerged: that the
public transport alternative was of greater net benefit to the community than the ele-
vation of the roads, and this option was indeed pursued in practice.

Ipswich: urban expansion (1970)
The issue here was whether the town should be expanded to the east or west. The latter was favoured by the farming community, since soil was sandy and not rich as in
the west, but there were urban advantages in going to the east. The dispute had been
pursued over for a considerable period with countervailing arguments, around consultant proposals that favoured the east. The PBSA analysis, which was different from
that carried out by the consultants, clearly came down to favour the east. As a result,
although only one month was spent on the PBSA study, the authorities were able to
take a decision in favour of the east without further delay.

---

2. The bibliographic references are in the Appendix, giving study locations and dates.

East Midlands: airport runway extension (1980)
The engineering options here were simplicity indeed. The one east–west runway needed to be lengthened to accommodate the expected increase of traffic. The options to either the east or west each raised local and regional planning issues. These made for complexity in the decision. The PBSA analysis was able to show that the overall community disadvantage of the expansion would be minimized by a compromise solution of partial expansion to the east and west. This was accepted by the Airport Committee and successfully fought through at the ensuing public inquiry, where the PBSA was presented in evidence.

Greater Manchester: Trafford Park (1983)
Within the three-year consultancy to Greater Manchester Transport, a detailed evaluation of optional bus services for Trafford Park was evaluated. The results were complex because of the large number of options, amounting to 11 in all. In this instance it was found that there were so many uncertainties about aspects of the CIE that a clear-cut conclusion could not be seen. But what emerged was the nature of the uncertainties, which were holding up the conclusions. From this it followed that further research on these particular uncertainties in the study (referred to as cycle 2) would enable clearer conclusions on the options to be drawn. And if such clear-cut conclusions were desirable then further resources on the investigation would be justified.

Greater Manchester: 3-year public transport plan (1984)
The public transport plan related in the main to buses and rail services. The plan envisaged minimal investment in transport facilities, but rather a rescheduling of services on existing routes. Many different packages of routes emerged as ten options. Faced with these the Greater Manchester Passenger Transport Executive had some nine evaluation criteria for choice (Table 13.1). Clearly, each criterion produced a difference in the preferred options. The CIE was able to evaluate the array of choices and options, and indicated its own preference for option 1. This enabled the GMPTE to consider its own weighting as between the varying criteria and to come up with its individual choice based on these weightings, which was option 4A.

Naples: regeneration/conservation of the historic centre (1988a)
The large historic centre of Naples is very rich in the cultural built heritage, and at the same time is sorely in need of revitalization and regeneration, within which conservation would be a major objective. A major planning study was commissioned to prepare such a plan by Italian specialists in conservation, urban planning, transportation, economics, physical resources, public health and so on (Di Stafano & Siola 1988). Each carried out its own sectoral study as contributions to the plan. Each provided rich material on the two prime options: do minimum or embark on a huge programme of investment for conservation/regeneration. It was to aid the choice that the community impact evaluation was directed. Its raw material were the sectoral studies of the specialists in the planning team. From these, conclusions were drawn that showed that while the investment in the conservation/regeneration option would be very large

for both the private and public sectors, the result in terms of financial, economic, social and cultural benefit would be sufficient to justify the programme.

8.4.2 Some guidelines for better comprehension

Such experience of the case studies can be distilled into some guidelines for comprehending the conclusions of community impact evaluation, from the admittedly complex display that emerges:

- The display is not simply in the tables that emerge from the boxes in the generic model, or its adaptations, but also in the textual backup for these tables which show how their predictions are founded on assumptions, facts and values. It is this total display that needs to be available as a basis for comprehending the implications of the analysis and the conclusions.
- For robust and justifiable conclusions to be drawn, the display should follow the rigours of the analysis as brought out in Chapter 7, which in turn must reflect the rigours of the two foundations of the analysis, namely impact assessment and cost–benefit analysis (Chs 5 and 6). Unless this be done, the interpretations from the display, and accordingly the conclusions, will not be reliable.
- But that in itself is not sufficient, for CIA is not simply the *adoption* of the two foundations as they exist, but is an *adaptation* of them for the purpose of evaluation in urban & regional planning. CIA has thus needed to develop its own theory and principles, largely in a pragmatic fashion from the case studies which have been carried out. These are summarized below (Ch. 10) and must also be respected in the rigours of the analysis.
- Of particular importance here is the comprehension of what is referred to below (Ch. 10) as the social accounts that describe the transactions of the producers and consumers within the community sectors, for these lead directly to the content of the critical Boxes G and H in the analysis.
- Within this it is also necessary to understand the role in these transactions of assumptions, predictions, facts and values; and also the differences between measurements in the transactions that are in money value (valuations) or in physical terms.
- Against this background, the search for the critical conclusions can be clarified by posing the questions involved in Box E, which specify the decision framework. As introduced above, these are of two kinds: the standard questions brought out in relation to Box H, and the specific questions introduced for each particular analysis. These, by definition, will vary from case to case and will be made up from the variety of sources and not simply from the analysts pursuing the methodology. The sharper and more pointed the questions, the more helpful they will be to drawings acceptable conclusions from the analysis.

In exemplification, we close this chapter by repeating from particular studies the particular questions that were framed as a basis for drawing the conclusions. They are taken from those case studies introduced above (§8.3).

8.4.3 Some typical specific questions from the case studies
Stevenage (1971)
A simple initial question was put by the Corporation when it commissioned the study: Should it facilitate the use of cars in Stevenage by building the elevated roadworks and accepting only a minimum supporting bus service, or should it encourage travellers to use public transport by supplying an enhanced, fast, modern, and convenient bus service, thus eliminating the need for the elevated roadworks? Our report gave a fairly firm indication that, on the hypotheses and data set out, the latter course would be preferable in the general community interest. This was amplified in the answers to the following four questions:
- Would the travelling public use either of the two bus systems which have been devised, to the extent predicted in the report?
- If the travelling public would use the bus system as forecast, what would be the differences in costs and benefits to all sectors of the community compared with the use of their cars?
- What prices would need to be charged to bus passengers to make the services financially self-supporting in the sense of balancing revenues and expenditures?
- If this charge were higher than bus passengers could be expected to pay, and so high as to counteract the attractions of the service, would there be the possibility of cross-subsidization from savings accruing to other sectors of the community from the enhanced bus service?

East Midlands (1980)
Having provided the community impact analysis, its findings were presented in response to the following questions:
- What will be the total array of impacts generated by the runway extension and associated works.
- Given this array, which are the candidates for amelioration and what degree of amelioration is necessary and, accordingly, what guidance can be given to both the Joint Committee and the County District Planning Authorities on operation and development control policies for the airport.
- From the array of impacts to be ameliorated and controlled, would there be aggregate net benefits to the community in extending the airport beyond what would have occurred on the runway as originally designed?
- From the array of impacts so ameliorated and controlled, will the distribution of costs and benefits (on sectors or geographically) be acceptable in terms of equity?
- Will more total net benefits accruing to the public from investment at EMA be greater than at another regional airport (e.g. W. Midlands Airport, Birmingham)?

Greater Manchester (1984)
Having presented the display showing preferences under each of the Executive's nine evaluation criteria (see Table 10.3), the implications for the decision were brought out in response to the following questions:

- What contribution can CIE make as one of the evaluation tests in the selection between plan options by Greater Manchester Transport and Greater Manchester Council?
- Which is preferred among the various options in terms of overall efficiency to the community?
- Which is the preferred option in regard to equity between different community groups, including their mobility and accessibility?
- How can efficiency and equity considerations be traded off against each other?
- What will be the return to the Greater Manchester Council for its revenue support for the adopted plan?
- What will be the impact of the public transport plan on the Council's highway system?
- How will the public transport plan affect County Structure Plan policies?
- What will be the effect of the Plan on overall government finance?
- What will be the effect of the Plan on central government finance?

Naples (1988)
The community impact evaluation itself was presented in the Consultant's Report. It is from that evaluation that answers were provided to specific questions aimed at bringing out the implications of the evaluation which were pertinent to the decisions facing Naples on the conservation/regeneration of the historic centre.
- Since the interventions can only be initiated by the investment of resources in conservation, what will be the financial cost and returns involved?
- Given the interventions which are proposed for conservation, which sectors of the community would be called upon to provide the inputs and which to receive the outputs?
- Given the intervention in the manner proposed, what will be the benefit that would flow, and what the cost, to each of these sectors?
- Given the enumeration of impacts just described, how are they categorized according to the kind which is of interest not only for the comprehensive evaluation but for the views of, and possible decisions to be made by, individual organizations and institutions in relation to the kind of impact that is of concern to them?
- Taking the impacts as a whole, how can they be distinguished further between those that are real in terms of the way of life of the people concerned, or are transfers in the sense of simply transferring wealth from one group to another, without affecting the way of life, except for the consequences of the redistribution?
- On the efficiency aspect, what judgements can be made as to whether the intervention is worthwhile in the sense of comparing net benefits with costs for the community as a whole?
- Can the distribution of costs and benefits between the sectors be regarded as equitable between them?
- How are these conclusions related to those in the specialist reports that have been prepared, which have been used as inputs to the evaluation?

- Having regard to the answers on the above, can recommendations be made for the adjustment of the project as so far described in terms of, for example:
  - modes of intervention
  - phasing of the intervention over time
  - amelioration of the disbenefits from the intervention, in the interests of greater equity.

8.4.4 Summary

In some case studies, the conclusions are reasonably certain. In others, the uncertainty persists and there was the need for cycle 2: further exploration and perhaps further investigation (Trafford Park 1983). In some instances this can be by sensitivity analysis. In others, it must take the form of clarification of intangibles and uncertainty by further measurement. In such situations the measurement justifies further expenditure, if a specific need for it has been shown to be necessary for improving the basis for a decision.

But even if the conclusions be uncertain and further exploration not carried out, experience in case studies shows the cycle 1 analysis alone to have various positive contributions. It enables a great deal to be learned about the strength and weaknesses of the various options, so that, should it be necessary to reach conclusions at that stage, a better basis for decision is provided than would otherwise exist. And where the decision is not to proceed because of the uncertainty, the analysis provides the basis for the generation of a different array of options which, by definition, would be superior to those used so far.

Thus, the least that can be said for the cycle 1 analysis is that it will have provided a basis for a judgement on the decision which is more comprehensible and defensible than would have been possible without the analysis. And as such it provides a basis for the further consideration necessary for improving the quality of decisions.

CHAPTER NINE

Simplification of the generic method

9.1 The occasions for the simplified method

The generic method has been streamlined for logical procedures and for potential computerization: by disaggregation into the discrete steps A–L with each step progressively articulated within itself. Its full application is certainly complex. While the complexity is no greater than in other applications of the rational method in planning (systems analysis, transportation modelling, population projection, employment and retailing predictions, etc.), it nonetheless presents a formidable face. Accordingly, for those situations in the planning process where the complexity is not warranted or practicable, something simpler (quick and dirty, short cut, rapid, crude) is needed. Similar situations arise in other forms of analysis. An example is cost–benefit analysis where rapid appraisal is called for to outline the key merits of a project at an early stage, with a view to deciding on whether or not to proceed (Bridger 1986: 4).

In CIE such situations arise where:
- There are resource constraints
  - in the background data readily available for the study
  - in the time and money available for making the background studies, or for the analysis
  - in the skills available for the background studies and the evaluation analysis.
- In plan-making (Lichfield et al. 1975, McAllister 1980: 272):
  - where the evaluation analyst is working as part of the planning team, there is the need for fairly rapid assessment of plans or part of plans in order to enable the team to advance its work to the next stage, or
  - where there is cyclical evaluation within the planning process in which the evaluation analysis is a step towards generating another option
  - where "crude" options have been generated for preliminary analysis, in order to distil them into a selected group of "fine" options for more careful evaluation.

These limitations are not of a uniform kind, so that the means of tackling the constraints would need to be considered *ad hoc* for each evaluation study, as brought out below in relation to study management (Ch. 20). For example, limitation in background data, which cannot be overcome, will affect the depth to which the evaluation analysis can penetrate; the limit in evaluation skills will affect the robustness of the evaluation method itself.

For these reasons the simplified methods introduced here must be treated as a variation from the "generic", with a view to its adaptation to the context of the study which is explored in terms of study management. But whatever the adaptation, "simplified" cannot mean "sloppy". The method must accord strictly with the basic theory, principles and application of CIA, for otherwise the conclusion will be unreliable and any subsequent refinement will be on unsound foundations, for example where there is the need to move from "crude" to "fine" options.

9.2 Which steps to simplify in the generic model

9.2.1 A review of their purpose and complexity

As seen in Chapter 7, the steps and boxes in the generic model vary in their purpose and are not uniform in their complexity. It is therefore pertinent here to make a review of the method as a whole, with an eye to its simplification.

Steps A–E: project description

Boxes 1, 2 and 3 These aim at defining the nature of the planning process involved, the project and the options. As such, they are an essential preliminary and, not being complex, should be carried out in full.

Boxes 4 and 5 Beyond indicating the concept, these play no part as such in the full generic method and the same will apply here.

Box 6 The aim here is to identify those variables in the plan (from implementation through to decommissioning) that will create the effects and impacts. It is thus useful for comprehension of the project and, as they are not complex, they should be attempted in full.

Box 7 This takes the clarification in Box 6 a stage further. It is not an essential step, except in complex cases, and could be reserved for them.

Box 8 This is a further aid to comprehension of the implications of the project for the system. As such it does not add directly to the analysis itself. But it is useful as a record of continuing changes in the system which would be introduced by projects, and as such could be useful in that context if not for the analysis itself.

Box 9 This is a critical step in the analysis, since it focuses on the questions that need to be addressed in the evaluation, be they standard questions, ones that were part of the initial terms of reference, or which have emerged in making the analysis. They are therefore an essential ingredient.

Steps F–G: analysis

Boxes 10, 11   These are the heart of the assessment and evaluation and, as such, are critical to the analysis. But by the same token they are the most complex and time consuming. They are therefore the main candidates for simplification and are addressed as such below (§9.3).

Step H–J: conclusion

Box 12   As the summerization of the two preceeding steps, this is both simple and essential for proceeding.

Box 13   From the viewpoint of the analyst, negotiations are not essential in order to reach some conclusion. But they are desirable from other viewpoints. They will help to test the analyst's assumptions *vis à vis* those who are consulted. They will contribute to the discharge of the decision-takers obligation for public participation. They will divert from later stages arguments that could have been disposed of earlier. They will be essential for the bargaining process. It is on these aspects that a view must be reached as to whether or not this step is taken.

Box 14: Evaluation report   Clearly a stage that cannot be avoided.

Steps J15/16: choice/decision and communication

Box 15   A raison d'être for the whole analysis and therefore essential.

Box 16   The logical end-product of the process which is thereby essential.

9.2.2 Conclusion

The preceding review explains how simplification would not be uniform throughout the method, but selective. In brief (referring to the Boxes):
- important for clarification of the issues and not complex in operation: 1, 2, 3, 6, 8 and 9.
- dispensable for the simplified method: 7, being a refinement for complex cases; 13, if the context of the evaluation process permits;
- essential for the method, but simplification called for because of complexity: 10, 11, 12, 14;
- essential, since it is the justification for the whole of the process: 15, 16.

9.3 Effect and impact assessment combined

This midway level of simplification is traced through by reference to the generic model.

Project description

Boxes 1–9   These steps in the generic model do not relate to the evaluation analysis itself but rather to the clarification of the subject matter of that evaluation. This is needed in any project, whether or not it is to be subject to rigorous evaluation analysis, if only to clarify to the analyst and explain to decision-takers and the public just what the project and options are and the questions being raised. And even if the subsequent analysis (§9.3) is to be only on a judgmental basis, the clarification will offer a better basis for informed judgement, by enabling the project itself to be better understood and to provide a firmer basis for any judgements required within the planning team, or for advice to the decision-takers and explanation to the public.

Effect assessment and impact evaluation

Box 10: Effect assessment   As indicated above, it is here that a major issue arises in respect of simplification. If the background studies on effect assessment are available, they can be simply summarized for transfer to this Box, to the level of detail available without further study. This was done in the Prospect Park case (see Ch. 14), which showed for each system element in turn the conclusions from the work of the environmental assessment team, from the baseline to the predicted effects following mitigation/enhancement. If the background studies are not available and cannot be provided to or by the evaluation team, within the resources available, then the simplified method would bypass this Box and proceed to Box G11.

Box 11: Impact evaluation   Such elimination of Box 10 and reliance on Box 11 is virtually a return to the method of the earlier case studies in planning balance-sheet analysis which did not specifically include the effect assessment (which was not known at the time) but went direct to the impact evaluation, in the equivalent of Box 11, then referred to as Table A (see Ch. 5). An example is given in Table 7.15. Except for the treatment of the effect assessments in Box 10, the simplified method would closely follow Box 11 for the analysis of the generic model (see Ch. 7). In brief: Box 11.1 identifies the checklist of community sectors; Box 11.2 brings down, as needed, the plan variables from Box 6; Box 11.3 presents the summary of impacts.

The impacts would be identified in relation to the project as a whole, or in cases of complexity to each of the plan variables in Box 11.2. The identification would proceed judgmentally by preparing background tables from the relevant plans and supporting material, as seen in Tables 7.5 and 7.6. Then would come the process of aggregation in relation to the community sectors (Ch. 7). This process was simplified in the Prospect Park study (§17.4.2) and further simplified in the Manchester Airport study (§17.4.3).

The remainder of Box 11 would then be followed as described in the generic model (Ch. 7). While accepting that the precision of the data would be less, there would be no short-cuts or diminution of rigour. A different approach was pursued in the Naples study (1988) where the team members did not pursue formal impact assessments but made judgmental assessments from their professional knowledge (economics, land economy, sociology, public health and traffic). These were enumerated by these topics and then aggregated by sector by the analyst.

Boxes 12–16   For the remainder of the process the generic method could be followed as described above, while bearing in mind that the data in Box G11 would be more judgmental than if the generic model had been pursued in full. In particular this would apply to Box 13–16, on consultation, recommendation, choice, decision and communication. Here there would need to be reference to the fact that the evaluation analysis had taken short cuts for simplification, so that any consequential limitations on the conclusions, recommendations and decisions could be taken into account.

In particular, there could be emphasis on any aspect that clearly depended on particular impacts, with the warning that, should they be considered particularly significant for the decision, it might be prudent to enter cycle 2, that is spend further research time on their examination, in order to test the sensitivity of the conclusions in relation to these impacts.

9.4 Applying the simplification

9.4.1 Negotiation and bargaining

The discussion, negotiation and bargaining that takes place prior to the submission of an application involves a process quite different from that undertaken by either the landowner/developer, in preparing his scheme prior to submission, or the local planning authority after the submission. It involves the exchange of views around the project involving either the two main parties alone, or these parties with others (e.g. consultees, third parties, public at large).

The essence of this process is not decision-taking as such, but rather the mutual exchange of views with the intention of mutual influence between the parties. This brings with it the exchange of information from each of the parties, in their support of particular viewpoints. This exchange can take many forms: at the one extreme, collaboration to evolve the scheme which satisfies the parties, and at the other strong advocacy without a meeting of minds. It is in this atmosphere that the question arises: how can the generic model help?

Although the dialogue could take place around a well thought-out analysis which is the basis of discussion (and as explained below it should do so), the typical situation would be one in which the full information was not available Then the aim of the discussion would be to build up a picture consistent with the generic model, step by step as need be. Pursuing this approach:
- Box 1: what is the aim of the process?
- Box 2: what is the project being put forward?

Should at this stage the project be well defined, then there could usefully also be a description of Box 3, the project options, and also 6, the plan variables, in order to understand the components of the project.

But in the typical case the discussion could revolve around a sketch scheme, with no options and only a sketchy indication of plan variables. Here the analysis could jump to Box 9, defining the questions to be answered, such as a mutual exploration of a scheme which is acceptable to the planning authority. From here, there would be another jump to Box 11, where the essentials to be established would be the community sectors (Box 11.1) in order to get the summary of the predicted impacts on those sectors (Box 11.3).

From this starting point, the discussion could lead to a succession of further options (Box 3) and then to the modification in Box 11.1 and 11.3.

The results would be a series of options, each with their comparative essential ingredients in Box 11. Scrutiny of these would lead in due course to an array of acceptable options, which would be ripe then for a more carefully prepared evaluation.

9.4.2 Development control: choice between options

The generic method of project appraisal is well suited to development applications, for each can be regarded as a proposed development project that needs permission to go ahead, so that refusal is another option. Development applications are also an important field simply because of the vast numbers that are processed in the planning system, even at low ebb some half million per annum, and also because there is considerable scope for improving the typical decision-making process. But there are, of course, great variations in the complexity of the projects which are the subject of the application. They verge from the simplest (for example, infilling in residential, office or shopping, where the surroundings dictate the obvious decision) to the very large, complex and controversial, which are the subject of appeal, and Secretary of State decision. Faced with this array, some guidelines are as follows.

Even at the simplest (excluding the trivial) it is the fundamental *approach* that matters. Having identified the planning process as development control (Box 1), and then the project (Box 2), it is important to identify the options (Box 3). These can range from the simplest (refusal, or permission with conditions) to variations in the project, of the kind typically brought about in the planning department, with or without discussion with the applicant.

From this preliminary statement, the analysis would then move directly to Box 11 in the simplified form. Starting with the checklist of community sectors (Box 11.1), predict by judgement the impact that would fall on each, that is the effects that would change their way of life (Box 11.3). The results would be placed in tabular form to strike the differences between the options (11.4), leading to an indication of sectoral objectives (Box 11.5) and then to conclusions on preferences (11.7).

Given the starting point of the two simple options, refuse or permit with conditions, the foundation is laid for inclusion in the simple analysis of further options which are generated via the consultation process, or the analysis in the planning department.

This would lead to the *balancing* that is essential in the report to the Planning Committee. But, in place of what is often an inspired leap to a conclusion from the balancing without supporting reasons, there would be a coherent basis for justification. In this there would be a great deal of intuitive judgement, not backed up by facts.

But that is superior to the intuitive judgement without a structured basis that often exists in the planning recommendations.

From this simple level there are clearly others according to the complexity, importance to the planning authority, resources available, and so on. It is here that the level of simplification is adjusted to the circumstances. At the other extreme, in the case that goes to appeal, in which the importance of the outcome to the applicant commands considerable resources, the full-blown generic method can be used.[1]

9.5 A conclusion: rigorous but flexible

From Chapters 7–9 a single conclusion emerges. The generic model is comprehensive, complex and rigorous. But its application is flexible according to the needs of the case in question. In this, however, it is possible and important to retain the rigour as far as it is practicable in the approach. And in this it is important to keep the relevant theory continuously in mind. To this we now turn.

---

1. As in Prospect Park (1992), East Midlands (1980), Brentford (1984), Manchester (1984), Brigg (1987).

CHAPTER TEN

Theory and principles of community impact evaluation

10.1 Theory and principles already covered

In Part I we saw how CIE was rooted in urban & regional planning, and in Part II how its theory and principles of were built up from two foundations: impact assessment and cost–benefit analysis. Accordingly, in presenting the theory and principles of community impact evaluation in this chapter, we rely upon what has been said about the theory and principles of these foundation elements, and here concentrate on those not already covered.

This we tackle by developing the theory that underlines the principles of the generic model (Ch. 7) step by step through the model itself. In this, Steps A–E, Boxes 1–9, have already been covered as follows:

- The planning process (Box 1)
- The nature of projects (Box 3, 5, 6 & 7)
- Projects in the urban & regional system (Boxes 2, 4, 8).
- The decision framework (Box 9)

We accordingly proceed directly to Steps F and G, which are the heart of the CIE.

10.2 Effect assessment: Box 10

10.2.1 Theory and principles derived from impact assessment (IA) in general

Our presentation on impact assessment above (Ch. 5) showed that, although it was possible to describe impact assessment in general, it was necessary to distinguish between different kinds of impact. In the assessment of these impacts in practice there were common elements accompanied by specific variations. These variations in the main have not been built around one common academic and professional discipline, but rather around the particular discipline that in practice makes the assessment itself; e.g. ecology, hydrology, Earth sciences, for the natural environment; sociology, economics, anthropology, for the socio-economic environment. From this it follows that the theory and principles that support practice are to be sought in a great variety of fields, some founded in the natural sciences, and some in the social sciences, together making up the body of academics and practitioners comprising either *natural or social environmental scientists*. Their work, which has been sketched out above (Ch. 5), enters as appropriate into the theory and principles of community impact assessment, namely: effects, impacts and environment (§1.1), the varying kinds of impacts (§1.3), prediction of impacts (§1.4), selection of impacts to be assessed (§1.5), assessing the significance of impacts (§1.5), devising the mitigation or amelioration of impacts (§1.5), measurement of impacts (§1.5).

10.2.2 Theory and principles derived from community impact evaluation

The impact chain

The prediction of impacts in CIE goes beyond that typically undertaken in IA in general, because of the importance given to two considerations introduced above (Box G11 in §7.5.2). First, there is the distinction between *effects* and *impacts*. Secondly, there is the attachment of the effects to individual community sectors, thus deriving the change in its way of life, which is the impact. These two form the basis of what was called the impact chain (Ch. 7). In passing it can be noted that the workings of this chain has echoes in the related concept in cost–benefit analysis which was introduced above (Ch. 6), in providing some clarification to the first, second and third-round effects of costs and benefits. From this description of the chain it is apparent that the linking of effects to people is a major departure from conventional effect/impact assessment, which generally speaking has tended to see the impacts as just scientific phenomena in themselves. Justification for this assertion can be illustrated by reference to the three methods of impact assessment which were amplified above (Ch. 5).

Impacts and people

Natural environment   Evidence here rests with the concentration on the natural environment in USA practice under NEP Act 1969 in the earlier days, with the human dimension being brought in later, primarily through social impact assessment. It rests also in the light reference to "human beings" in the practice which has emanated from the European Community.

Economic   The prediction of economic impacts proper relates to impacts of significance in relation to the national economy, such as, jobs, income and expenditures. Although these clearly relate to people, the concept is the totality of the population in the community under consideration, without normally attempting any identification of incidence on particular groups. Indeed, a limitation generally adopted is the definition of costs and benefits "on whomsoever they fall", which runs counter to any attempt to trace through sectoral impacts, as also does the insistence on the need to avoid double-counting; it is the ultimate, and not interim, impact which is explored.

Social   Although clearly relating to people, this could be only a somewhat narrow concern for the people as a whole. As examples: there is the attempt to limit the implications for the people in that they ". . . reside in the vicinity of the development site"; and also by the exclusion of "strictly economic perspectives".

Identification of relevant community and community sectors

"Community" is another one of those elusive words. It revolves around a Latin root, *communis,* namely *in common,* be it for example neighbourhood facilities, local administration, or an education system. But like other concepts borrowed from common-sense usage, it has been used with poetic licence. Indeed, in a literature review, some 94 meanings were identified in the research literature of the social sciences (Herbert & Raine 1976). In these, three elements tend to recur with a high degree of regularity: social interaction (sociological), common ties (human) and territory (geographical).

In community impact evaluation, the *common* relates not so much to these elements but to those that would arise from the aggregate of individuals affected by an injection into the urban & regional system of which they are part. Such impacts will not be uniform for each of the individuals or families. Any particular impact will be beneficial to some and adverse to others: short-term and reversible to some, or long-term and irreversible to others. This differentiation results in a delineation of "sub-communities", each of which shares a predicted impact "in common". But even such subdivisions are not homogeneous. Any individual can identify for himself a series of subsectors according to his activities and interests. These are multiplied in a family. The man has his work, social life, religious life centred on the church, recreational life centred on football, riding to hounds. The woman will share some of these interests (social or religious life) but have others: the daily nurturing of children, interaction with other mothers, educational interests centred on the Open University. The children will have their own interests (school, games, hobbies). From this it follows that a given injection into the system is likely to produce effects/impacts which are significant to particular families, but with varying implications for their way of life (as families or individuals).

Accordingly, when faced with an impact assessment on options, the individual or family cannot give a simple reaction of approval or disapproval to any particular one: the reaction, even for any individual, must be a mixed one. Each will need to weigh up the package of impacts on himself/herself; and each family will need to assess/evaluate the composite picture in terms of the family's "social decision". From this it follows that individuals or families cannot in reason ally themselves for or against particular projects in relation to particular impacts unless they have somehow resolved the balance of their individual and social choice. However, in practice, they probably react against the impacts they recognize to be adverse in particular options, in the hope of ameliorating them and thus advancing their expected net gain. And the chances are that they are not considering the opportunity costs in relation to the other impacts, should the options they resist be abandoned.

It might be thought from these wide possibilities that the enumeration of community sectors would vary significantly in different evaluation studies. But, as indicated above (Ch. 7, Box 10.3), this is not so. The experience from the case studies has been consistent: adopting the owners/producers/operators versus the consumers categorization, there emerges a reasonably standardized array of the sectors, which are built around their socio-economic functions. It is this characteristic that facilitates the preparation of the impact chain, in that a standard categorization of community sectors can be made provisionally (Box 10.3) as a basis for refinement when the effects are predicted more carefully.

It is from the delineation of sectors through impacts in the way described that we are then able to draw conclusions on the *relevant community* that experiences the impact. Sectors delineated this way will not aggregate into a single identifiable entity, but must be made up of overlapping geographical boundaries, many of which will not be tidy at all, and many of which can be seen to have the allegiance of particular individuals or families according to which of their different lives are impacted. From this it follows that the *relevant community* of those impacted by a particular project could have several meanings:

- the impacted communities, in the immediate locality and wider
- the functional community relating to particular impacts
- the administrative community, having legal and financial responsibilities for handling the impacts, compensation,amelioration, etc.
- the area of the planning authority within which the project falls, whose jurisdiction could be local, subregional, regional or national
- the national community, where a project is seen as having a particular array of costs and benefits which are of interest to national government and thereby can be given high priority (the British and French Governments on the Channel Tunnel compared with the local authorities, Kent or Amiens)
- a collection of governments, where the impacts are seen to be of importance beyond the boundaries (environmental pollution and the EEC)
- the public interest, which is a diffused expression of some or all of the above.

Insofar as these different communities are stakeholders, their decision frameworks will have a bearing. And insofar as any of these communities have overriding power (national government in certain instances in urban & regional planning), their decisions will have the power to prevail.

10.3 Impact evaluation: Box 11

10.3.1 Overview

The aim in Box 10, to extend effect assessment from scientific entities to incidence on people in a community, is still within the general approach of impact assessment (Ch. 5 above). In Box 11, however, we enter an area which is only touched upon in impact assessment proper but not significantly advanced. The aim in this box is to fill this gap: to move from impact assessment to impact evaluation (Lichfield & Marinov 1977, Lichfield 1985). It does so in Box 11.5 onwards, which is applied to information in Boxes 11.1–4, which we now take in turn.

10.3.2 The "valuation" of the impacts (Box 11.5)

The output of the effect/impact assessment is a measurement of the predicted effects/ impacts, with the measurement being in various dimensions (Box 10.3). If these are to be evaluated, they must be translated into the *values* of those community sectors who will experience them: how will these perceive and judge whether the predicted impact will advance or retard their predicted welfare. This step can be seen in terms of the theory of supply and demand relating to any goods or service (Lancaster 1977). These will have supply-side attributes intrinsic to the good or service itself, and can thus be described and measured objectively, in the way attempted in the measurement of predicted effects. Such supply-side attributes are perceived and judged subjectively by prospective consumers in terms of their expected utilities. They are thus *valued*, so setting up demand. In effect/impact assessment, it is the effects that are the *supply* and the human reaction to the impacts that represent the *demand*. Where the predicted and measured impacts are the kind exchanged in the marketplace – for example economic impacts such as jobs or incomes – then the value could be assessed from the starting point of market price. But where they cannot be so exchanged (being free goods, public goods or externalities), there will need to be mounted the surrogate valuations of benefits introduced above (§6.10). These could involve a major exercise. Thus, a short cut is necessary.

The short cut is pursued by postulating for each impact a "*sectoral objective*" (called "instrumental objective" in early case studies) by which each impacted community sector is assumed to perceive and judge their preferences (expected utilities) from the impacts in question. The judgement is simple in form: given the impact and the sectoral objective, would you consider that your welfare would be advanced or retarded (leading to a benefit or cost respectively) or be unaffected (leading to neutrality)?

The definition of sectoral objectives can be helped by some kind of "market research" in respect of the specific impact predicted. But in practice this will be practicable only if associated, for example, with a major study such as research on a planning inquiry. For one thing, the exercise would be laborious and costly; for another, in some cases it may not be practicable to address the question to the pertinent audience, e.g. future generations. For this reason some surrogate is needed. This could be attempted by informed expert opinion (estate agents experienced in buying/selling in the market; sociological or anthropological observation; the analyst guessing from everyday knowledge, with the help of experts in the Delphi method; the members of the planning team). The sectoral objectives approach here is clearly not *true* valuation of the impacts per se. In this we would be visualizing individuals who are each making a complex subjective valuation of the attributes of effects, whereby each individual would be presumed to differ from others in an unpredictable way. Rather it is visualizing a "notional sector of individuals" who are aggregated together for one particular impact, and who thus might be expected to have a more homogeneous view about their reaction to any particular attribute of an effect (e.g. traffic noise, atmospheric pollution). Thus, the sectoral objective must be seen as relating not to the *actual* people impacted, but to *notional* people for the particular type of impact; these are "notional" in the sense that they may be represented in different sectors for different effects.

10.3.3 Ranking options in costs and benefits from impacts (Box II.6)
The example demonstrated in the generic model was simplified for ease of presenta-

tion. It shows only four options (with one being the datum) from one plan variable. But in practice the array of costs and benefits could be considerably more complex, owing to the variety in the cases that could be tackled. Accordingly, we introduce here some generalization on the manner of handling the content of this sub-box, based upon experience in case studies:

Where the ranking is ordinal only, the question arises: does the ranking in fact convey which of the options would be most desirable? Some maintain that ordinal ranking would do so (e.g. Holmes 1975) and some in contradiction that the ranking is better in cardinal terms (Nijkamp et al. 1981).

Given the need for cardinal ordering, the question arises: can this be derived only from a ratio scale or is it practicable to make the translation from the ordinal to the cardinal? It is here that multi-criteria theory can be of help where the data are available (ibid.).

10.3.4 The role of objectives in CIE

Having introduced "*sectoral objectives*", it is relevant here to place their treatment in CIE within the long-standing controversy around the role of goals/objectives in the planning and evaluation literature. For this purpose we start with the classic formulation of the two major proponents in the controversy in the 1960s, Lichfield and Hill. Lichfield's stand was that the goals and objectives in planning must be geared to the normative planning process in the following way (Lichfield 1968):

- Formulate goals at the outset of the process (treating these as synonymous with ends, aims, objectives, values, policies) as *value statements.*
- Apply the goals to the whole or part of the area for which the plan is being prepared, with a view to recording whether they are already achieved in current conditions or, where they are not, by how much.
- Any shortfall would lead to a shopping list of specific *area* objectives, relating to issues that should be tackled in the plan. These can be defined as *planning problems* that exist currently, or could emerge in the future if conditions in the area follow the expected scenario. These *planning problems* can thus be defined as *gaps* that exist between current and predicted conditions and that *ought*, in the normative sense, not to exist, the normative clearly having some relationship to the setting of *planning standards.*
- These gaps are subjected to the conventional planning studies, having regard to *opportunities* in the areas which exist for their resolution (currently or in the future) subject to constraints of a physical, financial, economic, political nature. Such *opportunities* can be seen as *negative* problems.
- The result is the emergence of a series of *proposals* to tackle the *problems* and take advantage of opportunities. These can be translated into *programme targets* which are to be implemented in the planning process.
- The *targets* would be subject to the kinds of feasibility tests described above (Ch. 3).

Hill formulated his method of goals achievement matrix as a critical reaction to single-sector cost–benefit analysis and also Lichfield's balance sheet of development

(Hill 1968). His view of goals was much the same as Lichfield's: "An end to which a planned course of action is directed". But he has a different concept of the objectives flowing from the goals, namely that they are instrumentalized from the goals by making them more specific, as either means or ends: "For the purposes of the goals achievement matrix, goals should, as far as possible, be defined operationally: that is, they should be expressed as objectives. In this way the degree of achievement of the various objectives can be measured directly and the costs and benefits can be identified". And they are pre-weighted at the outset. They are thus not welded into the planning process and so made operational in it; it is the *achievement* of the initial goals/ objectives that provide the critical test of the plan's achievements.

Table 10.1   Goals achievement matrix (Rel. wt.: relative weight; C: costs; B: benefits).

Goals description Relative weight	a 2			b 3			c 5			d 4		
	Rel. wt	C	B	Rel. wt	C	B	Rel. wt	C	B	Rel. wt	C	B
Group a	1	A	D	5	E	–	1		N	1	Q	R
Group b	3	H	–	4	–	R	2		–	2	S	T
Group c	1	L	J	3	–	S	3	M	–	1	V	W
Group d	2	–		2	T	–	4		–	2	–	–
Group e	1	–	K	1		U	5		P	1	–	–
		E	E					E	E			

Source. Hill (1968).

Hill crystallized his approach in his adaptation of the planning balance sheet display into that of a goals achievement matrix as in Table 10.1. He sharpened his contrast with planning balance sheet in 1971 as follows (Lichfield et al. 1975: ch. 4):
- "Goals-achievement analysis emphasizes goals identification as the point of departure and the entire focus of this method is on the level of goals-achievement for the community as a whole and for sectors within it. The Planning Balance Sheet emphasizes the identification of sectoral effects as a point of departure . . .
- In GA analysis, community goals and sectoral goals are assumed, identified a priori, and provide the basis for both plan formulation and the determination of costs and benefits for purposes of plan evaluations. The Planning Balance Sheet emphasizes the identification of sectoral objectives in the course of the analysis of costs and benefits.
- GA analysis emphasizes the explicit derivation of objectives from higher-level goals.
- By determining how various objectives will be affected by proposed plans, the goals achievement matrix can determine the extent to which certain specified standards are being met. Is the transportation plan likely to meet minimum accessibility requirements and minimum standards of comfort and convenience?"

From this comparison of the classic approach in each method, it is seen that the difference can be simply expressed in answer to the question: is the planning process "goal" or "problem" orientated (Needham 1971)? The goals achievement matrix follows the first, and planning balance sheet/community impact analysis the second. More specifically, whereas the first seeks to test initial goals in terms of achievement, the second argues that any initial goals are relevant only in triggering off the planning process; and it is not the initial goals that should be tested but the predicted effects/impacts from the planning and programme targets. The "worthwhileness" which is being explored is the benefits versus the costs of the impacts rather than the degree of objectives achievement. So sharp is the distinction between these approaches that, whereas goals/objectives achievement cannot be pursued without the prior formulation and pre-weighting of the goals, the planning balance sheet analysis *could proceed* by confining the goals to initial value statements. In practice, this would require that the area study under step (2) of Lichfield's approach above (§6.3.4) would be carried out so as to generate objectives from problems, so bypassing the "value statements". Indeed there could be seen to be strength in this approach if the goals/objectives and initial aims/policies, etc. are too lightly and generally formulated. This can readily happen when those asked for the goals/objectives (decision-takers, planners or community interest) are detached from the discipline of feasibility, and then are not prepared to have them modified in the light of exercise of that discipline. Then the all too common situation can arise: a great deal of effort is spent on testing goals which in themselves have little foundation.

10.3.5 Difference between sectoral and planning objectives

Having identified the nature of *sectoral objectives*, we now indicate the difference between them and *planning objectives*. In essence, whereas the former provides a means of *valuing* the output of the planning process in terms of *effects* translated into *impacts* for those sectors that are impacted, the former relate in essence to the *input* to the planning process. These inputs commence with the statement of goals, explicitly or implicitly. One formulation is (Lichfield 1968):

> Goals are statements of directions in which planning or action is aimed. They derive from human values and as such are ethical, that is empirically untestable. They are fundamental in that they stem from the apparently insatiable wishes of the human species for greater self-fulfilment. As such they would not be suddenly changed or abandoned. They are ideals over a horizon which will never be attained, since progress towards them over time implies their reformulation in yet higher ideals.
>
> Objectives, on the other hand, are seen as specific steps towards the attainment of a goal, and thus although an end in itself also as the means of achieving a more distant goal. They are attainable, and thus factual as opposed to ethical in that the degree of attainment can be specified, measured and tested.

Although there is little dispute as to the nature of goals and their synonyms (aims, ends, ideals and, in certain circumstances, policies; Solesbury 1974), there is much about the nature and definition of objectives. In essence the differences relate not so much to the legitimacy of using objectives as inputs in the planning process as to the differences in the nature of their use for the purpose. As brought out above (§10.3.4) Lichfield sees the differences in meaning deriving from steps in the planning process itself (Lichfield 1968): There are three:

- As value statements, they may be *preservative* in the sense of wishing to maintain or retain as far as possible inherited and traditional aims; or they could be *acquisitive* in the sense of setting out to achieve something which does not yet exist (Ackoff 1962, 1963). It is inevitable that the value statements can be formulated only in general and perhaps vague terms. This arises because the aims relate to a complex of values about which individuals may not be certain, because of the indefinite time horizon, because particular goals can often be mutually conflicting and competing and thus act as constraints on each other, and perhaps be impossible of achievement within such constraints.

- The value statement is taken as a context for the area study that corresponds with the survey and analysis of the area being planned: its physical form and extent, the nature of the socio-economic activities being carried on, and the trends and expectations in these matters over the period for which the plan is being prepared. The study would show, for example, the deficiencies in physical form (inadequacy of road system) and the inefficiencies and disabilities of socio-economic activity (long journeys to work and incompatibility of shopping and motor vehicle activities in a shopping street). These stem from the nature of defects inherited from the past: clearly there must be objectives to remove them. But this alone of course is not sufficient, for the planners must also visualize the defects that could arise and opportunities that might be wasted in the evolution of the town over the future. For this purpose the forecasting studies in the planning process can be used to reveal where the town would be in say twenty years, and to set as an objective some desired deviation from this future state. The area studies thus lead to a shopping list of specific area objectives that emerge to be dealt with.

- The area objectives so defined are in essence statements of the *gaps* that exists between the conditions at the time of the area study and what, in a normative sense, *ought* to be found as an output of the planning process. As such they may not be realistic for the area in question. Some might be unobtainable (generous job opportunities for school leavers in a small town); some might be in conflict (proximity of residents to shops and the shops' economic viability) or beyond available resources (the need for redevelopment leading to a heavy burden on the local tax base). The area objective is therefore the *target* in the familiar process of the preparation of optional plans or strategies for the area. The result would be a plan or plans, which explicitly or implicitly contains a programme for change or resisting change, a programme which has been tested for viability in relation to internal consistency, feasibility, conformity with value statements, means of implementation, etc. This specific programme can then be regarded as the *programme targets*.

10.4 Social accounting in impact evaluation

10.4.1 Social accounting in Box 11.6

As seen above (Ch. 7) there are in the generic model a variety of background tables. From the amplification of Box 11 in Tables 7.15 and 7.16, it is seen that each linked or associated pair of owners/producers/operators and consumers are considered engaged in a "transaction" whereby the former produce services "for sale" to the latter. The producer's cost can thus be equated with the consumer's benefit, and the producer's benefit with the consumer's costs. These transactions are not confined to goods and services directly exchanged in the market. They extend as well to the indirect, for example the traffic noise "sold" to the occupier of an established house by the builders of a new traffic road nearby. Thereby it is ensured that, in the CIE transactions of concern in urban and regional planning, none be omitted, however intangible the services produced. Its existence, incidence and, where possible, order of magnitude, all help to fill out the picture. Thus, the Table presents a set of "social accounts" summarizing all "transactions" in the project, one account for each community sector, or subsector.

Given the number of transactions and options, and also the number of items that are not measured in money or physical units, there is clearly the need for some rigorous set of accountancy rules. The following have been found useful in the case studies (e.g. Table 7.16).

- Benefits and costs must either be kept separate (negative benefits are not treated as costs, nor negative costs as benefits) or treated uniformly as net benefits (positive benefits less positive costs).
- Capital and annual benefits and costs are kept separate, the former being those which occur once and for all, and the latter those that will recur.
- Where measurements of benefits or costs in money, time or physical terms are possible, figures are inserted in the Table. When a measurable entry could be measured but has not been, an "M" or "T" or "P" or "N" is used instead of figures, to indicate that money, time, physical units or number would be employed of a capital nature, or "m", "t", "p" or "n" to convey annual flows. An "I" (intangible) indicates that an entry is not measurable. Subscripts distinguish the different "M", "T", "P", "I" or "N" (or "m", "t", "p", "i" or "n") entries. The ordering of numbers progresses through each of the items if that entry is translated from capital to annual terms or vice versa, or is the same in a producers and consumers item.
- Where an increase or decrease in an "M", "T", "P", "I" or "N" item from a datum can be forecast, but not the amount of change, a (+) or (−) sign is used.
- No sign indicates positive (e.g. M) and a bar indicates negative (e.g. M̲).
- When the costs and benefits of certain producers and consumers are covered by those of others, an "E" (elsewhere) is introduced and cross references are shown in the remarks column. This avoids double counting.

Since each line for any item can contain costs and benefits that are not in figures and therefore cannot be arithmetically reduced, an algebraic reduction is needed. For example it will indicate which costs and benefits are smaller or greater (< and >), where they are equal (=), where the difference is not certain (N/C). The conclusion is shown in Box 11.7 by indicating which scheme has the net advantage for that particular entry, and where there is uncertainty (N/C). In all this, the important rule is followed: one particular scheme is always compared with the datum, so that differences always show in favour or against the datum, so leading to ordinal ranking.

Where any social account contains more than one entry, we "reduce" the account to its simplest terms by:
- ignoring E entries
- eliminating entries containing costs and benefits common to all alternatives
- adding the entries algebraically, offsetting negative against positive
- capitalizing annual entries or, more usually, converting asset values to their annual equivalents, to achieve a common form.

Where all entries are in money terms, the result is one figure for costs and one for benefits in each transaction. Even where the non-measured M, T, P entries are shown, it might be possible to see the direction in which differences lie, and perhaps to gauge the order.

10.4.2 Incidence in double counting

In the social accounts of the generic model regard must be had to the fact that the social accounting comprises "double entry" (as in financial accounts), in the sense that the benefits and costs are not netted out "on whomsoever they fall" as they are in economic accounts. On the conventions of the national income accounting, for example, it would be dubious to count both the savings to motorists through reductions in vehicle operating costs from a new bypass and also the increase in land values that could be credited to those very same savings. However, in community impact analysis, "double entry" is a necessary and deliberate feature, for it makes it possible to trace the *incidence* of the impacts on the various parties who are affected and to recognize the difference between the economic and transfer transactions. In this sense the accounts/transactions are not in fact double-counting. This becomes apparent in considering a "transaction" between producer and consumer. The former will have costs and benefits. The latter is the cost of the consumer, whose benefit is the utility/disutility from consumption.

10.4.3 Economic or socio-economic income

The question arises in relation to the social accounts just described: how can these supplement the conventional national income accounts on gross or net national product (GNP or NNP) these can be misleading (Pearce et al. 1989: ch. 4): "But GNP is constructed in a way that tends to divorce it from one of its underlying purposes: to indicate, broadly at least, the standard of living of the population. If pollution damages health, health care expenditures rise, that is an increase in GNP – a rise in the "standard of living" – not a decrease. If we use up natural resources then that is capital depreciation, just as if we had machines we count their depreciation as a cost to the nation. Yet depreciation on man-made capital is a cost while depreciation of environmental capital is not recorded at all".

By contrast, the social accounts in Box 11 include not only the kind of transaction which figures in the national income accounts but also those which do not, comprising sociological, cultural, natural environment and hazard costs and benefits. Thus, for the project option in question there is the possibility of considering the non-financial/ economic costs and benefits which will emerge, and thereby putting into better perspective the changes in standard and quality of life which the project will bring about.

What is not suggested here is that such project analysis can provide in itself the aggregate of adjusted national income accounts, which can chart the growth of the country in direct and indirect terms. But since such national accounts are in the end made up of a host of individual transactions in the economy during the year in question, what *is* suggested is that a balance sheet be prepared on the lines indicated for major projects so that in the decision-making their *true* contribution to the national income in these total terms can be derived, as a by-product of the community impact analysis, in both stock and flow terms. The double entry will enable incidence to be pursued, which is not possible in the national accounts. Although not entirely capable of numerical aggregation, the conclusion can at least give a lead to desirable practice on projects, plans and so on, which will for example be capable of recognizing the changes of growth which have insufficient regard to the environment. It will also enable the related questions to be considered: will there be sufficient economic growth from the project to sustain the environmental improvements that are sought; and will the project result in sustainable development (see Ch. 13)?

10.4.4 Nesting of the CBA family within CIA

We saw above (see Ch. 4) that CIE can be regarded as one of a family of cost–benefit methods. By comparison with the others, CIE embraces all pertinent community sectors, and thereby more, direct and indirect, costs and benefits. This leads to the possibility of CIE being able to act as a "supra" (meta) method of evaluation, in that within it can be "nested" the sectors and their costs and benefits related to the other modes of evaluation, such as financial appraisal.

The means of nesting in relation to transportation is demonstrated in Table 10.2. Column 1 lists a summary of community sectors. Column 2 presents various transport modes. For each of these is *tentatively* illustrated those community sectors in column 1 which must *necessarily* be reflected in the evaluation of that mode. From this display any decision-taker or stakeholder in column 1 can select for a particular mode that array of impacts on selected sectors (and the costs and benefits flowing from them) which are his *direct* concern in the evaluation method at the foot of the Table. And it also shows how any particular stakeholder, for example, landowners, government, travellers or public at large can, by tracing horizontally across the Table, make a comparative evaluation of any mode to answer questions relevant to his/her interests.

All the preceding is by definition embraced in column 3 which relates to integrated land use, transportation and environmental planning; it covers all the items in the evaluation reflected vertically or horizontally in column 2. To pursue the nesting it is necessary to reconcile the differences in definition of measurement and valuation techniques between the different members of the cost–benefit family. Notable examples relate to the treatment of common items in financial appraisal and economic

Table 10.2 Nesting principles in CIE for transportation.

	1. Community sector	2. Transport mode						3. Land use, transportation, environmental planning
No.	Description	Roads	Bus	Light rail	Rail	LRT	Air	planning
	Producers/operators							
1	Landowners	X					X	X
3	Infrastructure authority/company		X	X	X	X	X	X
5	Transport mode authority/company	X		X	X	X	X	X
7	Operator of mode		X	X	X	X		X
9	Economy:							
	local							X
	regional							X
	national							X
11	Government budget	X	X	X	X	X	X	X
	Consumers							
2	Occupiers	X						X
4,6,8	Users of transport mode:							
	travellers	X	X	X	X	X	X	
	pedestrians	X	X	X		X		
	facilities	X	X	X	X		X	
	Non-users of transport mode							
	natural environment						X	X
	social environment							X
	economic environment							X
	Workforce:							
	jobs							X
	incomes							X
	Public at large							X
	Evaluation method	FA COBA	FA SCBA	FA SCBA	FA CBA	FA SCBA	FA CBA CE	CIE Incorporating the others

Source: Lichfield (1992).

appraisal (Harrison & Mackie 1973, Sugden & Williams 1978). The former derives largely from accountancy and the latter from economics.

10.5 Decision analysis: Box 12

10.5.1 The standard questions

Efficiency
As brought out above (§8.4.6) Box 11 comprises a set of social accounts summarizing the transactions between the relevant sectors. In this the costs and benefits are the sum of the direct (private) and indirect (social, externalities), together making up the total

of costs and benefits. The division between the direct and indirect is somewhat arbitrary in that it follows established law at the time of making the analysis, the direct being defined by "property rights" and the indirect relating to impacts which have been called "amenity rights" (Mishan 1969). Should for example there be a change in law or practice, whereby residents had legal rights to privacy and quiet, and promoters would need to buy such rights in order to infringe against privacy and to cause noise, the boundary between the direct and indirect would change and so would the demarcation between private and social costs and benefits, with the total cost and benefit remaining the same. A similar situation arises in parallel where planning gain is present (Ch. 15), resulting in agreement for the private sector to bear infrastructure costs which are traditionally borne by the public sector. It is to the total costs and benefits that the three standard questions relating to efficiency should be applied (see §4.1) by means of the five criteria of value for money (§6.7.2). However, difficulties are immediately seen in doing so; in any particular analysis comparatively few of the transactions in costs and benefits are measured or valued. Thus, what can be done to squeeze from the data available the best possible *satisficing* answer to the standard questions?

There is little problem about applying the questions to those categories of costs and benefits which are typically tackled by financial analysis, social financial analysis, cost revenue analysis, cost–benefit analysis and social cost–benefit analysis (Table 4.1) for these are to be found by the "nesting" principle within the overall CIE display and can be extracted for the purpose (Table 10.2). For the CIE we are led to rely in the analysis on the ranking of options in terms of the *preference* by each of the sectors for the impacts that they will experience, the rankings being based on such measures and values as are available. It is the aggregation of such rankings that is transferred to Box 12 as a guide to the option which has the overall preference; it is to this box that we look to judge the preferred option. The determination is clear in those situations where none of the sectoral preferences is non-certain and the preference of the sectors is consistently for one particular option.[1] But where this simple outcome is not apparent (in the majority of cases) we have the following judgmental rules for the situations now described:

(a) No non-certainty among the sectors, but preference for the options is not uniform. If the sectors are assumed to be of *equal weight* for decision-taking, the preliminary conclusion is reached from adding up the number of sectors that prefer the particular option. But even this simple number is not conclusive, since the size and characteristics of cost and benefit for each of the sectors will not be the same. Here judgmental assumptions will need to be made for these variables, by scrutiny of the background material, leading to a conclusion in Box G11, as the basis for balancing the sectoral preferences for the options.

(b) As in (a) but *equal weights* between sectors cannot be assumed and varying weights must be applied to them. These should come from the decision-takers reflecting *their* "social values", leading to a "weighted preference", which can be taken as *the* conclusion for the decision. The conclusion could be illumi-

---

1. For example, the case study of Ipswich (1970), see the Appendix.

nated if the weighted preferences can be subject to a value sensitivity analysis, to explore for the decision-takers the implications of their weighting. A similar approach can be adopted if no weights are available from the decision-takers; the analyst here provides his own weights as assumptions, and subjects them to sensitivity analysis to reveal to the decision-takers the implications of a particular weight distribution, perhaps in terms of opportunity cost.

(c) As in (b) but with *non-certainty* about particular items being pursued as a basis for considering the implications of the non-certain items. By definition these could be capable of clarification with further research to probe for greater certainty in those items that are non-certain. But the need for such research might be avoided by pursuit of the question: is it likely that clarification of the non-certainty would alter the conclusions reached when ignoring the non-certain items? This question can be probed by sensitivity testing of the non-certain items to ask what would happen if they proved to favour options other than that brought out from the preliminary analysis. This might or might not lead to confirmation of the need for further research in order to obtain a reasonably clear conclusion.

The fact that the conclusions and recommendations from CIE cannot aim at *optimizing*, and must be content with *satisficing* and *second best* solutions, introduces considerable scope for judgement. This in turn suggests that the use of judgement will be facilitated if the ideal of pursuing the three standard questions of social cost–benefit analysis be abandoned in place of criteria that are less exacting. Some possibilities have been indicated above in relation to the evaluation of environmental impact (§5.4.2). Of these, three particular criteria would seem useful in that each is derived loosely from the cost–benefit family approach, but each avoids the rigours of the standard cost–benefit family criteria.

Equity of distribution
As brought out above (Ch. 6), unlike efficiency there is no clear staring point with guidelines from economic theory. But whatever the view taken on this question by economists (§6.3), there is no doubt that CIE *must* handle equity, for this is built into the prevailing decision rules for urban & regional planning. The test is not therefore yes/no but how. We pursue the answer under three heads.

Given that equity is to be handled, what rules to adopt?   Having a foundation in cost–benefit analysis, CIE embraces the answer to this question as described above (§6.3b). In addition, there are the special rules applicable to land-use planning.

The starting point is the classic view of the role of such planning, namely intervention in the market to achieve the *public* interest in the use and development of land. This can be interpreted as minimizing the divergence between the direct (private) and indirect (social, externality, costs and benefits), by bringing the indirect into the calculus as far as possible (Turvey 1963):

Where social cost exceeds private costs, private interests will sometimes do what is not in the social interest and there is a case for restrictive interference. Where social benefit exceeds private benefit, private interests will sometimes not develop and thus fail to do what is in the social interest. Here there is a divergence case for interference to encourage development. In other words there is a case for controlling development where the developer either does not bear all the costs of the development or does not receive all the benefit.

The calculus must clearly be carried out within the constraint that the redistributive effects are not such as significantly to undermine efficiency, namely output in relation to input (Cheung 1978).

In support of the equity approach, there is a prevailing ideology in urban & regional planning: that there should be a shift in distribution and that this should be progressive in favour of the community against the developer. More specifically, there is the prevailing recognition of the need to redress the contemporary balance in favour of minority groups, and the underprivileged, particularly the urban poor. Of the three criteria for doing so (§6.3b), there is a tendency to concentrate on needs, as opposed to rights or deserts.

This divergence and internalization is tackled every day in urban & regional planning through the plan implementation process, in particular of development control (see Ch. 13). In this the thrust is to seek to ameliorate the social cost and to enhance the social benefit on any development proposal.

One way of strengthening such regulatory controls is the taxation solution of Pigou or Viner (Ch. 6). But these in current practice are not readily available, because plan implementation and land policy measures do not generally provide for *ad hoc* taxation on projects, although "planning gain" could be seen as a rare instance (Ch. 15). However, proposals of a similar kind have been made in the field of environmental control, as a means of minimizing pollution by the use of the price mechanism (Pearce 1989: ch. 7). As such, considering the complementarity of the physical planning and environmental controls in practice, and the possibility of merging the two in respect of development control, it could be that such pricing will also be introduced in the urban & regional planning field, with a view to mitigating pollution costs, on the generally accepted "polluter pays" principle (PPP). In this the basic mechanisms are (Pearce 1989: ch. 7):
- *setting charges or taxes* on the polluting product or input
- setting *standards* the cost of which is initially borne by the producer
- setting a standard, issuing *pollution permits* in amounts consistent with the standard, and allowing those permits to be *traded*.

The estimation of the appropriate charge or tax level involves the valuation of environmental cost (Pearce 1989: chs 3, 12). But it is not only the estimation of the pollution costs that is complex and controversial, but also their incidence: in practice, will the polluter pay, or will the consumer of the product pay? If the consumer, will the amount of pollution be reduced if the polluter can pass on the price? And furthermore (Turvey 1962), ". . . the imposition of a tax upon the party imposing external diseconomies can be a very complicated matter, even in principle, so that the *priori* description of such a tax is unwise".

Difficulties such as these tend to discourage reliance on the market as against regulation, unless the price mechanism is simply used to encourage accordance with the regulation, under mechanism (b) above. It also argues against using mechanism (c), for pollution which might not otherwise take place is legalized through sale of rights to another party wishing to create the pollution.

Should compensation be paid for losses?   The payment of compensation is also implicit in the criterion of the Pareto improvement and the Hicks–Kaldor variations (§6.3c above). As indicated there, although the compensation principle is introduced there is in fact no provision for the compensation to be paid to the losers. Such provision does arise in the urban & regional planning system (Lichfield & Darin-Drabkin 1980: ch. 5), but from taxes and not the beneficiary. In Britain the pre- First World War statutory planning scheme established statutory development rights in accordance with the scheme, with the requirement for compensation to the landowner, should he be denied the use of those rights in the planning system. By contrast, in the immediate post-war development plan, development rights vested in the State under the 1947 TCP Act. But the unscrambling of the financial provisions of that Act in 1953 left the development rights to the State which could, under the planning permission system, grant such rights (for which no payment was asked) or deny their use (for which no compensation was paid). It is this grant of rights and denial of compensation that affects the magnitude of the benefits and costs that will flow from the implementation of the project.

Under the present formal financial provisions there is *generally* no compensation payable to landowners/developers who are denied development rights which potentially could impose social costs, and thereby undesirable impacts. Outside these formal provisions the question of compensation for injured parties does arise when the development is to be permitted and the amelioration, internalization, and so on, would still nonetheless result in damage to impacted sectors. It is here that provisions have been suggested for introducing the *polluter pays principle*, so that money adjustment is made between the polluter and those affected. This is different from using the principle to deter environmental pollution; by definition in this case the pollution is to be allowed, on the proposition that net benefits would flow to the community, so that unless compensation be introduced the injured parties have both the pollution and also are denied any financial payment (Lichfield 1989a). Instead of this reliance on regulations supplemented by market incentives, there is the other approach to externalities of making it possible for the parties, who either impose or suffer the external diseconomies, to negotiate financial settlement to their mutual advantage, whereby the diseconomy is suffered but compensated. This system calls for reducing transaction costs (Coase 1960). But *even* if the contractual procedures could be sufficiently simplified to minimize transaction costs; and so make the externalities more readily tradable in the market, we are still left with the harsh reality: the pollution and disutility will still prevail. All that has happened is that parties who would suffer would have a countervailing *financial* advantage. But the effect on the environment as a whole could be the same in *real* terms; and it is this overall environmental damage that needs to be reduced.

The social welfare function   From the above it is apparent that CIE can make its contributions towards improving equity in decision-making, simply through its display as a basis for action on the incidence of costs and benefits to different sectors in Box 11. Indeed this is one fundamental reason for its introduction. Furthermore, the availability of the display can help to firm up the use of a social welfare function, which sets up some *ad hoc* equity weighting principle to be employed in the decision. Being general principles, they clearly cannot take into account their implications for the different groups to whom they are addressed. From this it follows that the best possible intentions in equity redistribution could result in anomalies in distribution which were not apparent from the outset. By contrast, CIE displays the specific redistribution that would flow from any of the options under consideration, so that the equity rule could be adapted pragmatically as a result. From this it follows that, with a given set of guidelines, there would not be consistent or uniform redressing of the imbalance between sectors, but rather a series of adjustments based upon the individual case. Although this offers room for argument against the adoption of a clear principle, it nonetheless makes for a greater acceptability in terms of any particular case in hand: provided that the reasoning behind the equity adjustment is well understood and justifiable, and in itself does not produce inequity.

Trade-off between efficiency and equity   Since the efficiency tests would normally be numerical, and the equity test not, the combination is complex. In the absence of numerical solutions, a practicable approach would appear to be for the efficiency and equity to act as mutual constraints upon each other. Given the wide array of possible interpretations of the concept of both "sustainability" and "development" there will be the need for a specific definition to be adopted. This could come at various stages of the formal generic evaluation model, for example in the description of the planning and implementation process (Box 1), in the options (Box 3), in the decision framework (Box 9). The application of the specification to the project in hand will emerge in relation to Box 8, which describes the predicted change in the system as a result of the project, by reference to each of the system elements. In this way a picture can be obtained of the changes the project will produce in terms of: people (1a); nature (1b, 2b); man-made (1b, 2a); cultural built heritage (1b); the economy (1a, 2c). In relation to these elements will be predicted the impacts (Box 10) and the evaluation in terms of preferences (Box 11). Then comes the summary as a basis for the evaluation (Boxes 12 and 13) where the evaluation criterion are introduced.

Non-standard questions   By definition, non-standard questions may or may not be related to the standard ones. They would be, for example, if they merely introduced different efficiency criteria (such as minimum cost) or equity criteria (such as positive discrimination in favour of the underprivileged or low income groups). Then, by definition, the decision rules would be compatible with those in the standard questions.

Ranking   The same principles apply here as were introduced above in relation to Box 11, under ranking (§4.4.5 above).

Weighting At the heart of the evaluation is the derivation of preferences of the various sectors by reference to the relation of costs of benefits falling on them. In this derivation, certain implicit or explicit assumptions arise, as for example consumer sovereignty in choice; disregard for inequality of income and therefore marginal utility of money in making choice; money values or their surrogates as an expression of value. But whereas these assumptions must be rigorously retained in relation to benefits/costs for economic efficiency, they are typically disregarded in planning decisions, since at their root are the impacts on different sectors of the community and the need (for political or other reasons) to have regard to applying weights showing their relative importance in the mind of the decision-taker. It follows that this "weighting" is essentially a political decision. Given their indication by the decision-taker they can be applied by the analyst; or he can aid the decision-taker by illuminating the effect of different levels of weighting to the outcome; and further still by estimating the opportunity costs of particular weightings assumed by the decision-taker or adopted by him.

Weighing Having ranked and weighted, the stage is now set for weighing up opposing considerations. Where these can be expressed numerically, the conclusion is determinable. But where not, it is judgemental, in which matrices of options can help.

Social/collective as opposed to aggregated individual choice In the derivation of cost–benefit analysis from welfare economics theory, including the Paretian approach, the implicit assumption is that society is made up of an aggregate of individuals, each with differing valuations for utility and disutility, and that the cost–benefit analysis aims at an aggregation of the valuations of those individuals who are involved. But the decision-taker, in reaching his conclusion for choice, could depart from this assumption. He might take the view that the aggregation of the individual valuations does not reflect the true interests (as judged by himself, his Party or advisors) of the community as presently constituted, on whose behalf he is taking the decision; for future generations, whose welfare he must in planning clearly take into account, and whose individual valuations cannot be included into the aggregate valuations derived. This we call *social/collective* as opposed to individual choice.

This distinction is provided for in cost–benefit analysis through the above principles of "weighting", the adjustment of the efficiency measurements in the direction of social issues, such as distribution, the merits of different groups (children as against adults), the conservation of natural resources against economic growth or the cultural built heritage against demolition, strengthening national defence as against consumption (guns before butter), savings as against consumption (developing countries). The principles by which this is done (termed "social pricing") have been described above as "shadow or accounting prices" (Ch. 6).

10.6 The evaluation report (Box 14)

10.6.1 Presentation and discussion of the Report

The purpose of the evaluation report is to present the analysis along the lines of the generic model, in a form appropriate to communication and discussion, with a recommendation from the analyst. The evaluation report is thus the shop window in which can be displayed the relations between the analyst and others concerned for discussion. These others are:

- the remaining members of the professional team on the project. Their relations with the analyst and his work will be established at the outset (Box A1 on the planning context), and their data, information and opinions will have been reflected throughout. Insofar as they have views, these should have been ventilated and reconciled throughout the ensuing process, in which the presentation in the shop-window will give them occasion for comment. A likely instance will possibly be their preference for another recommendation, simply because their own "evaluation analysis" could be based upon their professional and academic background (e.g. economists or traffic engineers; see W. Midlands, Appendix).
- The decision-taker: The role here is quite different: since it is for the decision-taker to formulate the elements in the decision framework (Box 9) and for the analyst to present these conclusions in that form (Box 12) and in the evaluation report itself (Box 13). There is thus the interplay between fact and values in the analysis (see §10.8).
- Decision-takers/stakeholders: Although the main focus has been on the decision-taker, he/she in that decision must be sensitive to the large number of stakeholders whose conclusions in relation to their own subfields in the project will have a bearing on the decision-takers conclusions. Each of the stakeholder's contributions will act as some kind of a constraint on the project decision-taker, just as the project decision will act as a constraint on them.

Thus, the comprehensive analysis can be seen as comprising many sub-analyses, each relating to the decision context of a stakeholder. It thus would be practicable for the analyst to carry out these sub-analyses in a consistent way within the overarching analyses, if he were invited to do so. In effect, provided there is consistency in the rules in the various analyses, the sub-analyses can be said to "nest" within the overall analysis (Table 10.2).

But even if these sub-analyses do not enter into the overall exercise, the conclusions will emerge in the consultations, negotiations and discussions that will be taking place around the development project itself: informally during its preparation or more formally following its completion (when decisions are taken) and on formal presentation of the Report. Although it would not be the role of the analyst per se to be party to these negotiations, the conclusions from them would clearly have some relevance for the overall analysis, for they would illuminate the sectoral impacts, the costs and the benefits, and the criteria used.

10.7 Fact and value (Myrdal 1969)

From the generic model it can be seen that although evaluation is a technical process it is infused with value judgements, both of the analyst himself and also the decision-takers. The following illustrates:

- *The planning process (Box 1)*   What is the level and nature of the intervention envisaged in the planning process, both in plan-making and implementation.
- *Decision/space framework (Box 9)*   What are the inputs from the planners and analysts in terms of the options that are generated for evaluation, the constraints on them, the value content in the methodology. And what are the inputs of the decision-takers in terms of their constraints, questions posed and criteria for resolving them.
- *Effect assessment (Box 10)*   Which community sectors are presented by the analyst and how are they subdivided. And what have the stakeholders to say about sectors that are included or excluded?
- *Impact evaluation (Box 11)*   On what basis are the sectoral objectives formulated by the analyst and to what degree do the decision-takers seek to amend them in terms of imposing their political criteria in "social decisions".
- *Decision analysis (Box 12)*   What are the judgements introduced by the analyst in the preferences and ranking which are transferred from Box 11 to Box 12. And what are the inputs of the decision-takers in terms of equity criteria, weighting between sectors and social decision?

From this it follows that, in presenting the report to the decision-takers, the analyst needs to be conscious of the difference between his analysis, which is as objective as he can make it, and the subjective value judgements he has inserted. This consciousness would lead to emphasis in the report on the analyst's value judgements, on which it is reasonable for the decision-takers and the community to comment and challenge. If this consciousness is not followed, then the way is open for technocracy in the role of the planners and analysts in the evaluation. A related issue is whether or not the report itself should contain a recommendation or simply present conclusions upon which the decision-taker is to act. There is a danger in the latter course, since the analysis is sufficiently complex to justify careful handling if appropriate conclusions are to be drawn. For this reason, the analyst's role can usefully be extended to make recommendations. But if value judgements enter into the recommendation, they could usurp the political prerogatives of the decision-taker. This distinction comes to the fore if the analyst is faced with a decision-taker's conclusion which either cannot be objectively based upon the analysis and its conclusions, or does not follow the recommendation. Insofar as this is based upon the decision-taker's values, then the analyst must recognize the political prerogatives. But insofar as the decision can be said to be a distortion from the conclusions of the analysis that are not value laden, then the question arises: Should the analysts formulate an independent view? In one sense he should not. He cannot place himself in the position of the decision-taker and put forward some alternative decision, for that is not his prerogative, and he does not owe it to his professional conscience. It is also not his prerogative to suggest an alternative decision to that taken, where the difference lies essentially in values adopted by the decision-taker on, for example, the analyst's facts. But what he does owe to his professional conscience is some comment on decisions that do not legitimately reflect matters other than the decision-taker's values. His professional conscience could lead him to show the opportunity cost (in terms of community welfare) of the decision that is being taken against the decision that in his judgement *should* have been taken. In this sense he is postulating an additional client which may be different from that of the decision-taker, namely the community in general.

10.8 Value content in the cost–benefit family

10.8.1 Introducing value

In categorizing the above array of evaluation methods (see Ch. 3) we showed the recognition that grew up in the 1970s and 1980s of the need to distinguish between the methods on two counts (§4.3).

- the kind of planning/decision framework within which the method is being used, and thereby its objectives
- the kinds of questions that the decision-takers were posing, and thereby their criteria for choice.

This second distinction emerges in the criteria for evaluation used in the different methods (see Ch. 3), as for example:

- A1 checklist of criteria: subjective and somewhat arbitrary performance standards posed by the planners or decision-makers
- B1 unit cost: the cheapest for a given supply, where a value cannot be placed on the output
- C1 financial appraisal: financial rate of return, net present value or profitability to the particular promoter or investor.

Given this wide-ranging variety, the question arises: how is the choice to be made between evaluation methods? This question was explored above (§6.2) in relation to cost–benefit analysis. Since this has no standard approach, and each variant answers different implicit or explicit questions, each thereby can have its own underlying "value content" (Nash et al. 1975b: 86). From this it follows that in adopting any particular method for an evaluation there needs to be spelt out the underlying value set to which the evaluation method must be consistent. Furthermore, since the value set derives from moral notions, the value judgements which have been used should be made explicit, so as to be readily understood by those affected (the analyst, collaborating professionals, the public and the stakeholders). This being so, the question must necessarily arise: how would the conclusions from the different methods of analysis vary with different value judgements. This in itself could give rise to the need for value sensitivity analysis, just as there would be such analysis for, say, differing rates of discount. An instance of value sensitivity arises in the general presumption in cost–benefit analysis that the decision should be derived from an aggregation of individual preferences. This in turn requires supplementary value judgements in order to articulate the value set further, namely:

- Whose preferences should count?
- Which preferences should count?
- When should individual preferences count?
- How should individual preferences be aggregated as between (Ch. 6):
  - the Pareto Rule
  - the Hicks–Kaldor Rule
  - maximizing collective utility
  - market voting
  - management science.

This value content is either explicitly provided by the decision-taker, who is thereby pointing the analyst to a particular method which reflects that content; or by the analyst, who proposes a method for the particular situation, thereby implying the value content. From this it follows that if the method is not consistent with the value content in mind, the criteria for choice that are adopted will necessarily lead to a choice inconsistent with that value content; equally, if a method is adopted which does not specify the value content, the choice will implicitly be based upon that content.

10.8.2 Values influencing choice

Since the value content is based upon a moral, ethical or value judgement (that is, one which is not verifiable or falsifiable by fact), it follows that there will be variety in judgement between individuals, groups, government agencies; and there will be variety over time. This judgement is the privilege of the decision-taker who is to act on the evaluation, and it is the obligation of the analyst, in order to ensure that the method of evaluation is consistent with the decision-taker's value content, to point out any inconsistency between that content and that implicit in any particular mode of evaluation. It is also his obligation to draw attention to the opportunity cost of any decision that departs from that emerging from the analysis based on the indicated value content.

The influence of criteria for choice between a set of options, implying variations in value content, is brought out sharply in the Greater Manchester Transport study, which used an array of methods in one evaluation (Lichfield 1987b). The Transport Act of 1983 required a passenger transport executive to provide an annual rolling plan for the following three years. This was to conform to guidelines laid down by the Department of Transport and also the County Council as passenger transport authority, within the wider functions of the County Council in such matters as development and transportation planning, highways, responsibility for public sector expenditure, and so on. The first of such plans for Greater Manchester was prepared in 1983 and the second in 1984, and were submitted to the Department of Transport by the Transport Executive with comments by the County Council.

The plan was essentially a financial analysis with descriptive and analytical material regarding continuing bus and rail operations. It was made up of a series of options around the three variables of fares, level of service and subsidy. In devising an option, regard was had to the cost of the services to be provided, the level of demand anticipated and benefits to potential users. Table 10.3 presents in the columns the various evaluation methods that were used and in the rows the ten options evaluated. For each, where practicable, is shown the ranking in simple terms: best (b) or worst (w). Whereas the CIE pointed to option 1, the decision-takers weighing up the conclusions under the different methods chose 4B.

Table 10.3   Summary of evaluation criteria and conclusions: Greater Manchester.

Evaluation criteria	Options									
	1	2	3A	3B	4A	4B	5A	5B	6A	6B
Passenger transport exec. corporate objectives										
Objective trips per head per annum	b	b					w	w	w	w
Load factor	b	b		w				b		w
Vehicle-miles per standard hour	no significant results									
Cost per standard hour	b	b	b	b	b	b	w	w	w	w
Secretary of State revenue support	w	w						b	b	b	b
Economic evaluation			b	b	b	b	w	w	w	w
Strategic planning policies	b	b	b	b	b	b	w	w	w	w
Accessibility	b	b	b	b	b	b	w	w	w	w
Highways implications									w	w
Ratepayers										
Domestic			b							w
Other			w							b
Total cost of option (a) to community	no difference									
Community impact evaluation	b							w	w	w	w

Notes: Conclusions on ratepayers are based upon a comparison of Option 2 with Options 4B and 6B.
Community impact evaluation is based upon a comparison of the two "extremes": Option 1 with Options 5B and 6B.
Source: Lichfield (1987)

10.8.3 Comparing the value content of evaluation methods

Following this approach we are able to distinguish between the various evaluation methods in the dimension of value content. Some examples will illustrate:

- Conventional cost–benefit analysis, derived from welfare economics, measures the benefits by the aggregation of individual preferences, based on market prices (or surrogate market price) as an indication of willingness to pay, supplemented by consumers surplus. The value content here implies acceptance of the contemporary distribution of income and marginal utility of money for the individuals in question, and their individual assessment of the preferences of future generations. This clearly does not take into account any "social judgement" on the contemporary patterns of consumer demand, or of the "social" as opposed to individual discounting of current generations for their descendants.
- Planning balance sheet/community impact analysis: The value content here goes beyond that implicit in the aggregation of individualistic preferences in conventional cost–benefit analysis. For one thing, although individual preferences and their aggregation are respected as a starting point, there may be the need instead

to follow social preferences, even though these be judged to be somewhat pater-
nalistic. Examples are individual preferences reflected by market voting can
give undue weight to antisocial propensities (drugs, smoking); may not be sufficiently sensitive, or be oversensitive, to the needs of future generations through
adopting the current distribution of income and marginal utilities of money; or
if they are so sensitive, may adopt doubtful intergenerational weighting. For
another, it sets out to include all pertinent externalities, which by definition are
not covered by prices in the market. In brief, there is scope for evaluation that
introduces a "social decision" made by decision-takers on behalf of their relevant community, which would be a departure from the aggregation of individual
preferences derived from the market, including individual discounting by the
current generation for their descendants.

This general statement could be departed from in particular instances by the interplay of fact and value. Accordingly, each evaluation analysis has its own unique
statement of the values that are incorporated, just as each decision (whether based on
an evaluation or not) is a unique statement on the values that are implied. This calls
for the openness in value implication which was urged above (§8.3).

If this stance be followed, the content of the community impact evaluation will
have the common core presented above (see Ch. 8) with variations according to circumstances. One feature of this common core will be the "social accounts" that need
to be produced for each project, plan, etc. (see Ch. 7). Since these take account of all
the relevant impacts on the relevant community, which are translated into the relevant
costs and benefits (indirect as well as direct, measured as well as unmeasured), they
will not have the disadvantages attributed to an economy which is articulated only by
economic indicators relevant to the national incomes account. They will by definition
also take into account others, such as social and environmental indicators (Ekins
1986: ch. II.6).

10.9 From measurement to evaluation

10.9.1 Context

In this section we recapitulate certain aspects of the evaluation process presented
above which form an important stream throughout, namely the linking of *measurement* of effects in Box 10 with their *valuation* in Box 11 and their *evaluation* in Box
12. The nature of the stream can best be seen by briefly describing its flow in reverse.
The culmination of CIE is the evaluation of planning options as a basis for decision,
from the viewpoint of the welfare of the relevant sections of the community, that is
their greatest net satisfaction in terms of their needs, wants, desires and interests (Box
12). Each of the "notional individuals" is presumed to judge this satisfaction subjectively by applying his/her own individual *values* to the impacts that are predicted, in
terms of his/her own sectoral objectives (Box 11). Thus, the standards and principles
of worth to them of the predicted impacts are *human values*, for which purpose they
need to exercise their value judgement (Kaplan 1964: 370). These value judgements

reflect the *demand* of the notional individuals for the *supply* of the predicted impacts, whose attributes are physical, social, economic and environmental facts which can be measured, that is, objectively quantified (Lancaster 1971).

We now trace this stream in reverse in greater detail, starting with the *measurement* and proceeding through to the *evaluation*.

10.9.2 Measurement of effects

Measurement, in most general terms, can be regarded as the assignment of numbers to objects (or events or situations) in accord with some rule, which defines both the magnitude (or measurable attribute), and its measure. The *magnitude* is the property of the objects that determines the assignment of the numbers according to that rule. The *measure* is the number assigned to a particular object, giving the amount or degree of its *magnitude*. The measurement can be *fundamental* (presupposing no others) or *derived* (relating to fundamental measures) through calculation from them (ibid. 1964: 187). The rule of assignment, by which a measure is determined for any given *magnitude*, is a *scale* (also used to refer to the measuring instrument and sometimes even to the standard of measurement; ibid. 1964: 189).

As seen above (Ch. 5), the available *scales* of measurement vary according to the technology available for a particular effect. As we have also seen (see Ch. 7), in any planning analysis the *range* of effects to be included depends upon their relevance for the community in question. From this it follows that in such analysis we are faced with the need to introduce a *scale* for *all relevant effects*, even though the technology of measurement is underdeveloped for the purpose. In this, our quarry of available techniques of measurement is continuously expanding since systematic study is continuing (ibid. 1964: 213). ". . . by many devices which are less precise in strict quantitative measurement but nonetheless far better than unaided individual judgement. There is a direct line of logical continuity from qualitative classification to the most rigorous form of measurement, by way of intermediate devices, systematic rating, ranking scale, multidimensional classifications, typologies, and simple quantitative indices. In this wider sense of "measurement", social phenomena are being measured every day.

Indeed, ". . . if measurement is a mapping of objects into an abstract space, the range of possibilities is basically limited only by our imagination and ingenuity in constructing such spaces" (ibid. 1964: 206).

This continuing systematic study derives from the acceptance of the axiom that measurement of any phenomena advances our understanding of it. For the natural sciences the proposition seems clear enough (ibid. 1964: 172; quoting Lord Kelvin):

> When you can measure what you are speaking about and express it in numbers, you know something about it; but when you cannot measure it, when you cannot express it in numbers, your knowledge is of a meagre and unsatisfactory kind: it may be the beginning of knowledge, but you have scarcely, in your thoughts, advanced to the stage of science, whatever the matter may be.

However, in the social sciences, the validity of the proposition is less clear, for there the preoccupation is with human beings (as opposed to inert matter, flora and fauna) and thereby with *values* in addition to facts (Machlup 1994). Indeed it can be argued that, perhaps economics apart, measurement in social science represents not the end of the road in understanding but its beginning. More pungently, "if you can measure it, that ain't it" (Kaplan 1964: 206).

Thus, there is clearly the need to take measurement as far as practicable, in terms of the state of the art at the time of the analysis, to provide at least the *beginning* of defining and understanding. The current state of the art enables any effect to be measured by one or other of the scales of measurements presented above (Ch. 6) in order of increasing strength (that is, how much precision they can achieve). Accepting that measurement, however partial, is better than no measurement at all for advancing (albeit not completely) the understanding of phenomena, we thus have a useful array of tools for the purpose. But conclusions to be derived from the use of such scales must not be exaggerated or misinterpreted, for otherwise the understanding is not advanced (ibid. 1964: 198; citing M. Jahoda and others). "The basic question is always whether the measures have been so arrived at that they can serve effectively as means to the given end. The usual characterization of a valid measurement is that it is one which "measures" which it purports to measure". This leads to the conclusion: that there is the danger of "endowing the measures with all the meanings associated with the concept" (ibid. 1964: 199; quoting Coombs). "Too often, we ask how to measure something without raising the question of what we would do with the measurement if we had it. We want to know *how* without thinking of *why*".

Another way of being watchful in terms of false conclusions from measurement is to recognize not only what the scale does in fact convey but also what it leaves out in terms of "properties" and relations that are important in the conceptual frame within which we see the subject matter" (ibid. 1964: 208). It is here that very often the proponents of measurement are incautious (in omitting to underline what is not measured) and the opponents of measurement have their field day (in concentrating not on what the measure conveys but on what it does not).

10.9.3 Values

The measurement so far described (§5.9.2) relates to current and predictable facts, and as such can be said to be "objective". It is to these facts that human beings react "subjectively" in their selection of those that are relevant to them in their choice and decision-taking. This selection is made in terms of that elusive concept, *values*. To Kaplan these are the standards and principles of *worth* of the facts under consideration (ibid. 1964: 370). And to Kluckholm (1962) "... abstract standards that persist over time that identify what is right and proper for people in a society or group, which provides a framework that influences individual behaviour and affects social expectation ..."

The nature and scope of these values in any individual are complex. They derive from many different sources (upbringing, parents, society); they have different facets (likes and dislikes, religious or agnostic beliefs, social responsibility or otherwise); they can relate to the ultimate or terminal position (the purpose of life or the existence of God); or to intermediate or instrumental values aimed at more immediate purposes. Being subjective, values are incommensurable. This can raise problems within the mind of an individual, as for example where ultimate values are in conflict with short-term, or values derived from religion with those based on the everyday business of earning a living. But the difficulties are greater when the reactions to the facts are made by a group, be it a family, local community or the nation. It is here that the incommensurable trade-off between the values needs to be brought to the surface in discussion for choice and action. It is here that the "hard" or "wicked" choices need to be faced (see Ch. 2).

This incommensurability has been challenged (Dell 1989):

> There are two questions about commensurability; first, whether one can compare the various values that affect the quality of life of one person or animal; second, whether one can also bring into the comparison the values relevant to cases affecting more than one individual. It is our belief that both kinds of comparison are possible to the extent needed to resolve conflicts of values in environmental decisions.

Specific justifications put forward for this belief are (ibid.): First,". . . all the commensurability of values required is that one be able to make judgements such as "this is more valuable than that", "what we lose here is not as valuable as what we gain there", and so on . . .; and secondly, that ". . . governments do make such comparisons, and it is hard to deny that they make sense". And, accordingly:

> On the belief that values are commensurable, there is built up a mode of evaluation in relation to environmental issues, as an aid to sound environmental decision.

However, this begs the question. That values are commensurable in the mind of any particular individual, and that he can handle them, is a fact of daily life and is not at issue. What is at issue are the difficulties, theoretical and practical, of commensurability in groups (families, organizations, local authorities, governments). Furthermore, the fact that governments include values in their decision-taking, and make ordinal comparisons of value, is certainly true. But they sometimes need ordinal comparisons to give way to the cardinal; and they do not trade off necessarily in terms of open, explicit or acceptable values, being also interested in the implications of following particular less worthy values, such as ensuring the majority of voters are in their favour at the next opinion poll or election.

10.9.4 Value premises[2]

Value premises can take many forms; in each there is the attempt to explain the origins of values: is this to be found in the senses, emotions, experience, cultures, or elsewhere. Discussion about incommensurable values between individuals and groups,

---

2. Written communication from Robert Rapoport.

not to say cities and nations, can be assisted if there be exchange of information from all sides through debate about the *value premises* behind any particular value among the disputants. This could take the form of a public discourse in which value-positions are taken and groups conduct an internal dialogue, in which exhortation, resource mobilization and planning for action fill the agenda. In this way there is the chance of value clarification and reconciliation.

10.9.5 Value judgements

In order to express their views on values, individuals or groups make *value judgements* about the value. Since both the values and the judgements are personal and subjective, they are not verifiable or falsifiable in the way that facts are. But this does not mean that the person making the judgement cannot give a reason for it. This in essence relates to the *ground* of the values as seen by them (Kaplan 1964: 387–90): "Specifically, the ground of values consists, with suitable and important qualifications, in the satisfaction of human wants, needs, desires, interests and the like". The value judgements that derive from these grounds ". . . say something, in effect, about the intrinsic values empirically associated with what is being judged. It follows at once that every value judgement is contextual: it must specify, at least tacitly, under whom and what conditions there would be intrinsic value. In the concrete, a value judgement is an utterance, made by a particular person in particular circumstances".

Three sorts of circumstances can usefully be distinguished, as (ibid. : 290–391):

- *personal*: only the intrinsic value experienced then and there by the person expressing the valuation, which is a *value expression*;
- *standard*: "the judgement may be predicting that under certain conditions, not necessarily those prevailing in the circumstances of the prediction, certain people, not necessarily the speaker, will find intrinsic value". This leads to *value statements*;
- *ideal*: which allow for the widest possible range of predictions about intrinsic value, stating what can be expected on the whole, and in the long run. "These are *value judgements* in the *strict sense*".

10.9.6 Valuation

Value judgements are a means to an end, which is the achievement of the inherent values, i.e. the satisfaction of human wants, needs and desires (§5.9.5). In this, the particular value about which the judgement is made could be an instrumental means to achieving the ends of further values, in a *means—end* continuum (ibid. 1964: 395; quoting Dewey). Somewhere along the line of this continuum, value judgement will lead to choice, as a preliminary to decision or action (see Ch. 2).

For this purpose we need to introduce *valuation*, which is the process of estimating quantitatively the *worth* of the *ends* and also of the *means* of reaching those ends (ibid. 1964: 395). "Moreover, ends are appraised in terms of the means they call for, as well as providing a basis for the appraisal of means. We make our choices on the basis of costs, as well as on the basis of the satisfactions inherent in what we might buy.. . . There is always a price to be paid; actions have their conditions and consequences, and these inescapably reach beyond whatever end we might have in view as an end".

This formulation of *valuation* clearly offers direct and intriguing links with value theory in economics, as we now explore. Since Adam Smith (Pearce 1985: 456):

> Economists have traditionally separated the concepts of use value and value in exchange. Value in use is not an intrinsic quality of a commodity, but its capacity to satisfy human wants. Value in exchange is the worth of a commodity in terms of its capacity to be exchanged for another commodity. The exchange ratios of two commodities are the relative prices of those commodities expressed in terms of a constant money commodity.

Value in use   The capacity of a good or service to satisfy human wants is to be found in the bundle of its interrelated characteristics (Lancaster 1971). In some instances, as in a car, the attributes are material. In some, as in a house which is fixed in location, they are both internal (the number and size of rooms in a house) and external (view from the house or the neighbourhood in which it is located). Seen this way, any house can be precisely defined by its bundle of characteristics or attributes (Lichfield 1964b).

These attributes collectively have the capacity to satisfy human needs, wants and desires, and thus have the potential to make a favourable difference to their lives (Baier 1969, cited in Sinden & Worrell 1979). In economics this is known as their *utility*. How it can make a difference to such lives can be better understood by recognizing that the lives themselves are not a passive receptacle for sensation but are based upon motivations which the utilities can assist (Maslow 1970). By the same token there can be disutility: some element in the bundle of attributes can have an unfavourable difference on such lives, such as the poor view from a house or the pollution conveyed by an adjoining factory. But, not all the attributes of any particular commodity give equal utility or disutility. Therefore, the *use value* to the individual of any bundle equals *utility* from its use minus the *disutility* of doing so (Sinden & Worrell 1979). Utilities are thereby benefits (positive and negative); and disutilities are costs, namely sacrifices to be made for the achievement of the utilities. As such they are reminiscent of Bentham's polarization into pleasure (happiness, felicity) or pain (§10.7 above). Valuation is the process of making estimates about the quantities in this equation. The utility, positive or negative, cannot be measured directly since it is subjective. But what can be estimated is the individual's *willingness to pay* a certain price (WTP) for the positive utility; or the amount of compensation he will be *willing to accept* (WTA) for avoiding the costs of the negative utility if he had to incur it (Pearce & Nash 1981: 10). Since these are expressed in money terms, they also represent the opportunity costs in money to the individual concerned, that is the cost of the resources involved in the sacrifice to achieve these utilities or avoid the disutilities (Sinden & Worrell 1979).

Value in exchange   Measurability of utility in absolute terms was advanced by

Jevons in " 'Utils' in . . . which would be comparable to the units used for height, weight and distance" (Pearce 1985: 58). Later approaches, however, rejected such cardinal measurement and opted for ordinal measurement; that is, the comparison between utility levels could be expressed in intervals (differences) and not in relation to a real origin of zero (ibid.). The ordinal differences were to be derived from the indifference curve which measured marginal utility when faced with options. This marginal utility to both producer and consumer in the market led to the establishment of market price by the interaction of supply and demand, so establishing *value in exchange* which, by definition, is measured ordinally. However, in practice, there are large areas of the economy in which markets do not exist, as for example in relation to externalities, free goods or public goods (Ch. 6). For these the data can be provided only by devising surrogate markets (Pearce et al. 1989). We therefore need to consider the suitability of "market prices" for valuation under the following two headings.

WHERE THERE IS A MARKET   Generally speaking, a market is ". . . any context in which the sale and purchase of goods and services takes place" (Pearce 1985: 272). Such markets are not standard in any particular economy. The share market, for example, is capable of a worldwide almost instantaneous reaction to the forces of supply and demand through a network of telecommunications, and it produces an endless stream of market quotations which are recorded daily in the media. By contrast, the real estate market works sluggishly, owing to the long time that must contractually elapse between the initial deal (offer and acceptance) to the point when the legal investigation culminates in the exchange of contracts. The result is a slow flow of market prices which are not collected and communicated in an open way, although there have been great strides in recent years. However, whatever the nature of these differences in markets, they all pose a common question: how useful are they in indicating a net utility of the good or service in question? For a variety of reasons, as follows, the data must be treated with caution:

- in order to approximate net utility, the market needs to be "free" or "perfect", that is, offering conditions necessary for perfect competition where the following conditions hold: many firms each with an insubstantial share of the market; the firms produce a homogeneous product using identical production processes, and possess information; there is free entry to the industry (ibid.: 335). Such conditions are rarely found and imperfect competition is the generality, including where there is monopolistic competition, oligopoly and monopoly (ibid.: 196);
- even where the market prices are taken at face value, they cannot be said to rep resent net utility because, although they equate to what the buyer and seller agree as the exchange price, they may not precisely represent willingness to pay or accept (WTP and WTA). Of the totality of purchasers, some will be willing to pay more if the price were higher (consumer surplus); and of the producers, some would be willing to sell for less if the price were lower (producer surplus);
- by definition, the market price will not reflect all the externalities that are included in the bundle of attributes which are being bought and sold, precisely because the buyer is not compelled to pay for them (in that the seller can acquire

no title to them) and the seller cannot for the same reason charge for them since he cannot convey a title (Sinden & Worrell 1979: §2.5);

- where the market price covers goods and services which are not in themselves homogeneous in supply-side attributes (for example in real estate), the actual prices reflect the *compound* of the differing attributes, but may not be revealing on the *weight* of particular attributes.

In brief, therefore, although market data are of considerable significance in the economy, in being the only reliable data on exchange value, they need to be treated with caution as guides to utility, intrinsic value and value in use.

WHERE THERE IS NO MARKET   The absence of a market in goods and services is generally to be found where there are externalities or public goods of one kind or another (Pearce & Nash 1981: §8.1–3). This absence is taken to be an example of market failure, namely "The inability of a system of private markets to provide certain goods either at all or at the most desirable or optimal level" (Pearce 1983: 273). This does not mean that the absence of a market price indicates that the good or service has no value or utility; indeed, in the production of public goods or the need to make decisions about externalities, there is the need to make estimates of the ". . . kinds of information that also indicate the choices people would make if they had a chance to do so" (Sinden & Worrell 1979: 422). The goods and services in this general category are referred to as "unpriced values" or "values without prices" (Ch. 6).

With the growing search for efficient investment in the public sector and in relation to the environment, there has also been a growing exploration of tools for estimating unpriced values of the kind described (Sinden & Worrell 1979: pt II). These are not uniform in their approach. Some will aim at estimates of total utility or disutility, others of opportunity cost, and others of willingness to pay. Since it is the latter which is at the heart of the estimates of utility, we conclude with an indication of such methods which have been aimed at estimating utility in that most complex of all "unpriced values": environmental externalities (Pearce et al. 1989: ch. 3; Winpenny 1991):

- Preventive expenditure (PE): the value that people place on preserving their environment can be inferred from what they are prepared to pay (WTP) to prevent its degradation. The extra costs incurred for amelioration of the potential adverse impacts is an example.
- Replacement costs (RC): a special case of preventive expenditure is the relocation of the source of the expected degradation, including where the potential victim would prefer to move away from the source. A simple example is double-glazing against aircraft noise or, where this would be inadequate, the substitution elsewhere of the dwellings themselves.
- Travel cost (TC): where people would travel to enjoy some environmental good (such as a national park), it can be assumed that the benefit of so doing is at least greater than the cost of reaching the national park and the entry price (WTP).
- Hedonic price (HP): the use of known prices in the market to assess, through regression analysis, the utility of selected attributes of the property. A typical example is the valuation, by regression of property prices, for noise nuisance

near an airport, in the search for the willingness to pay (WTP) to avoid the damage that would be caused by aircraft noise; or the willingness to accept (WTA) compensation for submitting to such noise.

- Contingent valuation (CV): posing hypothetical questions to respondents to indicate their valuation of particular factors (e.g. loss of wilderness or of recreational facility) in terms of willingness to pay taxes or charges to safeguard these attributes.

10.9.7 Evaluation

We now close the circle by reference to our starting point, evaluation. The task here is to *choose* between the options as a basis for decision and action, having regard to the *intrinsic value* of the end: human welfare. For this purpose we need to predict the impact on people, in terms of the resulting utilities and disutilities, estimated in terms of the value of the costs and benefits to the people impacted, according to their instrumental values (that is means to an end), which we have called their sectoral objectives. This is explored in Boxes 10–12 of the generic model.

10.10 CIE and utilitarianism[3]

The previous two sections are an attempt to show how community impact analysis can contribute to social decision-making, with the prime objective of advancing community welfare. As such it has clear echoes in the theories of utilitarianism, which can be regarded as its ethical basis. Utilitarianism has rich antecedents in the eighteenth century and possibly with the Greeks (Halevy 1972: ch. 1). But it was Jeremy Bentham who arranged the earlier thinking into a philosophical and practical doctrine (Sabine 1963: ch. 31): "Nature has placed mankind under the governance of two sovereign masters; *pain and pleasure*. It is for them alone to point out what we ought to do, as well as to determine what we shall do. On the one hand the standard of right and wrong, on the other the chain of causes and effects, are fastened to their theme" and "the greatest happiness of all those whose interest is in question is the right and proper, and the only right and proper and universally desirable end of human action". In essence government actions should aim to achieve the "greatest happiness for the greatest number" by seeking to enhance pleasure and minimize pain, which could be objectively measured in terms of intensity, duration, uncertainty, propinquity or remoteness, fecundity and purity. This was the famous felicific calculus.

This apparently simplistic formulation has attracted criticism and been the subject of controversy ever since. Some find it so riddled with philosophical difficulty as to make it simple-minded and therefore to be dismissed (Williams 1972: 149):

> Simple-mindedness consists in having too few thoughts and feelings to match the world as it really is. In private life and the field of personal morality it is often possible to survive in that state ... But the demands of political reality

---

3. For considerable amplification, see Moroni (1995).

and the complexities of political thought are obstinately what they are, and in place of them the simple-mindedness of utilitarianism disqualifies it totally. The important issues that utilitarianism raises should be discussed in contexts more rewarding than that of utilitarianism itself. The day cannot be too far off in which we hear no more of it.

But the funeral has not yet arrived. Support and defence came not only from the contemporary J. S. Mill (Mill 1974), but also from our own contemporaries of various disciplines, for example social science (Braybrooke & Lindblom 1963: 205–206):

> Utilitarianism has not collapsed beyond repair but it does need rehabilitating . . . Why should we focus on utilitarianism to the exclusion of other provinces of traditional ethical theory? We do so because utilitarianism, at least in the English-speaking world, is the school towards which most social scientists are inclined, if they are inclined towards any. There are historical reasons for this inclination: important branches of social science, among them economics and sociology, grew out of utilitarian preoccupations. There is also a natural convergence in preoccupations between utilitarianism and social science. Utilitarianism, after all, insists more strongly than any other ethical theory on forcing moral judgements to the test of facts – the facts of social science.

Utilitarianism also has its contemporary supporters in philosophy, who aim to make it operational in action. Of particular relevance here is Smart, who distinguishes between (Smart & Williams 1973: 12–42):

> Act-utilitarianism is the view that the rightness or wrongness of an action is to be judged by the consequences, good or bad, of the action itself. Rule-utilitarianism is the view that the rightness or wrongness of an action is to be judged by the goodness and badness of the consequences of a rule that everyone should perform in like circumstances. There are two subvarieties of rule-utilitarianism according to whether one construes "rule" here as "actual rule" or "possible rule.

The reliance on *consequences* in act-utilitarianism clearly has affinities with impact assessment and thereby impact evaluation. The *consequences* echo effects. The *greatest happiness for the greatest number* echoes *impacts* and their *size*. Their *characteristics* are echoed in their intensity, duration, certainty, propinquity or remoteness, fecundity or purity. The goodness or badness echoes the *sectoral objectives* and *criteria of welfare*. But even so, Smart continues: "But is he to judge the goodness and badness of the consequences of an action solely by their pleasantness and unpleasantness?"

It is here that the controversy flourishes. And it is here that the aim of utilitarianism is laid bare (Braybrooke & Lindblom 1963: 206). "Utilitarianism, like other ethical theories, undertakes to identify and explicate the criteria for intelligible moral judgements. Having identified these criteria, it has also endeavoured to provide procedures for applying them to disputed cases. Utilitarian authors have differed in the weight that they have given to these two undertakings. Hume, confining himself to treating what he considered to be established judgements (that such and such traits are virtues and such others vices), was preoccupied with the first; Bentham relying – too confidently – on his idea of the felicific calculus, with the second". On the latter, Braybrooke & Lindblom go on to say (p. 225):

> It is now time to call into question the intelligibility of the felicific calculus. Bentham's suggestion, realized or not, has never gone out of fashion. Ethical theorists and economists – the two classes of people who are worried most about the problem – have always conceived that the problem of making utilitarianism perfectly definite and intelligible was a problem of carrying out Bentham's suggestion of a calculus. Attempts to carry it out continue in our own day. In matters of controversy, utilitarian considerations seldom fall wholly on the side of one policy and against every other with which it is compared. If one is to obtain definite results for utilitarian considerations in such controversies – if he is to show, for example, that observations of utility vindicate some pre-emptory rules but not others that at first sight seem as useful – then either the felicific calculus must be made practical or some other set of procedures must be found.

Braybrooke & Lindblom go on to question the feasibility of the calculus and to raise problems about the concept of the consequences and identify suitable reference groups, the consequences for which are to determine the rankings of policies (Braybrooke & Lindblom 1963: 226–44). Their conclusion is that their strategy of disjointed incrementalism can fill the gap. By comparison, here we also recognize the difficulties of the felicific calculus but suggest that community impact evaluation can be considered as a contemporary *approach* to the original – even if simplistic – formulation.

But this is not to say that CIE would be accepted as orthodox utilitarianism when tested against its three requirements (Moroni 1995):

- *Consequentialism:* judging actions or policies in terms of their consequences for individual human beings.
- *Welfarism:* judging states of affairs of policies in terms of the level of satisfaction achieved, which is identified with utility.
- *Sum ranking:* the technical term for the summation of everyone's satisfaction to give a good verbal utility total.

On consequentialism, the CIE is certainly orthodox utilitarianism. On welfarism, CIE is different, whereas utilitarian welfare comprises an aggregation of individual preferences, CIE starts there but also has room for the decision-taker introducing overriding social preference. On sum ranking, CIE is completely different, because there is no attempt to aggregate all the preference without consideration also of the distribution of costs and benefits among different groups. In brief, CIE is not orthodox utilitarianism as an approach, but has some elements similar to it.

This approach is illuminated by the work of Allison (1975), a contemporary political philosopher who, although quite critical of Bentham's original formulation, considers that it can be modified to make a contribution to contemporary problems in modern environmental planning

> Following a rejection of the felicific calculus and other Benthamite technical devices, attempts to re-establish utilitarianism have moved in two opposite directions that I call "economic" and "ethical" utilitarianism. Economic utilitarianism consists of a parallel technical construction to the felicific calculus based on the elements of preference theory. Here is a refined model of rational entities with transitive, connected, preferences whose welfare can be maximised using the calculations of cost–benefit analysis, which generate monetary measurements of wellbeing through the application of the Hicks–Kaldor compensation principle.

> Allison calls his appraisal "modern Benthamism". Perhaps CIE can be called "contemporary economic consequential or preference utilitarianism". As such, it appears to be in the nineteenth-century tradition of the attempt of ". . . a small but determined group of philosophical radicals to introduce the enlightened principles of their utilitarian programme into the sphere of town planning" (Hyde 1947).[4]

---

4. My thanks to Gordon Stephenson (1995) for this link.

CHAPTER ELEVEN

CIE in democratic planning

11.1 Is the planning process democratic?

It is generally accepted that there are strong elements of democracy in the standard planning process in Britain, as the following pointers show. The basic aim is planning for people; the ultimate planning powers are in the hands of the people's democratically elected representatives, at local and central level; the remit of the representatives is the relevant community, at least in the administrative sense; government has stated that the planning is in "the public interest" (DoE 1992); and throughout the process the public are involved through participation.

But it is also argued that there is considerable room for improvement. For example, how to define the public interest? Is this really being promoted? Is public participation more than a formal charade? Does not the government often act like an elected dictator? Are the local representatives really representative of the community they serve or rather of sectional interests, political or otherwise? Is there not inequality of power between the development industry, the local planning authorities and the people who are impacted (as evidenced by the respective legal and professional talents at so many an inquiry)?

We thus have a situation where the aim is clear and it attracts support; there are very few who would openly disdain democracy in favour of, for example, technocracy, totalitarianism, regressive qualification of voting rights, etc. But there is difficulty in achieving the aim. Part of the difficulty is that there are so many differing and often conflicting concepts of what is implied in democracy (that simple but elusive term). It is simple in that it requires that the people have their say in governmental decision-making, and it soon becomes apparent in any particular instance whether or not this has taken place. It is elusive since, despite discussions on the nature of democracy in the centuries of literature since the Greeks (Sabine 1963, Halevy 1992, Miller 1976), what is clear from what has been called the "noblest offerings of history" is that there can be no simple formulation of rules and criteria for the democratic process. Should we follow Plato or Aristotle, the utilitarians or the idealists, Hayek or Rawls, Kropotkin or Marx (McAllister 1980: ch. 3)?

This obstacle, the lack of consensus on the meaning of the term, is in itself the cause of a yet greater obstacle: the very looseness of definition means that it is not difficult for opposing groups both to espouse democracy and to claim that what they are doing is fully democratic, even though to others their proposals are an offence.

That this is so can be seen where proponents of particular views will argue fiercely that theirs and theirs alone are democratically based, simply because each is basing its presumption on differing concepts and formulae, often without being specific on what the democratic process truly involves, and often proposing a formula that is democratic in name only just because it suits their ends. Some random examples from newspaper reports on recent British politics will illustrate.

Margaret Thatcher, the longest-serving Prime Minister of this century, and wielding unusual levels of personal power for a British Prime Minister, has had only about 40% of the voters in four elections in her favour. Her determination to use that power for her own ideology and programme was not diminished when, in the British elections of 1989 for the European Parliament, a greater number of Labour Members were elected than Conservatives. In no way was this taken as a guideline for British European policies. More specifically, she raised strong opposition to the social chapter supposed by the other European Countries unanimously. Gerry Adams, leader of Sinn Fein, at a conference following the IRA bombing of the Brighton Hotel, at which were staying many the British Cabinet, including the Prime Minister, stated that "contrary to being against democracy it was a blow for democracy". He added following the cease-fire in 1994 that a referendum in Northern Ireland would be undemocratic. The struggles within the Labour Party to amend its constitution in relation to constituency nominations for MPs was a clear illustration of how different rules could be promoted in the name of democracy, which were clearly intended to serve the power structure of those that proposed them. The struggle to introduce proportional representation in Britain is both advanced and rejected on the slogan of democracy. Although the majority of the Scots MPs support a National Assembly, the Conservative Party oppose it, as being against the interests of the union of the two countries, yet favour subsidiarity in Europe on behalf of Britain.

In this situation, how is the democratic process in planning to be advanced? We review certain possibilities for reaching a conclusion on an approach.

A mode familiar in planning is to adopt a set of goals, objectives and principles to be followed as guidelines to specific decisions. But there would hardly be conclusive in democratic discussion, since they could be numerous on any particular issue, often in conflict and having different weights in people's minds.

One possibility is a referendum for selected planning issues. But although this could be practicable on occasions where the issues are clear cut, it would hardly be workable either for the multitude of development control decisions nor for the minority that are sufficiently contentious to go to appeal. The balloting rules would be too difficult, and the conclusions not convincing.

It could be accepted from the outset that, once the facts of a particular issue are known and agreed, the disputation will be on values, and it is these that should be reconciled. Some think they could be, on the proposition that they are *commensurable* provided that the judgement is ordinal, stating that "this is more valuable than that" (Dell 1989: 43). This proposition is disputed by others who maintain that values, being subjective to individuals, are incommensurable (Cohen & Ben Arie 1993). Even though many would accept that individual values are incommensurable, they would argue that social values (attributable to a particular community or society) are not. For example, Rawls has advocated a particular approach to distributive justice (Rawls 1972). He derives his rules from the social arrangements individuals would choose if they were uninfluenced by their own purely personal vested interests (their personal values). If placed in this "original position", they would choose a set of institutions for society with certain characteristics that should become the adopted rules.

However sound the Rawlsian argument, it is unlikely that individuals in particular circumstances will adopt the value of a hypothetical "original state" derived in this abstract way. Indeed, from the great variety in these "noblest offerings of history" it is unlikely for any contemporary community to reach agreement on the particular democracy of the past that they would espouse. And even if they were to reach such agreement, it is unlikely that they would all persist in following the accepted principle through thick and thin on all issues and decisions, when adverse implications of the particular case in terms of their own values were made only too apparent. They would then be sacrificing not an original imaginary position but a current one. In this it is the predicted impacts on themselves and their families that will be in the forefront of their minds; and it is these impacts to which they would apply their individual values and value judgements.

In the event, although the values of individuals, both personal and in relation to society, can and should be discussed at great length (for any effort to reach mutual understanding of the foundations of value, their formulation and dynamic is all to the good; McAllister 1980: ch. 2), it is the perceived facts on potential impacts that are uppermost in the individual and collective mind as a basis for the values and value judgements that are exercised (Ch. 10). In effect, the trade-off in such minds can be: how much of an explicit or implicitly stated value should be given up to enjoy beneficial impacts, or pursued to resist adverse impacts? Or how would the different values, implicit or explicit, lead to choice amongst the optional impacts that are posed?

It is in this way that the reconciliation can be made between the use by different individuals, families, religions and communities, which hold to apparently similar values and yet adopt differing actions in practice, on given options of potential impacts. Having adopted and stated a value, the choice between the options would be on that which would advance most the particular stated value or mixture of values; a test is made of the "value of the value" to the person or group concerned. In this way, the goats can be sorted out from the sheep.

A contemporary example rests in the almost universally held value of reducing atmospheric emissions that would advance global warming. From newspaper reports, for Britain the Thatcher Government put forward a programme that fell short of reducing the impact from Britain to desirable levels on the grounds that a faster rate would increase costs to the British economy and therefore reduce its competitiveness; the Labour Party, holding the same value, introduced a constraint that was even more narrow: the apprehension of reducing job prospects in the coal-mining industry (strong supporters of the Labour Party) which would follow a move away from the burning of coal as fossil fuel.

11.2 Advancing the democratic process in planning

The example of equity just given is an instance of generally advancing democracy in planning, which derives from the following emphasis. Planning is a continuing process, with cyclical feedback, with each stage of the process (plan-making, etc.) being capable of articulation into a series of steps. Thus, instead of thinking of democracy in the electoral model, whereby it is exercised with great energy as a pet animal by the political parties concerned on particular election dates, it is seen as penetrating throughout the process at each possible step. If the penetration is sufficient, the democratic process must be enhanced, as the following examples show:

The planning process operates in a complex social context, in that there are many interested parties: the planning authorities at different levels; the landowners, developers and others who wish to implement change; the sectional interests who would be impacted by the change (e.g. nature conservation, chambers of commerce, conservation pressure groups, etc.); the immediate neighbours to the changes in question; the public at large, both in the plan area and beyond. From this it follows that the planning process should proceed in some form of consultation and bargaining between the interests. Where any of these cannot be present at the table (e.g. unborn generations or tourists from abroad), it is for the professional planners and politicians to ensure that their views are represented and not overlooked.

Given this array of interests, which are bound to be conflicting and therefore not all capable of being satisfied in a particular planning solution, the aim of the planning is to find that solution with the maximum net benefit, or at least the minimum net damage, to all concerned. Because of these conflicts, it cannot be expected that all interests will be happy, but at least those who would suffer from any given proposal could have the comfort of realizing that, although they would suffer, there would be overall net gain, particularly if some means could be found of compensating the losers from the gains of the winners. In this sense there is an attempt to reach a "consensus". But it is not a consensus based upon the non-worldly presumption of absence of conflict (Simmie 1974). Rather there is the recognition that in planning you cannot please all the people all the time, so that some must suffer for the greater good, in the "public interest". The hope is that for this reason many if not all would be prepared to swallow a decision adverse to themselves.

In seeking the view of those concerned on optional proposals, it is necessary to recognize that it is *their* objective to which consideration needs to be given, and not necessarily those of the planners or decision-makers. But it does not follow that the conclusions drawn from following the aggregation of such objectives are necessarily those to be followed by the governmental decision-takers. In such situations the government is entitled and empowered to take the view that there are "social objectives" which could override the aggregation of individual objectives of the people impacted. A clear example here is sustainable development, for example on the need to ensure that natural resources, historic buildings, and so on are conserved for future generations in situations where conformity with the objectives of the current generation could mean their erosion.

11.3 How can CIE help?

11.3.1 In general

It is therefore relevant to enquire: how could the wider introduction of community impact evaluation help to strengthen the rôle of democracy in the planning process? We present here the view that it would do so

CIE is built on the following propositions: planning is carried out *for* the people; it recognizes that people are not homogeneous but must be seen as sectors with conflicting interests in any project proposal or plan; the sectors cannot all be net beneficiaries, since some must lose; the gains and losses to people must be seen from the viewpoint of *their* objectives and not those of the planners and politicians; planning therefore aims not at a consensus solution but at one which does the maximum good or the least harm. That would serve the public interest.

From all that is said above (Ch. 4) it is apparent that CIE has been specifically devised from the outset to address itself to these requirements of planning. For this purpose it adapted the theory and principles of cost–benefit analysis towards this end (Lichfield 1960, 1964, 1968) because the basic search is for the benefits minus costs falling on impacted people, with a view to assessing the outcome in terms of *their* perception of *their welfare* (see Ch. 5). It is with this in mind that the costs and benefits are derived from the sectoral objectives in relation to the predicted *impacts* on those people (Chs 5 and 6).

Furthermore, it attempts to define a "relevant" community on whose behalf the assessment is and should be made, and so side-steps those decisions that might be made by a planning authority in terms of its own administration (to the neglect of others outside) instead of those who are functionally linked to the town in question; or by regional or central government, who tend to think of the national interest without full regard to the local, including English interests without full regard to Wales, Scotland or Northern Ireland. Furthermore, the careful distinction can be attempted throughout between the planner/analyst as a technocrat and the political decision-takers/stakeholders as representatives of the values they hold or for which they were elected. And the evaluation and its conclusions are seen as statements for discussion, negotiation and bargaining between the sectoral interests that are affected.

11.3.2 Decision-takers and stakeholders

Although in the planning process there are key decisions throughout taken by decision-takers, these are in turn dependent on those of the stakeholders as decision-makers for their own interests. Accordingly, the taking of a decision itself implies some framework for collaborative discussion at the critical point in the process between all those who are directly concerned with the decision.

11.3.3 Politicians and professionals

In a democracy, of whatever kind, the ultimate responsibility for the decision must rest with the politicians, who are elected or nominated for the purpose. In discharging their responsibilities, they typically need two kinds of help: from the professionals who advise on decisions; and from management/executive/administration responsible for carrying out the decisions. This tripartite relationship is not standardized for all situations; indeed in some situations it is in the building of appropriate planning institutions that the respective rôles are defined for continuing use.

In any such institution there is a clear division of responsibility, even though it be blurred at the edges. Since the politicians have been elected/nominated to represent the people, it is their privilege and responsibility to decide values which will inform the discharge of those responsibilities; it is these values that the managers must take as guidelines; and it is these values that the professional planners must reflect in their work.

In this context, the pursuit of democracy in planning is a major value. But, as we have seen in §4.1, how the concept of democracy informs this value is very perplexing. But it must be resolved in the evaluation process. One impressive approach here is that of McAllister (1980: ch. 3). He initially reviewed from the literature *his* selection from "the noblest offerings of history" in Aristotle, Hobbes, Locke, Montesquien, Rousseau, Jay and Malyan, and Toqueville. From them he sought the definition of democracy. This led to the enumeration of a series of values which in his view *should* guide evaluation (McAllister 1980: ch. 3).

This approach suffers from two difficulties. The shortlist of principles can only be valid in evaluation if they are in fact subscribed to by the politicians in question; but any particular distillation from history may not command general support *per se* as guiding principles in contemporary political situations, even in countries that are recognizably democracies. From this it follows that if McAllister's enumeration of principles is used by the planners without sanction, they could be arrogating to themselves the values of long-dead theoreticians in place of those contemporary elected representatives. This corresponds to the economist who, following Mishan, insists on a cost–benefit analysis of adopting a "value content" derived from economic theory as opposed to the values of the decision-taker (see Chs 5 and 8).

The alternative to this approach emerges above (§4.1). Even if particular values are specified, although they are critical in guiding decisions they cannot automatically be applied in any particular instance, for they then obscure the implications of insisting on the particular values. In certain instances the achievement of a particular value can result in little opportunity cost, whereas in others the same value would imply the use of resources in a manner that imposed heavy costs. That would hardly make for robustness or confidence in the advancement of democracy in planning.

11.3.4 Fact, value and value judgement

The preceding suggests that a tidy distinction can be maintained between the politicians' values and the planners' facts. But whereas it is important in practice constantly to have in mind distinctions between fact, value and value judgement (§6.6.2), it is very difficult to avoid overlap in practice. Politicians become aware of the substance of the planning and evaluation process and cannot be constrained in exercising their views; and even where the professional respects the prerogative of the politician in deciding on values, he cannot but reflect his own values in the professional contri-

bution; in a sense he is arguing for a modification o
when he urges a change in decision through demonst'
politician's inclinations. And since there is no hom
arguments for change will be diverse. Furthermore
to modify the stance of each, as they progressivel
ning and evaluation process.

11.3.5 Public participation

Within these arrangements for communication, CIE can play a p
participation. Effective public participation requires involvement with
various stages in the planning process. For this purpose there needs to be sum
communication. Generally speaking, the work of the planning process in its conventional dress hardly meets this bill, for it tends to be dressed up in the technical language of the planning professionals (surveys, statements, policies, plans, statistical
tables, graphs). But what the public are concerned with in the ultimate is the impact
of the proposal on their standard and quality of life, both as they exist and as they
might be. It is this that can be conveyed as a distillation of the content of the evaluation in Steps F10–I13, for example in the impacts that arise, the community sectors
affected, the distribution of the beneficial and adverse impacts between the community sectors, the assumptions made on the sectoral objectives, the degree to which the
objectives will be advanced or retarded, the ranking that emerges, the weighting put
on the sectors by the decision-makers in relation to the rankings, and so on. These are
matters that are readily communicable and so can form the basis of direct discussion.

The results of the participation could generate comment on any of the matters
raised, and in certain respects will lead to a reformulation of the analysis, as for example a restatement of the sectoral objectives and what flows from them. But that apart,
it is not to be expected that there will be general agreement, simply because of the
underlying conflicts of interests that must exist in relation to any project or plan.
These in themselves could be reduced through the introduction of ameliorative measures in the plan proposals, which is a strong feature of the environmental assessment
process (Ch. 5). But even if such measures do remove some of the objections, they
cannot remove all of them. Clearly, "one man's meat is another man's poison". But
what can emerge to the losers is the conclusion that their loss is to be seen in the perspective of wider gains to the community in general; and indeed it could be that to
any individual the reconciliation is offered between the adverse impacts on himself
and his family and the beneficial impacts on others (the destruction of rural amenities
in a village that will result in new dwellings which import people, and so give benefits
of larger numbers to support the local services, etc.).

In this sense it is puzzling to understand the consistent charge that town planning
aims at "consensus" (Simmie 1974). How can it possibly aim to do so in the light of
the obvious underlying conflicts? What planning does aim to do in this regard is to
seek a consensus on what should be offered as acceptable to the community in the
public interest, in offering maximum net benefit overall, or the least minimum damage overall, or damage with equalizing compensation.

tiating, mediation and bargaining

out in §14.6, planning can be seen as a process of complex bargaining a large variety of actors. Such bargaining proceeds whatever the level of com-ation. In essence it often amounts to sectoral views being expressed against (and y for) any proposed change, be it by local residents against intrusion into a resi-ntial area or by statutory consultees in respect of water, drainage, roads, and so on.

Whatever the stage for the bargaining in the planning process (comment on planning application, presentation of planning application, public inquiry, public participation generally), the bargaining will achieve greater efficiency and clarity if it takes place around a central statement that lends itself to the purpose, for otherwise the discussion and argument typically result in a mass of conflicting statements, points of view, values, and so on, through which it is difficult to see a satisfactory conclusion. Such a central statement could be the community impact analysis, assessment and evaluation in a simplified form, which portrays the change in the standard and quality of life to different sectors that will emerge. It is the impacts and their consequences that are of relevance and interest to people. The use of this vehicle for communication will be of value whatever the stage setting for the bargaining process itself.

For example, it would be relevant in the resolution of impacts through their mutual trading-off (Susskind et al. 1978); in the *in camera* discussion on analysis between representatives of the decision-makers and stakeholders (Phillips 1987); at a public inquiry where, instead of the conventional testing of the evidence by cross-examination, which often results in mutual rubbishing on all sides, the CIE could be tabled in advance in full, as an *independent* statement in which the technical and quantitative inputs have been agreed as far as possible between the opposing parties, as a basis for the disputation on the differences.

11.3.7 Weakness for democracy of the grand index

The preceding discussion can be sharpened by reverting to the critique of the grand index as a conclusion for evaluation (see Ch. 9). Although understandable from the viewpoint of the analyst who wishes to sharpen his conclusions as far as possible in numerical form, it is inimical to any democratic process in planning. It presumes to have taken into account many of the issues discussed above, as for example introduction of the democratic process, the distinctive rôle of the politicians and professions, the distinction between facts, values and value judgements, the involvement of the public, and the place of negotiating, mediation and bargaining. And, by the same token, it does not typically lay out the value content of the analysis in a way that will enable decision-takers, stakeholders and the impacted public to see how their interests are involved. We are thus led back to the conclusion reached in relation to cost–benefit analysis (Nash et al. 1975a,b Nash 1992): that although an apparently cumbersome display, the planning balance sheet (and thereby CIE) does offer raw material to encourage the participatory process in a democratic way (see also McAllister 1980: ch. 14.2).

CHAPTER TWELVE

The planning process

12.1 The role of evaluation in the planning process

As brought out in §12.4.1 evaluation is not a discrete but an integral part of the planning process. However, this is a comparatively recent innovation. Thus, it was possible to say in a research study in 1975 that (Lichfield et al. 1975: para. xv):

> Evaluation is frequently treated in practice as a discrete activity, functionally separate from other plan-making activities, with those responsible for undertaking the evaluation work having little or no influence over the nature of preceding work. Alternatively, evaluation is left until too late in the study for it to make an effective contribution to subsequent decision-making procedures. A variety of difficulties may result.

In order to overcome the difficulties, and to enable evaluation to be better integrated in the planning process, the research critically explored case studies, in the use of evaluation in the process, by the main methods that were then currently practised, namely threshold analysis and planning balance-sheet analysis in regional studies, goals achievement matrix in subregional and urban studies, linear programming for a new town, and social cost–benefit analysis for a major airport (Lichfield et al. 1975: part II).

These studies showed that the particular planning process itself influenced the method, and vice versa. This has also affected our treatment here of the rôle of evaluation in the planning process: it has been necessary to adapt the evaluation process to the planning process within which it is to take place. For our purpose we have selected the development planning practice of Britain, post Second World War (Cullingworth 1994). However, as shown in Chapter 1, this label needs interpretation, having regard to the changes since its inception in the late 1940s. Its essence can be grasped from the following:
- having the form and content associated with physical planning, with which is identified the social and economic activity implicit in the physical fabric
- operated through vertical integration in the national, regional and local levels, including budgeting and policy
- having the instruments of both regulatory and allocative planning
- having its plan-making in the comprehensive rational, analytical mode, with

implementation decisions being taken against the background of the plan, but not necessarily in accord with it.

But, even within the one evaluation method and the one planning system, there is no one model for application, just as there would not be for other inputs to the planning process, for example in survey and analysis. As brought out in §8.4, this was found to be so from practice where, in a large array of case studies in planning balance-sheet analysis/community impact evaluation, variations in the basic method were adapted to reflect some six variables.

It was just because this variation between the case studies denied the use of a *standard* method that it has been necessary to formulate a *generic* method (see Ch. 7) which can then be adapted as necessary to the variables found in the particular circumstances in any particular planning situation in any particular country. In this chapter we show in demonstration how the generic method for project evaluation can be adapted to the plan-making, implementation and review phases of the British style of development planning process (see Ch. 1). Evaluation is relevant in each phase, albeit for different purposes and to answer different questions. And since the three phases are themselves interrelated in a cyclical sequence, with feedback from the later to the earlier, the respective evaluation analyses are also so linked. We now consider each phase in turn.

12.2 Plan-making

12.2.1 Subplans

The typical urban or regional plan is prepared for a specific area (the plan area) which is defined by a commissioning body. In content, the plan will typically show proposed changes within the context of areas of stability, that is "no planned change". In the plan-making process it is necessary to carry out studies in wider or narrower study areas that are functionally related to the topics in question, and thus not necessarily coincident with the plan area.

For the areas of change, there will be within any plan interrelated "subplans" that typically cover such matters as the following: policies, proposals, projects, programmes of projects, and strategies. We now describe them in turn.

- *Policies:* A policy is a statement of intent, for the whole or part of a plan area, which will guide the making of decisions that arise on interpreting or implementing the plan (Solesbury 1974). The conditions in which the policy is used could be formal (e.g. development control) or less formal (e.g. planning briefs to project planners within the area). The use of policy as such comes within the field of "policy science", which aims to improve the quality of public policy-making (Dror 1971). As seen there, policy and plan-making have strong affinities to the use of the rational method. For example, evaluation in policy-making looks for inputs and outputs, which are measured as far as practicable as a basis for the evaluation.
- *Proposals:* In contrast to a policy, a proposal relates to a site or area, as for

example in use zoning. Thus, although more specific than a policy, it is less specific than a project.

- *Projects:* Private and public sector schemes for specific sites, which imply the allocation of resources for the transformation of the site within a visualized period, by means of specific agencies using specific implementation means.
- *Programmes of projects:* Projects throughout the area which are necessarily interlinked in space and time, since they have been visualized in the same planning framework. Examples are sectoral programmes (such as roads, housing, schools and shopping).
- *Strategies:* Compared with the other elements, the meaning of strategy is more elusive: it is widely used but difficult to define. Indeed, dozens of different definitions have been developed, but there is no real consensus concerning which one is best (Schwenk 1988: 4). We must therefore devise a meaning that is relevant to urban & regional planning. But although there is a common approach to forward looking in different fields, the field itself requires adaptation of the approach. So it is here. The strategy must be devised having regard to features such as:
  - ends which are fundamentally social, economic, environmental and physical
  - a mixture of developing and operating agencies, under various degrees of control by the planning authority
  - a long timescale for implementation of the "battle", which therefore implies programming for both the short term and middle term;

  Against this background, strategy is the art and science of employing resources as a means (policies, proposals, projects and programmes of projects) towards achieving the ends implicit in the plan, in the recognition that those ends are in themselves the means in the chain for achieving the socio-economic values in the plan.

12.2.2 Adaptation of evaluation method to subplan

Having reviewed the different kinds of subplan, we now explore how the generic method of community impact evaluation presented above for projects can be adapted to each. It will be relatively straightforward for "programmes of projects", since these can be seen as one major project with interrelated subdivisions in space and time. But the adaptation will be more difficult in respect of, say, regulatory proposals, for although the input will be apparent it will often be difficult to predict the output of impacts with any precision in terms of development or timing. In passing it might be added that, although this is clearly so in the British "flexible/discretionary" situation, it will be less so in a plan in Western Europe, where it is the plan itself as designed which is to be implemented (Davies et al. 1986). And the adaptation will also be difficult for policies, for these are merely statements of intention, often conflicting, of the context in which the planning decisions are to be made, in particular in relation to development control. This situation can be overcome to a degree if scenarios are predicted as a guide to the intent behind the policies, for then it is the scenario that can be evaluated on the assumption that it will be realized (Khakee 1991).

Whatever the subplan and the nature of the evaluation used for it, a distinction needs to be drawn between the "linear" and "cyclical" approach to the evaluation (Lichfield et al. 1975: ch. 12). In the former the evaluation process produces some conclusions that are then used as the basis for choice, decision and action. In the latter, the conclusion of the evaluation is not choice but the beginning of another stage in the evaluation process. For example, the conclusion could be that none of the options should be chosen as a basis for decision, but instead a new option devised, taking the strengths and avoiding the weaknesses of those examined, by definition leading to an option which should be superior to those under review. Or the conclusion could be that the next step is not to devise a new option based on those already examined, but to take advantage of the learning process that the earlier evaluation has provided, to start afresh on the devising of new options.

Because of the nature of the cyclical approach, one rule needs to be maintained: consistency in choice criteria between the evaluation frameworks during the various stages of the cycle. If there is not such consistency, it follows that at the end of the process the best possible option may not have been found. This does not mean that the depth of analysis for each phase needs to be the same. On the contrary, with the cyclical approach the initial round could be full and vigorous, with the succeeding rounds comparing only the significant variations between the options. Alternatively, the preliminary evaluation rounds could be more generalized, appertaining to key factors which are of importance in the initial selection, with the later analysis going into detail on these and other elements.

12.3 Plan implementation

The aim of *ex ante* evaluation for *plan implementation* is to aid the choice between options in respect of the *means* of implementation that are proposed in the various subplans. For this it is necessary at the outset to be clear on various questions:

• Are the proposals intended for implementation, for if not then there will be little point in carrying out the evaluation

Enough has been said above about the different kinds of planning mode to show that these imply differing degrees of intention to implement, and thereby of relationship between the making of the plan and its implementability. In illustration, in "Utopian" planning there is only a tenuous link, since the intention is to influence ideas and aspirations, but not necessarily events on the ground, although there are some brilliant instances of the latter also coming true, as in Ebenezer Howard's Garden Cities. By contrast, development planning and resource planning are certainly intended to influence subsequent physical development and socio-economic activities on the ground; they are *designed* for "implementation".

• Whatever the intention, are the proposals feasible or practicable, for otherwise again there is little point in the evaluation itself.

The prospects of feasibility are heightened if the intended implementation process itself is borne in mind during the plan-making and not left to the end, as is so common. For this it would be useful to have a rational implementation model comparable with that used for plan-making (see Ch. 1). As yet, none is available. Pending that, one approach is to carry out feasibility tests at each step in the plan-making process, the test being adjusted to that step. For this purpose a checklist of tests, adapted to the particular plan-making model, is helpful (Lichfield & Darin-Drabkin 1980: ch. 2).

- If intended for implementation, what is the meaning attached to the term in the plan in question?

There are differing connotations. For example, in initiatory or project planning there is the clear purpose to execute the project (Morris & Hough 1987). In regulatory planning, the evaluation could relate to the degree in which the execution has or has not followed the plan itself (Alterman & Hill 1978). In flexible/discretionary development planning of the British kind, the evaluation could relate to an outcome permitted under development control which was not in fact in accord with that plan; the evaluation test is whether or not this would be a better outcome in terms of impact on the community compared with what was proposed in the plan (Lichfield & Lichfield 1992). To some the outcome is not as important as the management of the process of implementation (Barrett & Fudge 1981).

In these circumstances, in different cases the implementation measures themselves will be of varied kinds, some being more susceptible to evaluation than others. This is brought by recalling the typical plan implementation measures presented in §1.6.4 and noting the degree to which effects can be predicted for the purpose of evaluation.

- general influence: by definition this is relatively uncertain as to the impact on implementation, and thereby in the outputs it would produce
- intelligence and information: as above
- organization: the outcome would be seen in improvements in implementation efficiency
- fiscal: where too general the outcome is not clear for particular areas, but where specific the impacts can be predicted
- direct: projects can be handled directly by the generic method but the outcome in development control would be more uncertain
- direction as to home or work: while happily of no relevance in non-totalitarian countries, it is mentioned as an example of where evaluation of implementation would be quite distinct.

In terms of the generic model, this contrast relates to the plan variables that are categorized as "implementation" (Box 6), for it is by these measures that the implementation is to be secured. The evaluation will compare the difference in outcome from the different implementation options that could be employed.

12.4 Plan review

It is a commonplace today (although it has not always been) that plan-making is a continuing process, so that the plan itself is monitored with an eye to review, and change as necessary. The nature of the review will again differ with the mode of planning. Where the statutory planning process is strongly linked with the maintenance and defence of property rights (as under the British TCP Act 1932), it is the common experience that plan reviews take a considerable time, simply because of the need to take into account the possibilities of objections property owners leading to compensation for diminution of their rights in the review. By contrast, the process is much more speedy, indeed in some instances continuous, in the contemporary British scene, where entitlement to development rights on any property does not accrue from the plan itself but instead from the *ad hoc* planning permit which is given. The speedier process facilitates the aim of keeping the plan continually up to date, in order to make it more effective as an instrument for guiding development.

Whatever its nature, the review addresses itself to the changes required in the plan that was adopted for implementation. As such, the review not only considers the changes flowing from the forecasts in the plan-making (population, employment, traffic, etc.), but also compares the actual implementation with what was visualized, and the reasons for variation.

A distinctive feature arises where the plan was accompanied by an *ex ante* evaluation as an aid to choice and decision, based upon the facts, assumptions and thinking at the time of the plan-making. Accordingly, what can also be evaluated under review is the *ex ante* evaluation itself, with a view to assisting in the refining through experience of the evaluation methodology, by offering greater insight into the process on the next round. For example, were the plan variables (Box 6) the ones that in fact triggered off the change, or were there others? How did the actual change in the urban system compare with what was predicted (Box 8)? Were the predicted effects reasonably accurate (Box 10)? Were the predicted community sectors in fact impacted and how (Box 11)? Should additional community sectors and subsectors have been included (Box 10 and 11). Were the sectoral objectives chosen for the evaluation those that the community sectors would still stand by (Box 11; Lichfield 1995)?

12.5 Relationship of plan review to other kinds of review

Whereas the monitoring and review in the planning process have their roots in the planning system, their principles and practice are closely related to review in three other related fields in what is generally called "programme evaluation", in each of which there is borrowing from the other.

Programme evaluation
The growth of the Welfare State has seen the introduction of social action programmes, in the areas of health, education, employment, social security payments, etc. (Dennis et al. 1960). With this introduction has gone the need for *ex post* evaluation of the purpose and effectiveness of the changes that have been introduced, even where the programme may or may not be derived from a plan (Weiss 1972). In essence, the evaluation seeks to answer four questions (ibid.: 54; quoted by Suchman 1967):
 • The kind of changes desired.

- By what means is this change brought about?
- What is the evidence that the changes observed are due to the means employed?
- What is the meaning of the changes found?

Protection of the natural environment
With the growth of concern for environmental pollution, whether it be from continuing sources or new, the need has grown to keep the pollutants and their effects under review. This is processed through (Bisset & Tomlinson 1988):
- monitoring: an activity undertaken to provide specific information on the characteristics and functioning of environmental and social variables in space and time "in order to see whether an impact occurred as a consequence of the project and to estimate its magnitude to see if it is of significance"; and
- auditing: comparing the impacts predicted in an environmental assessment with those that actually occur after implementation, in order to assess whether the impact prediction process performs satisfactorily.

These processes are designed to meet three basic objectives (Sadler 1988):
- in project regulation, to ensure that activities conform with operating conditions previously established following an environmental impact assessment;
- facilitating impact management, by providing an opportunity to manage the unanticipated effects, from modifications to mitigating measures to project design;
- aid development in improving the practice and procedures of environmental assessment and its supporting processes.

Social audit/accounting
Closely linked with the process just described is that brought out above under this heading (Haughton 1988). In essence, it is the same process as that described for the natural environment, but it can relate to current practices in any field, with the emphasis being on social responsibilities rather than those of the individual or company concerned. The subject matter of the review can thus be varied, so that the process itself will need corresponding adjustment to the particular field.

12.6 Relationship of evaluation analyst to the planning team

Whether we are concerned with the generic method designed for projects, or an adaptation to one or other of the situations described, the application of the evaluation needs to be adjusted for the way in which the evaluation analyst is related to the planning team. In this, major variables arise. Is the evaluation (Lichfield et al. 1975):
- a discrete step, whereby the analysts receive information from the planning team and report on conclusions, or is it
- integrated into the planning process, whereby the evaluation analyst is part of the team.

Within either of these there are specific situations that have a bearing on the evaluator's terms of reference, such as:

- in the design phase: in-house or not to the planning team
- in the decision-making phase: direct access or not to the commissioning body
- following the evaluation: in helping to generate further options as part of the cyclical process, where the evaluation conclusions are the beginning of another planning design input, rather than being completed by an output.

CHAPTER THIRTEEN

Roads: a case study comparing COBA, framework appraisal and CIE

13.1 Review of highway evaluation methodology

In (§4.5) we presented 30 methods of evaluation in urban and regional planning. Of these, some 11 are known to be used in transport evaluation. These are shown in Table 13.1, where against the costs and benefits for each sector involved in transport are shown the 11 methods with the relevant costs and benefits they would employ in their analysis. For ease of cross reference, for each column the category of evaluation method from §4.5.1 is also shown.

Within this array of transport evaluation methods, the Department of Transport has used economic appraisal only since 1967, adopting the then-contemporary method of cost–benefit analysis (COBA) in 1973 (ACTRA 1977: ch. 4). The economic measures for COBA are traditionally narrow: capital and maintenance costs of the road compared with savings to vehicle users (in time spent travelling; direct costs of fuel, vehicle maintenance and depreciation; direct and indirect costs from road accidents). This narrowness was criticized by ACTRA 1977, which proposed a new method of framework appraisal alongside the economic assessment (ACTRA: ch. 20). This recommendation was adopted by the Department of Transport, who then experimented with various forms of Framework Analysis. This experience was crystallized by the Leitch Committee into one form in 1979 (SACTRA 1979), which was then put into effect by the Department of Transport in 1982 for trunk roads, in association with its Environmental Appraisal Manual (DoT 1983, revised 1993). In this there was some modification of the interest groups in the Framework Appraisal of the 1977 Report as follows (SACTRA 1979):
- Group 1 Travellers
- Group 2 Occupiers
- Group 3 Users of facilities
- Group 4 Policies for conserving and enhancing the area
- Group 5 Transport, development and economic policies
- Group 6 Finance implications.

The extension of this work to urban roads was recommended in the SACTRA Report of 1986. It recommended that a wide array of environmental and social impacts should also be reflected in the appraisal for urban roads (ch. 11) and demon-

Table 13.1   Costs and benefits of community sectors in transport evaluation methods.

Sectors, with their costs and benefits	1	2	3	4	5	6	7	8	9	10	11
Producers/operators											
Project											
Costs: capital and operating					X	X	X	X	X	X	X
Benefits: value and output				X		X	X	X	X	X	X
Transport operators											
Costs											
Capital	X				X	X	X	X	X	X	X
Operating	X				X	X	X	X	X	X	X
Benefits											
Fares					X	X	X				X
Output				X	X	X					X
Landowners: displaced and non-displaced											
Costs: net difference in land value								X		X	X
Government: central and local											
Costs: grant and subsidy								X			X
tax and rate income											X
Community at large											
Benefit: net difference in GDP											X
Consumers											
Passengers – any modes											
Costs: fare, time (access and trip)				X			X	X	X	X	X
Quality of service		X		X							X
Safety				X							X
Benefits: trip end											X
Residents, workforce, visitors											
Costs: environmental impact			X	X						X	X
Benefits: accessibility					X						X
Community at large											
Costs: taxes and rates				X					X		X
heritage										X	X
Benefits: socio-economic activity											X

Notes: 1: unit costs (CM); 2: quality of service (CE); 3: environmental impact assessment; 4: goal achievement; 5: corporate objectives achievement; 6: passenger miles per £; 7: financial cost and return; 8: social financial analysis; 9: cost–benefit analysis; 10: framework appraisal; 11: community impact analysis.

strated by case study how these could be applied (chapter 14). This wider approach was accepted with reservations on detail by the DoT (DoT 1986). In its reservations the DoT rejected the recommendation that the Framework Appraisal method should give way to a new format to be called the Assessment Summary Report. It also rejected the similar recommendations in the SACTRA 1992 in their response of 1992 (DoT 1992f: para. 16.28).

strated by case study how these could be applied (chapter 14). This wider approach was accepted with reservations on detail by the DOT (DOT 1986). In its reservations the DOT rejected the recommendation that the Framework Appraisal method should give way to a new format to be called the Assessment Summary Report. It also rejected the similar recommendations in the SACTRA 1992 in their response of 1992 (DOT 1992f: para. 16.28).

13.2 Context for the case study

Brigg is a market town of some 5000 population within Glanford Borough, County of Humberside. As part of the Brigg Local Plan, Glanford Borough Council proposed an Inner Relief Road (IRR) to facilitate among other things a proposed pedestrianization, following the removal of traffic, of Market Place and Wrawby Street (Glanford Borough Council 1987; Fig. 13.1). This pedestrianization would differentiate Brigg from most other towns and villages in the area, and would so draw in additional population to live, work and shop there.

That public opinion supported the proposal is seen from a random sample of 530 passers-by in Brigg Town Centre between Thursday, February 12 to Saturday, February 14 1987 inclusive. This survey showed pedestrianization as the improvement people would most like to see, followed by environmental improvement, an Inner Relief Road, improving shopping facilities, improving entertainment facilities and the creation of additional employment opportunities.

In October 1986 Glanford Borough Council commissioned Nathaniel Lichfield & Partners to prepare a study of potential economic development in Brigg. As part of the study, the Consultants were asked to consider whether the town centre proposals, including an inner relief road could form a good basis for Brigg's regeneration. This

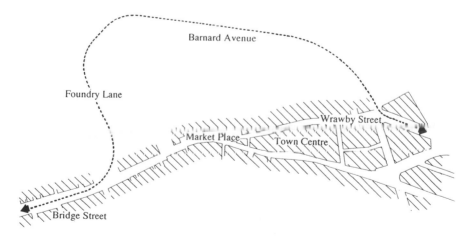

Figure 13.1   Brigg Central Area By-pass (*source:* adapted from Glanford BC 1987).

included, among other things: an assessment of the contribution the inner relief road would make to economic regeneration in Brigg, and a critique using CIE of the economic evaluation of the Inner Relief Road by the County as Highway Authority, prepared with the help of the Glanford Borough Council Officers.

This critique was included in evidence by Dalia Lichfield at the Local Plan Inquiry (Lichfield 1987b). It is this critique that is presented below (§13.5) for comparison with the county's economic evaluation (§13.3) and an evaluation using the DoT urban road appraisal (§13.4).

13.3 Highway authority economic evaluation of IRR

In order for the pedestrianization to go ahead, the traffic that went along Wrawby Street and Market Place would have to be diverted. Humberside County Council (1986) looked at two options for the alignment of an inner relief road. The first ("do minimum") visualized the continuation of the use of Market Place and Wrawby Street for traffic, with traffic management schemes (Fig. 13.1). Although this option involved only minimum capital costs, its impacts would not necessarily be limited: traffic disbenefits would go on worsening as traffic continued and perhaps increased. The second was the preferred option for the inner relief road, namely Bridge Street/ Foundry Lane/Barnard Avenue/Wrawby Street (Fig. 13.1). It would cost some £1.7 million, towards which there would be no County grant. Humberside County Council (1986) recognized the objectives of the Borough Council in seeking the inner relief road; considered in detail the two alternatives for the road; concluded in favour of the Foundary Lane, and made a highway economic evaluation of it.

The economic evaluation itself is stated to be ". . . a crude analysis to give an indication of the level of benefits and disbenefits of the proposals" (§5.4). Its components are confined to:
- *Costs:* capital costs of property acquisition and the road works (§5.2).
- *Benefits:* savings to motorists in terms of money value of time and vehicle operating costs for an average vehicle, capitalized over 30 years, and discounted to 1979.

From this analysis there is the conclusion "Overall there are no benefits *to the motorists* if the inner relief road was built. The time and distance savings for traffic terminating in Brigg are almost exactly balanced by the out-of-the-way travel caused to through-traffic. In crude terms, the scheme would be worth building if it did not cost anything". (§8.1)

The Report, of course, recognized that this economic evaluation took no account of the other benefits which led the Borough Council to formulate the proposal, including the environmental issues: "Although the existing roads can carry the current and predicted traffic flows, it is unpleasant to shop in a narrow street with the continual passing of heavy traffic" (§1.2); but neither the environmental gains nor the wider economic gains are included in the highway economic evaluation. Although it is acknowledged that "the desire and need for pedestrianization in Brigg is the essential feature of the local plan and there appears to be general support for this element of the plan, the means to achieve this aim – constructing the inner relief road – is costly and unrewarding *in conventional highway economic terms*" (§8.2) [our emphasis].

The result was this. Although the County agreed that the IRR was a good and desirable scheme, and that its benefits were not reflected in the highway economic evaluation, the COBA analysis did not enable them to offer priority for the Brigg inner relief road within their current programme. As a result, if Glanford Borough wished to proceed on environmental/regeneration grounds, they would need to bear all the costs.

13.4 Urban road appraisal of Brigg inner relief road

13.4.1 Description of appraisal method

The appraisal method was that recommended by SACTRA 1986 (as an *Assessment Summary Report*). The full description of the method is contained in Chapters 8–13 and demonstrated in the worked case study in Chapter 14. In essence:

1. define the scheme objectives (§14.12, Fig. 14.3)
2. define the groups affected by the objectives (§14.13, Fig. 14.4)
3. define the problems associated with these objectives and groups and their measures (§14.4, Fig. 14.4)
4. make an assessment of the way in which each option is expected to contribute to the achievement of each scheme's objective (§14.15, Fig. 14.7)
5. present traffic and economic data in relation to the above (§14.20–14.41)
6. present environmental and social impacts in relation to the above (§14.41–14.80)
7. present a summary of the achievement of objectives by the options, as reflected in the data
8. compare the achievements of the options and indicate the preference (§14.81–14.83 and Fig. 15.30).

13.4.2 Application of the method

This method is now applied to the options compared in the County Report, namely the DM and the IRR schemes (Fig. 13.1). For simplicity of presentation it does not follow all the individual tables of *Urban road appraisal* (SACTRA 1986), but condenses into one table (13.2) the essentials from Chapter 14, in Figures 15.3, 15.4, 15.7 and 15.30.

Column 1 shows the scheme objectives, which were essentially the focus of the appraisal. These objectives derive from discussion on the Brigg Local Plan in the context of national objectives for trunk roads and major local authority roads, and local objectives from the Structure Plan. In Column 2 are shown the current ". . . problems which relate to the objectives which are reflected nationally and in local development plans" (SACTRA 1986). Column 3 lists the community groups that are affected by these problems, and whose condition will be ameliorated or aggravated in respect of the problems under each of the options. Column 4 shows as a datum the "do minimum" option. In column 5 a judgement is made on the amelioration of the problems through the adoption of the IRR option, compared with the position under "do minimum" in column 4. The changes are indicated descriptively but could in practice be quantified.[1] Column 6 compares the objective achievement in column 5 with the datum in column 4. In all cases the provision of the IRR compared to DM (column 4) would assist the scheme objectives enumerated in column 1.

13.4.3 Conclusion

From Table 13.2 the conclusion is clear. The provision of the IRR compared with the DM will assist in all the scheme objectives for traffic movement (both vehicles and pedestrians) which are enumerated in column 1, and by so doing give benefits to the groups affected in column 3. This is what would be obtained for the expenditure of £1.7 million by Glanford Borough on the inner relief road.

13.4.4 Comment

Whereas COBA showed the benefits to vehicle users in money value of time and vehicle running costs, the Assessment Summary brought out benefits to other groups (established local commerce and industry, owners and occupiers of new sites, pedestrians, adjoining shop occupiers). The benefits were not measured or valued but presented in terms of "assisting in their objective achievement". As such they are difficult to compare with the cost in order to reach a conclusion in cardinal as opposed to ordinal terms. It is therefore not easy to strike a balance.

This difficulty is also seen in the case study in the SACTRA Report on *Urban road appraisal*. The "Summary and preferred option" (p. 168) is in judgmental terms without the benefit of a "balance sheet". This was surprising for SACTRA (ch. 8.2) endorsed one of the six criteria for assessment which were presented in the ACTRA 1977 Report namely:

"f. Balance costs and benefits, however described, in a rational manner".

Having regard to this criterion it was also surprising that they did not follow up the ACTRA 1977 Report (§20.28–30.35) which recommended the use of "multi-criteria appraisal techniques" for the achievement of the "balance of the assessment" on which they said (ACTRA 1977: §20.32):

"In particular, we were impressed by the work of R. Travers Morgan & Partners for the Urban Motorways Committee (and subsequently in Scheme Appraisal) [chapter 1.45, Cost–Benefit Matrix] and by the Planning Balance Sheet developed by Professor Lichfield. These approaches can provide a summary of the effects relevant to a decision".

---

1. From the data in the Brigg Local Plan (1987) and the County Technical Report, no. 152 (1986).

Table 13.2 Brigg Inner Relief Road: comparison of options by assessment summary.

			The options		
Scheme objectives	Problem measures	Group affected	Do Minimum	IRR scheme compared with Do Minimum	Objective achievement (col. 5 − col. 4)
1	2	3	4	5	6
(a) Aid Economic regeneration Development Environmental impact By removing through-traffic from town centre	Noise Air pollution Travel flow Accidents	All vehicle users Established local commerce and industry New commerce and industry	–	Increase in travel costs Remove traffic nuisance Improve accessibility	Assisted
(b) Help movement of industrial and commercial traffic and avoid congestion	Journey time Vehicle operating costs Accidents	Vehicle users	–	Traffic movement facilitated	Assisted
(c) Improve access to new development sites	Diversion of traffic	Owners and occupiers of new sites	–	Access provided	Assisted
(d) Improve road safety	Personal injury Accidents	Vehicle users Pedestrians	–	Less vehicle/pedestrian conflict	Assisted
(e) Enhance conditions for pedestrians	Pedestrian delay Accidents Trip quality	Pedestrians	–	Better conditions for pedestrians	Assisted
(f) Pedestrianized shopping streets	Free flow for pedestrians	Pedestrians Adjoining shop owners	–	Better conditions for pedestrians	Assisted

Source: Brigg Inner Relief Road (1987).

13.5 CIE of Brigg inner relief road

13.5.1 Description of method

For ease of presentation as evidence, a simplified CIE method was adopted. In essence this provided (see Table 13.4) the equivalent of Box 11 of the generic model supported by a textual description. The analysis identified the sectors that would be affected (column 2 and in part their size in column 3), described the kind of impact on them (column 4), defined the sector's objectives (column 5), and noted the unit of measurement and valuation of the impact, should these be possible (column 6). It then compared the IRR with a DM datum option (column 8) to show which of the two would give more net benefit and therefore preference to each sector (column 9). This prediction is supported in column 7 by analysis of representations on the IRR at the Local Plan Inquiry.

13.5.2 Conclusion

Since Table 13.4 is complex, Table 13.3 facilitates a grasp of the overall picture by summary. This table shows a list of the sectors involved (column 1) and which of the two options they prefer (columns 2 and 3). The table shows a predominant sectoral support of the IRR, with only three sectors in support of the DM option, all producers on site:

Table 13.3   Brigg Inner Relief Road: summary of community impact evaluation.

Column 1	DM Column 2	IRR Column 3
Producers		
On site		
1.0 Landowners and occupiers displaced	3	
3.0 County council as highway authority	3	
5.0 Borough council as highway authority	3	
Off site		
7.0 Landowners and occupiers not displaced		3
9.0 Landowners – developable sites		3
11.0 Borough council as rating authority		3
13.0 Brigg as economic entity		3
15.0 Regional cultural heritage		3
Consumers		
2.0 Occupiers displaced (in 1.0)	E	E
4.0 Vehicular traffic		3
6.0 Occupiers – not displaced (in 7.0)	E	E
8.0 New occupiers in developable sites		3
10.0 Brigg taxpayers		3
12.0 Brigg residents and visitors		3

Notes: 3 shows option favoured by sector; E shows conclusion included elsewhere.
Source: D. Lichfield (1987).

- 1.0 Displaced landowners and occupiers
- 2.0 The County Council as Highway Authority
- 3.0 The District Council as Highway Authority

If the sectors are regarded as of even weight, then the IRR is clearly preferred. But they are not. The question therefore arises: is the cost to the 3 on site sectors of the IRR greater than the benefit to the 5 offsite producers and 4 consumer sectors. From the discussion above (§2.5) this hardly seems so. Put another way, the benefit to the five producer and four consumer sectors would seem to be a good return for the cost to the three onsite producers; and the weight of these three sectors cannot be judged to overcome the weight of the remaining nine sectors which show preference for the IRR.

13.6 Is the cost of the IRR worthwhile

The County Council as highway authority would clearly be faced with the major costs (in land acquisition, construction and maintenance of the inner relief road). To justify that cost they must consider its benefits. But in accordance with "conventional highway economic terms" (Technical Report (TR), para. 8.2) the only benefit they take into account are the economic savings to vehicle users, and furthermore these are confined to savings in time and vehicle operating costs (TR Appendix 1). As such, their cost–benefit analysis does not take into account in this admittedly "crude analysis" (TR 5.4): other benefits to traffic which they recognize, namely accident relief and benefits to pedestrians (TR: para. 5.1); environmental benefits which should be accepted in terms of the Assessment Summary Report in *Urban road appraisal*; economic benefits that relate to the Brigg community as a whole, as brought out in the community impact evaluation.

It is because these benefits were ignored that, based on their narrow interpretation of costs and benefits, the County as Highway Authority dismissed the return to the County Council investment in the inner relief road as inadequate: "Overall there are no benefits to the *motorist* if the inner relief road was built. Time and distance savings for traffic terminating in Brigg are almost exactly balanced by the out of way travel caused to through-traffic. In crude terms, the scheme would be worth building if it did not cost anything" (TR: para. 8.1).

Their conclusion took no account even of the other gains they recognized in TR 152, for example improvement in shopping environment (§5.3), from new development (§5.4), and pedestrianization (§7.1 and §8.1). But these benefits did weigh with the County as Planning Authority in the terms put forward in the Brigg Local Plan (TR: Appendix 3).

"From a strategic planning viewpoint the scheme seems worthy of support. It would surely benefit the town centre and would help to attract more commercial investments".

But under conventional highway economic analysis it did not rank for grant aid, because of the low benefit in time savings and vehicle operating costs to vehicle

Table 13.4 Brigg Inner Relief Road: comparison of options by community impact evaluation.

Community sector		Number in sector	Impact Kind	Sectoral objective	Unit of measure	Majority number	IRR cf. with DM	Sectoral preference
1	2	3	4	5	6	7	8	9
Producers/operators								
On site								
1.0	Landowners and occupiers displaced	4	Property loss	Minimize loss	£	1	–	DM
3.0	Highway authority (CC)	1	Construction and maintenance costs	Minimize costs	£	1	–	DM
5.0	Highway authority (BC)		Maintenance costs	Minimize costs	£	1	–	DM
Off site								
7.0	Landowners and occupiers not displaced		Change in value from vibration, noise, air pollution, hazard		£			
7.1	Along the IRR	75		Maximize gain	£	19	+	IRR
7.2	In the town centre Balance 5.0	160		Maximize gain	£	110	+	IRR
								IRR
9.0	Landowners: developable sites (9.1–9.3)		Change in development value	Maximize gain				IRR
11.0	Glanford BC as rating authority		Change of rate assessment in shopping street and developable sites	Maximize rate base	£		+	IRR
13.0	Brigg as economic entity (13.1–13.3)		Change in economy	Maximize gain to economy	Economic indicators		+	IRR
15.0	Regional cultural heritage		Enhancement	Maximize	1		+	IRR

Table 13.4 Brigg Inner Relief Road: comparison of options by community impact evaluation.

Community sector		Number in sector	Impact Kind	Sectoral objective	Unit of measure	Majority number	IRR cf. with DM	Sectoral preference
1	2	3	4	5	6	7	8	9
Consumers								IRR
On site								
2.0	Occupiers displaced (included in 1.0)							
4.0	Vehicular traffic							
4.1	General		Change in: time vehicle operating costs accidents trip quality driver stress	Minimize adverse change	£ £ £ I I		+} +} +} +} +}	IRR
4.2	Service		Difficulties of access for loading, etc.	Minimize	£			
	Balance							IRR
4.3	Pedestrians: town centre outside IRR		Improvement in wellbeing Need to cross IRR	Maximize Minimize	I I		+ =	IRR IRR/DM
	Balance							IRR
Off site								
6.0	Occupiers: non-displaced (already in 7.0 above)							
8.0	New occupiers on developable sites		Gain livelihood Gain new public facilities	Maximize Maximize	£ I		+ +	IRR IRR
	Balance							IRR
10.0	Brigg ratepayers		Reduced rate assessments	Maximize	£		+	IRR

Table 13.4 Brigg Inner Relief Road: comparison of options by community impact evaluation.

Community sector	Number in sector	Impact Kind	Sectoral objective	Unit of measure	Majority number	IRR cf. with DM	Sectoral prefer-ence
1 2	3	4	5	6	7	8	9
12.0 Brigg residents and visitors		Improved prosperity & well-being	Maximize	I		+}	IRR
		Improved cultural heritage	Maximize	I		+}	DM
		Disruption from construction	Maximize	I		–	
Balance							IRR

Notes: IRR = Inner Relief Road option; DM = Do Minimum; + = IRR better; – = IRR worse; "=" = IRR/DM equal.
Source: Brigg Inner Relief Road 1987.

users, even though, as shown in the economic evaluation in TR 15.2 (Appendix 1), the vehicle users themselves would prefer the inner relief road as against the do nothing; the Urban Road Appraisal showed that the Brigg Local Plan objectives would be advanced by the inner relief road (§2.2); the community impact evaluation shows that the preponderance of the community sectors affected by the impacts of the two options would prefer the inner relief road; and Glanford Borough Council clearly considered it a worthwhile investment having regard to the wider impacts.

Put more specifically, why in these circumstances should the rate of return on a highway investment, and the ranking of priority in a County programme, be limited just to the benefits to motorists and not take into account benefits to other sectors?

CHAPTER FOURTEEN
Development control

14.1 Development control: a key element in plan implementation

In introducing plan implementation above (§1.6.4), we isolated development control as one of its measures. It is a particularly significant measure for several reasons. Unlike others, it bites specifically and deeply into the continuing physical development process and thereby has the greatest potential of all measures for influencing development on the ground. Although the control is "negative" (in that it relies upon the power to refuse permission), it is capable in practice of being "positive", in that it can be used creatively to steer and stimulate the market to operate with greater regard for development in the public interest.

14.2 Development control is conditioned by the statutory planning process

Development control becomes possible when, in the planning process, there is the legal requirement to obtain a permit for the carrying out of development, so enabling the planning authority to exercise its influence on what is proposed. Just how this influence can be exercised is therefore conditioned by the statutory and administrative processes of the country in question This is exemplified in a review of development control in Western Europe (Denmark, France, West Germany and The Netherlands; Davies et al. 1989). Although there are variations, in very general terms in each the plan is legally binding, so that development in accord cannot be refused, whereas development not in accord cannot be granted. This produces a quite different development control system from that in Britain, with its discretionary approach, to which we now turn, in order to provide a basis for the discussion below.

14.3 The development control process in England and Wales[1]

14.3.1 The legal basis for control

The statutory basis for development control in England and Wales stems from a simple requirement, which has remained unchanged from the TCP Act 1947 until the PC Act 1991: that planning permission needs to be obtained for carrying out development, except where some alternative control is introduced outside the planning system (e.g. trunk roads). But although this simple requirement remained stable, its context has changed over the years under a series of statutes and regulations, supplemented by administrative and policy guidance of the relevant departments of government, particularly the Department of the Environment in its Circulars, Planning Policy Guidance, departmental decisions on planning appeals and judicial case law. As such, the context presents a complex and formidable thicket of statutory, administrative and judicial controls (Grant 1982).

For our purpose in this section, it is necessary only to recognize the main principles on which this superstructure has been erected. They are now set out by reference to the sequence of steps taken to secure the planning permission needed to embark upon development.

For what development is permission required?

The development for which permission is required has remained constant over the period. It is either construction (the carrying out of building, engineering, mining or other operations in, on, over or under land) or the making of a material change in the use of any buildings or land. However not all such development is subject to control, there being exclusions for minutiae, some of which can be important in the local context.

The planning application

A formal application for planning permission can be made without any prior approach to the local planning authority (LPA). But a prior informal approach is common and desirable, for obtaining information, consultation, pre-application discussion, negotiations and bargaining. There are two kinds of permission:
- outline, for the erection only of new buildings, where the LPA agree that the application can be considered to establish the principle of whether or not development would be permitted;
- full planning permission in other circumstances.

In support of an outline application the minimum is a plan showing the site and particulars on a standard form, but there could also be an explanatory letter or memorandum that describes the scheme and argues the case for planning permission, and in addition illustrative plans, technical reports and an environmental assessment if required (see Ch. 17). Application for the full permission needs fully worked-out plans.

---

1. The planning system in Scotland and (to a lesser extent) Northern Ireland is broadly similar to that of England and Wales (Young & Rowan-Robinson 1985).

Consideration of the application

Until 1991 (PCA 1991), the law had remained the same since the TCPA 1947. In deciding whether or not to give permission, the local planning authority, and also the Secretary of State and his Inspectors when involved (in respect of call-in applications and appeals), were guided and constrained by the TCPA, 1990, Sec.70. They: ". . . shall have regard to the provisions of the Development Plan, so far as material to the application, and to any other material considerations, and . . . may grant planning permission, either unconditionally or subject to such conditions as they think fit; or may refuse planning permission".

During the late 1980s, DoE policy placed increasing weight on the role of the development plan. This principle was enacted in the PCA (Sec. 54A): that ". . . applications shall be [determined] in accordance with the plan unless material considerations indicate otherwise". In consequence "An up to date and relevant development plan could normally be given considerable weight, and strong contrary planning grounds would need to be shown, to justify a proposal which conflicted with it" (§14.6.1).

In striking this balance, the Authority needs to exercise its discretion. In contrast to the system in other countries (Davies et al. 1986, Patricios 1986), this discretionary power in the determination of planning applications constitutes an area of the greatest flexibility in British administrative law (Grant 1982). Given this degree of discretion, there have been, as can be expected, many occasions for recourse to the Courts for interpretation. Thus, it is in case law that the boundaries of the discretionary power have been delineated and, as happens with judicial law, often in an untidy fashion.

The planning decision

The Authority is under obligation to give a decision within two months of registration of the application, unless they secure agreement for an extension of the decision time.

Conditions

In granting planning permissions, authorities take advantage of their opportunity to impose conditions "as they think fit", which in effect allow permissions to be granted that otherwise would be refused. Conditions can be imposed on the grant of an outline planning permission or on a "full" application. In imposing conditions, the guideline "as they think fit" implies considerable discretion. This has been circumscribed by the Courts (for example, conditions must generally relate to the use and development of land, and more specifically to the development permitted by the permissions) and have been codified in criteria (DoE 1985).

Planning agreements

Planning authorities have long since been empowered to make agreements for restricting the planning, development or use of land outside the provisions of the planning statutes. Previously, agreement required prior ministerial consent. Following relaxation of the requirement, the practice expanded, as did similar agreements under other statutes (Local Government Act 1972, Highways Act 1980 and local Private Acts). It is these agreements that provide for "planning gain" (Ch. 15).

Planning appeals
Having completed its consideration, the planning authority will either refuse (giving its reasons), or agree, subject to stated conditions, which in themselves are again supported by reasons. Should there be a refusal, or the imposition of conditions that are unacceptable, the applicant has a right of appeal within six months. Should he decide to go ahead, he will present the formal appeal (with grounds) following which the appeal machinery is set in motion. The appeal itself can be dealt with by written representations, an informal hearing or a local public inquiry for the purpose of informing the Minister. For these a programme will be set, with dates.

Appeal decision
Following the appeal, in any of its forms, there will be a decision letter. In the majority of cases the decision is taken by the Inspector; in the minority by the Secretary of State against the background of the Inspector's report.

Appeals to courts
Should there be a refusal (by the Inspector or Secretary of State) there may be grounds – on legal points only – for an application to the High Court to overturn the refusal. If this is not pursued, or it fails, the proposal must be either discarded or followed by a new application. Should there be permission with acceptable conditions, then the developer proceeds to negotiate the outstanding reserved matters in the permission and in the planning agreement if necessary.

Enforcement
Should the developer proceed without permission, or proceed on a permission but without conforming to that permission or its conditions, the planning authority have the powers of enforcement against breach of planning control. To this the developer must either comply, subject to penalties for non-compliance, or he can appeal against the enforcement, which will invoke much the same procedures as in an appeal against a refusal of permission.

14.4 Project planning and development control

The control of development just described is a statutory supra-imposition on the project planning process referred to above (see Ch. 7). We now explore the two in combination by reference to Table 14.1.[2] This shows at the head the five main parties involved (landowner/developer, local authority, consultees, Secretary of State and the courts) and under each shows where they could be active in each of five stages.

    1. *Project plan*
        The developer and his team can elect either to proceed directly to the planning

---

2. The figure is an approximation. For detail see Mynors (1987), and Salt (1991) for a succession of flow charts.

Table 14.1   Relation between project planning and development control.

Landowner/developer	Local authority	Consultees and third parties	Secretary of State	Courts
Project planning	Development control	Devt control	Appeals	Appeals & judicial reviews

1 *Project plan*
Informal consultation
Environmental assessment
 	
Feasibility: planning
Feasibility: financial
Evaluations of options,
   decision, choice
	Informal consultation
	Development plan and
	   policies
	Planning brief
		X		

2 *Planning application*
Informal consultation
   (as needed)
Amend
Abandon
Proceed
	Formal & informal
	   consultations
	Development plan/policies
	Other material considerations
	Decision:
	   grant without conditions
	   grant with conditions
	   refuse
		X		

3 *Appeal* Abandon Proceed	Contest appeal	X	X	X

4 If *permission* proceed with
   development team
Application for reserved
   matters
	Decision on reserved matters			

5 On completion: Occupy (commission), operate and manage Dispose				

application, or to work up proposals by informally consulting all relevant parties, and negotiating and bargaining with the planning authority and its officers. They will need formal consultation if an environmental assessment is involved (see Ch. 14). The approach to the local planning authority could be at different levels of intensity. At the simplest, the developer and his planning team would refer to the development plan documentation available and have meetings (with the planning department, utility undertakers, consultees, etc.). Or there may be an exchange between the two sides over all the relevant considerations, which will amount to negotiation and bargaining (RTPI Development Control Working Party 1976). Linked with this there could be informal consultations with private or public bodies who could be involved with an application; and presentations by the developer/planning team to the public likely to be affected. Should they be consulted, the Authority will have regard to the development plan and also to other considerations, possible conditions on the permission if granted, the need for a planning agreement, the possibility of planning gain (subsequently planning obligation) and the need for an environmental assessment (Chs 15 and 17). Faced with the results of the consultation/negotiation/bargaining/public presentation, the developer and his team will finalize their project plan for submission of a planning application. This process will include some kind of evaluation of optional project designs as a basis for choice and decision. In that choice there could be four major issues:

- financial feasibility/viability: which of the alternatives would show an acceptable financial return to the promoters, having regard to private costs and benefits, and those externalities that need to be internalized as a result of the planning conditions, agreement and planning gain. This test would be provided by financial analysis or social financial analysis (where the fortunes of the landowner, financier, consumer, etc. are also presented) (Table 4.1);
- planning feasibility: having regard to its knowledge of the policies and attitude of the planning authority and DoE, the likelihood of the options getting planning permission, either from the planning authority or, if need be, on appeal to the Secretary of State;
- combination: given a matrix of ranking of the answers to the preceding two questions, which would be *best*, bearing in mind the possible need to accept a lower financial rate of return in order to have a higher level of planning feasibility.
- evaluation of options, choice and decision.

2. *Planning application*

Having made this decision, the landowner/developer will submit the planning application, which will be considered by the planning authority as a basis for its decision. Faced with the decisions they can then decide whether to abandon or proceed.

3. *Appeal*

Should the decisions be adverse and they still wish to proceed, the applicant will set in train the appeal proceedings. If adverse, they can then consider judicial review.

4. *Development*

Given a favourable consent, the developer/landowner can then proceed with the development team. During this stage they will apply to the planning authorities for reserved matters on an outline application, and to the local authority on building regulations, bylaws, etc.

5. *Completion*

On completion of the project, it would be ready for occupation, operation and management; or disposal.

14.5 Considerations in deciding a planning application or appeal

14.5.1 Context

As seen above (§14.3), authorities have considerable discretion in making planning decisions. This would be relatively straightforward if the issues that needed to be considered under the twin blades of "development plan" and "other material considerations" were simple. But they are each very rich indeed and complex in content, so extending the limits of the flexibility and discretion. For this reason they have been a legal minefield in the forty years since the 1947 Act. Despite the shift of weight towards the Plan, they could well continue to be so (DoE 1992, Gatenby & Williams 1992, Lichfield 1992d).

14.5.2 What is a development plan? (DoE 1992b)

The answer to this seemingly simple question is quite complex:
- the development plan is not a single document but a hierarchy of plans (structure to local plans) and perhaps a hierarchy of authorities (county, district);
- the statutory development plan is supported by other plans and policies of differing status: non-statutory; under some other legislation (urban development corporation or special planning zones); informal local plan or policy framework; planning briefs;
- the development plan, and other plans and policies, are constantly evolving under a time-consuming review, so that inevitably parts will be out of date when they are finally published, and therefore of less significance than the more recent ones;
- there could be gaps in the development plan and other policies relating to the application under consideration; and, if not, there could be several policies, possibly conflicting, which are pertinent;
- in the nature of the notation for development plans, and the policies and explanatory statements, it is not always easy to identify whether or not the application is in accord with the plan.

14.5.3 What are other material considerations (Mynors 1987)

There is general agreement that (following *Stringer vs MOHLG* 1971):

> In principle any consideration which relates to the use and development of land is capable of being a planning consideration. Whether a particular consideration falling within that broad class is material in any given case depends on the circumstances.

In practice, the array of planning considerations is very wide and diverse, possibly covering (Mynors 1987): policies, standards, departmental policy (e.g. in Circulars, PPG Notes, appeal decisions, views of other government departments, regional policy), planning history of the site, development within curtilage, the proposed development in principle and detail, strategic issues, site-specific considerations, policies, general relationship to the surrounding area, impact of the proposal on the site, design, density, problems of construction and operation, access and layout, existing precedent, creation of a new precedent, alternative sites, non-planning factors, financial consideration, personal circumstances.

However, even so, they may not be "material" planning considerations (see Loughlin 1980, Layfield 1990).

14.6 Guidelines for balancing the considerations

14.6.1 Official guidelines

Since development control is such an important part of the planning process, and since it has been applied year-in year-out since 1948 in many millions of cases, it might be reasonably thought that official guidelines on weighing up these multi-faceted considerations, with a view to leading to a justifiable decision, might have been more mature by now. But this is far from the situation. Let us look to the one source where central government guidelines have been pouring out on all facets of planning since the passing of the 1947 Act: the Ministry of Town and Country Planning (1943–51) and its successors leading to the contemporary Department of the Environment. The guidance is given in:
- Circulars (1943 to date) (being phased out);
- Development Control Policy Notes (being phased out);
- Planning Policy Guidance Notes (1988 to date) containing a series of guidelines on how planning considerations should be employed in certain substantive areas.
- Advisory Bulletins
- Research publications on good practice.

From these we concentrate on official guidance on the principles of weighing up planning considerations as a basis for the decision, which is our primary concern here.

Purpose of the planning system
Planning Policy Guidance Note No. 1 (DOE 1992) emphasizes that "the town and country planning system "is designed to regulate the development and use of land in the public interest" (para. 2).

But the guidelines are difficult to follow. The term "public interest" is notoriously elusive (Schubert 1961, Friedrich 1962, Friedmann 1973), and because of that can be merrily claimed by advocates of viewpoints that are quite clearly *not* in the public interest. And it is particularly difficult to establish in urban and regional planning, where definition of the "relevant public" is not easy to determine and there are an abundance of conflicting interests.

Changing role of development plans
In the Second Reading Debate on the Planning and Compensation Bill that introduced Section 54A of the TCPA 1990, the Minister, Sir George Young, noted the change of emphasis (Young 1991):

We have always regarded the development plan as important, but Circular 14/ 85 appeared to downgrade it by referring to it as only one of the material considerations. Those days are well behind us. Today's debate should leave no doubt about the importance of the plan-led approach.

This however did not mean that the other material considerations had no weight:

But it would still allow appropriate weight to be given to all material considerations. . . . One material consideration would, of course, be the extent to which the development plan was up to date . . . and consistent with national and regional policies as well as relevant to the proposal in question.

In further amplification the Minister added:

. . . But it would be wrong, within the planning law framework that has survived for more than 40 years, to change the status of a plan into a prescriptive document. We do not operate zoning mechanisms such as are found across the Atlantic. In our system, it is important that each planning application should continue to be considered on its merits and that all the material considerations should be weighed in the balance in reaching development control decisions.

We must be careful in our amendment not to remove the flexibility in our system that allows the weights to be attached to the various considerations to be determined by the relevance and significance of those considerations.

Presumption in favour of development or development plan
This presumption was introduced early in the planning system, when it was necessary to offer the following advice (MOTCP 1949):

The Minister fully appreciates the difficulty with which the local planning authorities are faced in dealing with applications in which the needs of the individual and the claims of the community appear to be evenly balanced, but he is of the opinion that the necessity for appeal could in many cases be avoided if full weight were given to the considerations set out in this Circular.

The considerations referred to have a contemporary ring, in terms of, for example, pre-application discussions, giving precise reasons for refusal, and being careful to limit conditions to legitimate concerns of planning. That the advice was not being fully effective was clear from a Circular soon after, which repeated the essence of that in 1949. Again there is a contemporary ring (MOHLG 1953):

Where there is no clear and specific reason for refusing permission or attaching conditions, the applicant should be given "the benefit of the doubt" . . . Development should always be encouraged unless it would cause demonstrable harm to an interest of acknowledged importance.

Indeed, this 1953 formulation, with its specific encouragement for development, seems fitted to the 1980s, which is associated with deregulation (Ch. 19):

> The planning system, however, fails in its function whenever it prevents, inhibits or delays development which could reasonably have been permitted. There is therefore always a presumption in favour of allowing applications for development, having regard to all material considerations, unless that development causes demonstrable harm to interests of acknowledged importance.

This is certainly consistent in the changing seas of planning. But despite the consistency, it is striking that over the intervening forty years there has been so little success in establishing an authoritative definition of the ruling caveat just mentioned. It is here that CIE can clearly play a role. It can demonstrate harm (cost or disbenefit), identify interests (groups or policies), or bring out their importance by reference to options forgone (opportunity cost). The CIE also has this role in the new circumstances since the introduction of Section 54A/1990 via the PCA 1991. As a result ". . . this introduces a presumption in favour of development proposals which are in accordance with the development plans" (DoE 1992a).

Onus to give reasons
Whenever an Authority refuses planning permission, it is required to give reasons that (DoE 1985a): ". . . must be precise, specific and relevant to the application: they must demonstrate clearly why, in the local planning authority's view, the proposed development cannot be permitted". This is emphasized elsewhere (DoE 1985):

> Except in the case of inappropriate development in the Green Belt the developer is not required to prove the case for development he proposes to carry out;
> if the planning authority considers it necessary to refuse permission, the onus is on them to demonstrate clearly why the development cannot be permitted.

There is certainly a clear logic for this presumption. Landowners and developers have the legal right to put forward how they wish to use their land to pursue their own objectives: and it is a fair presumption that in doing so they will have prepared carefully, with the aid of advisers, and examined the potential possibilities with greater care than could the planning authority, whose responsibilities are spread over *all* the land in that area, usually without adequate resources for their discharge. But the special case of inappropriate development in the green belt" has been widened as a result of Section 54A (DoE 1992a).

"An applicant who proposes a development which is clearly in conflict with the development plan would clearly need to produce convincing reasons to demonstrate why the plan should not prevail".

Summary
These guidelines clearly indicate that balancing/weighing up are needed. But they could hardly be said to provide a coherent approach or method to the issue raised at the head of this section: how to weigh up multifaceted planning considerations. This in itself would not be so serious for a practitioner in the development planning process, on either side, if there had been over the years authoritative and accepted guidance elsewhere. But that is not so, as we now see.

14.6.2 Unofficial guidelines

The Dobry Reports (Dobry 1974, 1975)
The terms of reference for this inquiry, commissioned by the DoE, did not invite views on the topic, but rather on the procedural streamlining of the development control system. But although containing no guidance on the topic, they did anticipate the use in development control of impact studies that would ". . . describe the proposed development in detail and explain its likely effect on its surroundings" (Dobry 1975: sec. V.18).

PAG Report (Planning Advisory Group 1965)
Although set up by the DoE to make a "general review of the planning system", the Group interpreted their remit as concern ". . . for the broad structure of the planning system and our attention has therefore been concentrated on development plans which are the key feature of the system. This is the main subject of our present report" (Introduction, paras 1 and 2).

As regards development control:

> we have examined the scope of development control to see whether it should be tightened or relaxed. On balance we recommend no change . . . As regards procedures, our general conclusion is that the present system of development control is basically sound and works efficiently . . . we consider that attention should be concentrated on management control and measures designed to tone up the system and improve the general level of performance . . . In our view, however, the main defect of the present planning system lies not in the methods of control but in the development plans on which they are based and which they are intended to implement. (Planning Advisory Group 1965: paras 110, 111, 112, 115).

The Royal Town Planning Institute
Following an earlier Report that virtually ignored the topic (RTPI Development Control Working Party 1976), a second report (ibid. 1979) saw the ". . . fundamental aims of development control to be the achievement of the intention set out in development plans and policies . . .". But, although recognizing the role of other material considerations, the working party said little on their balancing.

The Nuffield Report

This non-governmental but authoritative inquiry into the planning system (Committee of Inquiry 1986), had a wide terms of reference: "To inquire into the assumptions and purposes of the Town and Country Planning system, its past and present performance and its proper role in the future; and to report" (Committee of Inquiry, Appendix 1, para. 1). But, having described the legislative framework for development control (Ch. 2), they had little to recommend on its future practice except an endorsement of the moves towards its greater efficiency (ibid.: paras 9.22–24).

House of Commons Committees

Another possible source of improvement in planning control was the weighty deliberation on three occasions on planning procedures (House of Commons 1976–7, 1977–8, 1986). Although these reports contained quarries of significant information from those submitting evidence, the opportunity was not taken to tackle the issue of balancing that is raised here.

14.6.3 Empirical guidelines

Even in the absence of clear guidelines, the proof of the pudding could be in the eating in the form of the actual decisions that the local planning authorities make on planning applications, and the Departmental Inspectors and Secretaries of State on planning appeals. Here the conclusions are disappointing. On the former, the typical planning officer's report to committee tends to comprise a recital of the facts, a statement of the policies, other material considerations and the consultations. But it is often not possible to see a clear link between these and the final conclusion and recommendation in the report. In general terms, Inspector's reports, even in complex cases, are often masterpieces of condensation of a mass of background material in a structured form. But here again it is often difficult to see how the *balancing* is carried out. This situation is compounded in the Secretary of State's decision letters, which may disagree with the Inspector's conclusion and recommendations, but often with a somewhat tenuous link between the report and the actual ministerial decision.

We finally come to another possible source of clarification on decisions: the appeals to the Courts on points of law arising from decisions by Inspectors or Secretaries of State, or from judicial review of local authority decisions. Following legal requirements, the concentration here is on whether or not the law (statute, regulation or cases) has been followed. There is accordingly the introduction of logic to balancing *legal issues* on particular cases. But, since the Courts in general avoid giving views of matters of *planning judgement*, they rarely give guidance on how to go about the balancing of the conflicting considerations in decisions.

14.6.4 Conclusions

In brief, whereas in plan-making there are volumes on logical analysis leading to the plans themselves, in development control the field, which should have been ploughed with millions of planning decisions, is disappointingly barren on the logic of balancing multifaceted and often conflicting planning considerations in a particular case. On this we now offer an approach based on CIE.

14.7 How should the balance be struck?

14.7.1 The approach: planning for people as though they really mattered[3]

The varied planning considerations resemble the "apples and pears" commonly used in facing a comparable problem in economics: how can an economic calculus be made in purchasing, investment, and so on, when the decision-taker has to weigh up and balance variegated supply-side attributes in the object under consideration against the value to himself of spending resources on those particular attributes as opposed to other opportunities available. In economics, the short answer is to use the "common measuring rod of money", which can balance in one dimension the values of the attributes against the costs of the resources. It is around this common dimension that, in economic life, the involved parties discuss, negotiate, bargain, agree and exchange.

The use of money for this purpose clearly cannot be pressed in planning, since so many of the considerations are outside the market process. However, one common measuring rod can be identified: that all the considerations in question can be construed as having their impact on the basic resource for which, it is accepted, the planning is carried out, namely *people*. Thus, if the discussions, negotiations and bargaining that make up the plan-making/-implementation process can be transmuted into a dialogue around the impacts on people, we have the common measuring rod for planning decisions. How this is to be done is seen by recognizing the approach that is implied by the various parties involved in negotiating and bargaining on development projects that are under review by the planning authorities.

When considering a planning application an authority has two prime options: permission subject to conditions, or refusal. If the authority is disposed to approve, subject to conditions aimed at amelioration of any potential undesirable costs, it is in effect saying that the conditioned proposal would accord with its policies, and therefore would be in the public interest. If it wishes to refuse, it is saying in effect that there are no material considerations that justify departure from plan and policy. To these options can be added many suboptions which will arise in discussions, negotiations and bargaining of the parties, both during the preparation of the project prior to application, and subsequently, leading to permission or refusal. In this, each party would be comparing the disadvantages (costs) and advantages (benefits) that would fall upon him under the options. In this comparison, each party would, implicitly or explicitly, use the standard approach to the appraisal of development projects, namely that of cost–benefit analysis, albeit obviously not necessarily the standard method in all its rigour.

The individual cost–benefit *approach* for the parties would not be uniform, for each would vary and certainly simplify the standard method according to its role in the project and criteria for decision on it. These variants can be regarded as members of the *family* of cost–benefit analysis, each adapting the generic approach to accommodate the costs and benefits with which he is concerned and their geographical

---

3. With deference to the late E. F. Schumacher, with a switch from "economics" to "planning". This section amplifies Lichfield (1992).

spread (on site, off site, etc.). Some typical instances are as follows, the parentheses enclosing a description of the variant method:
 (a) the user: what service will I obtain and at what direct cost (cost minimization);
 (b) the landowner/developer: what will be my direct financial costs and returns? (financial/investment appraisal);
 (c) the local authority: what will be our direct costs and direct revenue in terms of tax expenditure and revenue, etc.? (cost–revenue analysis); or direct/indirect economic costs and benefits (cost–benefit analysis);
 (d) official consultees: what will be the repercussions on my statutory interests (cost–revenue or cost–benefit analysis);
 (e) the impacted public including third parties: what will be my indirect costs and benefits (cost–benefit analysis);
 (f) the government: what will be my direct costs and tax revenue (cost-revenue analysis); and the indirect benefits generated with the costs (social cost–benefit analysis);
 (g) the local planning authority: what will be the direct and indirect (externality) costs and benefits on the various community groups who are impacted, leading to answers to the questions: given acceptance of the proposal, who would suffer, who would gain, who would pay and who would be compensated? (community impact analysis).

In the negotiations/bargaining, even where the developer accepts that his proposal is not in accord with plan, he could still try and show that, having regard to other material considerations, its net benefit to the community would be greater than adhering to policy. For this he could not rely on his own financial appraisal (that is net benefit to *himself* in (b) above) but would need to adopt that of the planning authority in (g), in order to show that his proposal for the use and development of the land is *more* in the *public interest*. To meet this criterion, both parties would need to ask: would the net benefit to the impacted people (i.e. the relevant groups in the relevant community) from the proposal which infringed policy be greater or less than the net benefit from following policy.

This criterion would be the *common denominator* for the various options that would arise in the discussions. To test the criterion the options should be compared by the "balance sheet approach". For this the "community impact analysis approach" would be used, to a depth and formality appropriate to the case. We now go on to show how.

14.7.2 From planning considerations to impacts on people

We showed above (§3.5) that there are in everyday use a very wide and diverse array of planning considerations that need to be taken into account in deciding a planning application or appeal. Because of the diversity, there is little hope of finding a common denominator among them. But since they are so well embedded in the planning literature and practice, it is necessary for them to be translated into our common denominator, namely *impacts on people*. We now propose to show how this can be done. For this purpose we will not use the considerations described above (§14.5.3.) but instead the results of a research study into the relationship between development plans and development control; or, more specifically, which considerations would the development control officer or inspector have to take into account, and where lay the authority for doing so, in the development plan or elsewhere (Davies 1986)?

The considerations were found to be in four groups:

- The need to have regard to development plan policies that ". . . define the objectives, or lay down a course of action that guides the process of decision-making".
- The need to take account of central government policy, based on White Papers, Departmental Circulars and so on.
- Within the guidance of the development plan package ". . . the characteristics of the development proposal which need to be taken into account in assessing the application for planning permission".
- The views from other actors, comprising the statutory or major consultees: informal or minor consultees; objectors; neighbours; public at large.

Within the framework of these four groups, the review found that there was a ". . . broad consensus about the range and scope of development control as expressed through the medium of planning considerations". This consensus was summarized in a checklist which, stemming from the application characteristics, is the raw material that needs to be handled by the planning department in the consultation and decision process, in the light of the development plan and policies (Table 14.2). From the table it is seen that there are six first-tier features that subdivide into 14 second tier and 87 third tier.

The planning considerations resembled those in appeal decisions except that there are ". . . issues identified at inquiries by Inspectors in their decision letters as matters that either were mentioned by parties to the appeal (including matters of policy and other considerations) or influencing the Inspector in arriving at a decision or recommendation on appeal". As such, the appeal considerations are shown in Table 14.3. They were not found to be identical to the planning considerations, but they resemble them in general.

Whether or not the sample of authorities and inspectors in their view just described is representative of development control and appeal practice throughout the country, the review makes the major points of concern here. In the consideration of any application or any appeal, there is a welter of variegated and disparate information, of different weights and from different sources, which somehow has to be integrated to lead to a particular planning decision. Furthermore, this needs to be done conscientiously, in protection against a possible challenge to the decision (on appeal to the Minister or to the courts on judicial review), if irrelevant considerations are taken into account or relevant ones are not dealt with adequately. This richness and variety in planning considerations has led to the view (Davies 1986) that development control is "potentially a form of open government which requires:

The resolution of conflict: between strategic requirements and their local impacts; between old established residents and newcomers bringing change; between conservationists and developers; between a massive, but diffused and

Table 14.2 Checklist of planning considerations used in development control, by land use (*source*: Davies et al. 1986: 15).

1st Tier	2nd Tier		3rd Tier		a	b	c	d	e	f
	201	Site Characteristics	301	Topography	●	●	●	●	●	●
AMENITY (INCLUDING APPEARANCE)			302	Landscape Features: vegetation, water etc.	●	●	●	●	●	●
			303	Impact on Historic etc Buildings		●	●	●	●	●
			304	Archaeological Site	●●●	●	●	●	●	●
	202	Design (visual quality)	305	Issues of Architectural Style	●●●	●	●	●	●	●
			306	Intrinsic Architectural Merit: scale, mass etc.	●	●	●	●	●	●
			307	Relationship with Surroundings	●	●	●	●	●	●
			308	Treatment of External Spaces		●	●	●	●	●
			309	Density/Plot Ratio		●				●
	203	Physical Impact/ Quality	310	Daylight	●	●	●		●	●
			311	Sunlight	●	●	●		●	●
			312	Protection from Noise	●	●				
			313	Visual Privacy		●				
			314	Orientation		●				
			315	Outlook	●	●				
	204	Operation/Amenity (effect on amenity)	316	Noxious/Hazardous Uses			●	●		
			317	Hours of Operation			●	●		
			318	Effects of Construction				●		
			319	Litter				●		
			320	Obstruction (i.e. "comings and goings")	●					
	205	Relationship to Surroundings (i.e. off-site)	321	Impact on Historic etc. Buildings		●	●	●	●	●
			322	Impact on Protected Land - Amenity		●	●	●	●	●
			323	Non-conforming Use - Removal		●	●	●	●	●
ARRANGEMENT	206	On-Site Layout	324	Roads - Layout		●	●	●	●	●
			325	Roads - Capacity		●	●	●	●	●
			326	Parking - Layout		●	●	●	●	●
			327	Parking - Capacity		●	●	●	●	●
			328	Pedestrian Movement: gradients etc	●	●	●	●	●	●
			329	Creation of New Pedestrian Routes		●	●	●	●	●
			330	Cyclists						
			331	Disabled Persons					●	
			332	Refuse Collection					●	●
			333	Children's Play Space - Layout		●	▨	▨	▨	
			334	Children's Play Space - Space Provision		●	▨	▨	▨	
			335	Residential Internal Accom. - Layout					▨	
			336	Residential Internal Accom. - Provision/Facilities					▨	
			337	Residential Private Open Space - Layout		●	▨	▨	▨	
			338	Residential Private Open Space - Provision		●	▨	▨	▨	
			339	Residential Amenity/Open Space - Layout		●	▨	▨	▨	
			340	Residential Amenity/Open Space - Provision		●	▨	▨	▨	
			341	Non-Residential Ancillary Accom. - Layout	▨	▨				●
			342	Non-Residential Ancillary Accom. - Provision			●			
			343	Back-Land Development		●	●	●	●	●
			344	Security e.g. defensible spaces		●	●	●	●	●

Section		No.	Criterion
207	Off-Site Relationships	345	Compatible/Related Uses & Activities - Proximity
		346	Compatible/Related Uses & Activities - Capacity
		347	Proximity to Incompatible Uses and Activities
		348	Highway Network - Proximity
		349	Highway Network - Capacity
		350	Public Transport - Proximity/Access
		351	Public Transport - Capacity
		352	Utilities - Proximity
		353	Utilities - Capacity
103 EFFICIENCY	208 Resources: on-site	354	Sub-surface Conditions
		355	Condition of Buildings
		356	Conversion Potential of Buildings
		357	Vacant Land/Buildings
	209 Off-site	358	Loss of Natural Resources: agricultural land etc
		359	Blight - Physical
104 CO-ORDINATION	210 Phasing: on-site	360	Linkages in Mixed Use Schemes
		361	Interim Measures
	211 Phasing: off-site	362	Other Linked Proposals: Physical
		363	Other Linked Proposals: Socio/Economic
		364	Phasing - by Quantity
		365	Phasing - by Areas
	212 Operation/Time	366	Temporary Uses
105 QUANTITY AND DISTRIBUTION	213 Quantity	367	Loss of Existing Use
		368	Addition or Increase in Use
		369	Residential Mix/Non-Residential Unit
		370	Ancillary Uses
		371	Special Categories (Residential)
		372	Special Categories (Non-Residential)
		373	Expansion of Existing Premises
		374	Employment Generation
		375	Impact on existing uses - off-site
	214 Distribution/ Location	376	By Sub-Areas
		377	By Groups
106 OTHER CONSIDERATIONS	No 2nd Tier	378	Planning Gain
		379	Agencies
		380	Prejudicial to another (preferred) scheme
		381	Competition
		382	Applicants' Needs (including personal occupancy)
		383	Site Assembly
		384	Financial Viability
		385	Impact on Existing Occupier - on-site
		386	Impact on Existing Occupier - off-site
		387	Precedent

14.7 HOW SHOULD THE BALANCE BE STRUCK?

Group		Appeal Consideration		Equivalent Planning Considerations
1.	AMENITY	01	Visual Amenity	301-9, 321-3
		02	Amenities (Residential)	310-20
		03	Pollutants	312, 6, 8, 9, 347
		04	Unsocial Hours	317
		05	Pedestrian and Vehicular Activity	320
		06	Privacy	313
		07	Light Loss	310-1
		08	Outlook Loss	315
		09	Amenity Areas	333-4, 337-9
2.	DESIGN	10	Design of Buildings	305-6
		11	Relationship with Surroundings	307, 321-2
		12	Site Layout	314, 331, 341
		13	Existing Buildings On-Site	303, 355-6
		14	Density/Plot Ratio	309
		15	Internal Space Provision & Arrangement	335-6, 342
		16	Site Characteristics	301-2, 308, 354
3.	INFRASTRUCTURE	17	Access to and Adequacy of Road Networks	343, 348-9
		18	Road/Footpath Pattern - On-Site	324-5, 328-30
		19	Off-Street (On-Site Parking)	326-7
		20	Off-Site Parking (Includes On-Street)	326-7
		21	Public Transport	350-1
		22	Utilities	352-3
		23	Social Infrastructure	345-6
4.	STRATEGIC	24	Physical Resources	376
		25	Social/Ecological Resources	304
		26	Agricultural Resources	358
		27	Loss of Landscape/Townscape	376
		28	Containment	376
		29	Strategic Location	379
		30	Phasing	360, 362-5
		31	Target or Land Availability	367-8, 370-3
		32	Social Balance	369, 377
5.	SOCIAL AND ECONOMIC CONSIDERATIONS	33	Public Opinion	none
		34	Moral/Religious/Ethnic/Cultural	none
		35	Security and Public Order	344
		36	Employment	374
		37	Financial Viability	384
		38	Appellant's Circumstances	382
		39	Local Needs	377
		40	Prejudicial to another Preferred Scheme	380
		41	Use of Resources	357, 359
		42	Temporary Uses/Interim Measures	361, 366
		43	Competition/Impact on Existing Occupiers (Off-Site)	375, 381, 386
		44	Impact on Existing Occupiers Off-Site	385
		45	Property Values	none
6.	LEGAL	46	Create Precedent	387
		47	Existing Precedent	387
		48	Citation	none
		49	Planning Gain	378
		50	Natural Justice	none

Table 14.3 Checklist of appeal considerations by relationship with planning considerations land use (*source:* Davies et al. 1986: 45).

political action generalized benefit and an intense, but local, cost. The point is
that is that the calculus of cost–benefit and the objective method of rational
decision-making break down in these circumstances and conflict can be
resolved not by reference to a predetermined plan but through political action.

Although the final sentence in this quotation reflects reality, it accepts that there
is no identifiable bridge between the planning considerations and rational decision-
making, and as result assigns the resolution of the conflicts mentioned to the politicians. It thereby fails to find a way of helping the politicians in this admittedly intrac-
table task. It is here that we precisely argue that the help can be offered by translating
the planning considerations to impacts on people.

A perusal of the planning considerations, in development control and on appeal,
shows the tendency to describe such considerations in "physical terms", for this is the
dimension in everyday use in "physical land use planning". But although this is
understandable, stemming from the definition of development in the Acts, physical
planning is really a shorthand for embracing also the activities of people on the
ground in relation to the physical attributes. Clearly, the implications behind the term
relate to people, even though not specifically spelled out. This is clearly so in the
category of "social and economic considerations" in the checklist of appeal considerations (Table 14.3). It is also so in the more physical category, that relating to
"design", as the following indicates. In each case it is possible to show what the
impact will be on the way of life of *people*.

10 Design of buildings: within the building, on residents, passers-by, etc.
11 Relationship with surroundings: people in the surroundings.
12 Site layout: people moving around the area in question.
13 Existing buildings on site: on those owning and using those buildings.
14 Density/plot ratio: in respect of residents' amenities.

This shows that, although the typical form of communication and jargon traditionally used in planning emphasizes the *physical*, they are nonetheless readily capable
of being seen in terms of impacts, on people in their family, social, economic and
other lives.

14.7.3 From people to community sectors

It will be readily accepted that "planning is for people", but in practice it is recognized that a community is not homogeneous but in fact is composed of different sectors or interest groups. This is brought out in the above checklist of planning
considerations used in development control (Table 14.2) in relation, for example, to
the category "arrangement-on site layout". Here the different modes of traffic
(motorists, pedestrians, cyclists, disabled persons, children, residents, etc.) are considered separately. Such sectors are also recognized in the fact that the negotiation
and bargaining that takes place around planning applications and appeals is made by
interest groups (as for example the landowners, objectors, neighbours, etc.), with
each of the actors having their own values and interests.[4] And even where the interests of certain of the actors (notably the planning authority) are not expressed this

way, but rather in terms of the planning principles and policies (as for example protection of the green belt, relief of congestion), these are inherently aimed at the welfare of particular groups of people.

14.7.4 A demonstration

The approach in §14.7.1 is now demonstrated by reference to the planning considerations reviewed above. Table 14.4 allocates the considerations in Table 14.2 to their relevant community sectors, subdivided into producers/consumers and operators in accordance with Box 11.3 of Table 7.2. From the Table it is seen that the bulk of the considerations can be allocated to community sectors, accepting that the allocation is generalized and may not be accurate in relation to the specific considerations. Only a few could not be allocated, namely those that were essentially descriptive of the project options (366 and 380) and those that had to do with precedence (378). These would not be relevant to Box 11.3, but respectively to Box 3 and Box 4. A striking feature of Table 14.4 is the many characteristics relating to consumers as opposed to producers/developers. This may have to do with a disinclination to "double count"; or it may relate to genuine omissions. In brief, planning considerations can be readily turned from *abstraction* to *impacts on people* in various community sectors. In this way the weighing up and balancing can be carried out much more readily.

Table 14.5 has been drawn up in a similar way for Table 14.3 with corresponding sectors. The majority of the items find a place in the Table. The exceptions relate to how the proposal would affect the implementation of any other scheme proposed for the site (540) and planning gain (649) which in effect introduce a comparison of options; five legal considerations/creating precedence (646); existing precedent (647); previous legal decisions (648); and natural justice (650).

In summary, the tests show that the bulk of the two kinds of consideration can be allocated as impacts to community sectors of the kind introduced in CIE. The remainder are not so relevant to the weighing up as to the decision context.

14.7.5 Conclusion: from planning considerations to community impacts

The search for improved decision-making in development control must build on but go beyond the traditional judicial interpretation of the statutes and the administrative interpretation of Circulars and policy guidance. In effect, the aim is to move beyond decision-making by quasi-legal rules to decision-making based on judgements relating to specific impacts from the planning application. In this, the clue is to recognize that, on any case, all sides of the pertinent discussion, negotiations, and so on, should have a common denominator around which the case revolves; in planning, this is the *prediction of the impacts on the different groups of people* who would be affected by the decision, be it as a result of adhering to the plan or of other material considerations. By bringing this array of impacts into a common framework, the necessary balancing and weighing up can be more rigorous and better understood and defensible.

This process would be eased if all planning and appeal considerations were to be

---

4. See Bruton (1983), for studies structured around this kind of analysis.

Table 14.4 Planning considerations of local authorities, allocated to relevant community sectors.

	Producers/operators			Consumers	
Item	Description	Plan con.	Item	Description	Plan con.
1	2	3	4	5	6
1	Property owners, on site		2	Occupiers, on site	
			2.1	current	3.63 3.85
			2.2	potential	3.01 3.63 3.86
3	Property owners, off site		4	Occupiers, off site residents visitors	
			4.3	neighbours	
			4.4	Passers by	
5	Landowner/developer	3.84 3.79	8.8	Users of limited urban facilities	
		3.82 3.83 10.3	8.1	Utilities	3.62 3.82 3.82
7	Limited urban facilities		8.8	Private transport facilities	3.62
7.2	vehicular transport		8.21	roads	3.48
7.2.1	roads				3.49
7.2.2	car parking		8.22	car parking	
			7.3	Public transport	3.50 3.51
7.3	Public transport		8.2	Residents	
7.4	urban uses			Schoolchildren	3.46
7.4.1	shopkeepers		8.3	Non-resident	
7.4.2	business		8.4	Employers	
9	National resources			Shoppers	3.75
9.1	agricultural land		8.8	Townspeople generally	3.59
				(balance)	3.64
11	National heritage				3.65
11.1	urban	3.03			3.67
		3.04			3.68
		3.21			3.69
11.2	rural	3.22			3.70
11.3	agricultural	3.58			3.71
					3.72
					3.73
					3.74
					3.76
					3.77

Source: adapted from Davies et al. (1986: fig. 2.1 and Appendix A).

Table 14.5 Appeal considerations identified by Inspectors, allocated to relevant community sectors.

Producers/operators			Consumers		
Item	Description	App. con.	Item	Description	App. con.
1	2	3	4	5	6
1	Property owners, on site		2	Occupiers, on site	5.4.2
1.1	current		2.1	current	5.4.4
1.2	potential		2.2	potential	
3	Property owners, off site current potential		4	Occupiers, off site residents visitors	
5	Landowner/developer	5.3.7	4.3	neighbours	
		5.3.8	4.4	passers by	
7	. . . urban facilities		8	Users of . . . urban facilities	
7.1	Utilities		8.1	Utilities	
7.2	Private vehicle transport		8.8	Private vehicle transport	
7.2.1	roads		8.2.1	roads	
7.2.2	car parking		8.2.2	car parking	
7.3	Bus/rail transport		8.3	Bus/rail transport	
7.4	Urban uses		8.4	Users of urban . . .	4.30
					4.31
					4.31
7.4.1	shopkeepers		8.4.1	Residents	
7.4.2	businessmen		8.4.2	Schoolchildren	
			8.4.3	Non-residents	
9	National resources		8.4.4	Employees	8.3.6
	urban land	5.41			
	agricultural land		8.4.5	Shoppers	
	minerals	4.2.4			
	water resources	4.2.4			
11	National heritage		8.5	Town and rural residents	5.3.3
11.1	urban	2.11		generally	5.3.9
		2.13			5.4.3
		2.21			5.4.5
11.2	rural	2.11			5.4.3
	ecological	2.25			5.4.5
11.3	agricultural	2.26			5.4.5

Notes: Application considerations are issues identified by Inspectors in their decision papers as influencing the inspector in arriving at the decision or recommendation.
Source: adapted from Davies et al. (1986: fig. 2.1 and Appendix B).

couched at the outset in terms of the impact chain (Fig. 7.3) whereby effects of injection into the urban & regional system could be translated to impacts on people in communities. But although desirable in the interests of better decisions on projects, it is asking too much for the planning system to switch in the *short term* from its well ingrained habits of dealing in planning and appeal considerations to effects and impacts. But in the longer term, where evaluation of alternatives is called for in a rational manner, this switch can be readily made as illustrated above (§14.7.4).

Having demonstrated the means of translation from planning considerations to impacts on people in communities, we can now proceed to show how the generic model can help in development control.

14.8 Application of the generic model to development control

14.8.1 Planning

- Box 1: The project planning and development control system as described above (§2.1–2.4).

14.8.2 Project in the system

- Box 2: The project as visualized in the development application, be it formal or informal, as in project planning prior to the application. The description will need to be in the detail necessary to bring out the project characteristics of the generic method (see Ch. 7, Box 2).
- Box 3: Once the application has been submitted the options are reasonably clear: outright permission with no conditions (which is rare); permission with specified conditions; outright refusal.
- Boxes 4–5: The content of these boxes are not needed for the analysis. However, whereas in the generic model it is not *essential* to identify the pre-project and post-project urban & regional planning system, this information is *valuable* when available. It is generally available in relation to a planning application: the current system being that shown in the current land use and other surveys, and the prospective system in the development plans themselves. Thus, the change implicit in the planning application can be put in context.

14.8.3 Options specification by plan variable

- Box 6: Plan variables would be considered in turn in complex cases.
- Box 7: For the simple options described above (Box 3) the context would be as stated there.

14.8.4 Effect assessment

- Box 10.1: The plan variables would be translated from Box 6.
- Box 10.2: Where the application is accompanied by an environmental statement, formal or informal, the summary (in Box 10.3) will relate to its content. But where it is not, then judgement on effects will need to be used as in the simplified method (Ch. 9).

14.8.5 Impact evaluation

- Box 11: Impact evaluation of options by plan variable: In the simplest and perhaps the majority of cases, the conclusions will be fairly clear. But where analysis is needed, to reach a conclusion and justify the recommendation, the impact evaluation will need to be carried through. In the simpler cases, this will follow the simplified and/or quick and dirty method thus not requiring the effect assessment under Box F10. In the more complex cases, the fuller method will need to be embarked upon.

- Box 13: Evaluation, consultation, negotiation and bargaining: It is the analysis so far described that would be used as the basis for these steps.
- Box 14: Evaluation, report and recommendation: As indicated, the evaluation report will be part of the Planning Officer's report to Committee, following the typical content of description, consultation, and so on. It is here that the recommendation for decision is spelt out, based on the preceding analysis. This step is clearly essential, however simplified the analysis itself. It will be important, as necessary, to spell out any assumptions made in the analysis, so that the robustness of the decision can be understood.
- Box 16: Decision communication: This will be typically embraced in the decision letter from the authority, with its reasons.

14.8.6 Application of the generic model to negotiation and bargaining on a planning application

Having identified the nature of the evaluation process on the planning application itself, we now turn to the more complex situation: the negotiation and bargaining that precedes a planning application by the two major parties: landowner/developer and his project, and the planning authority's regulatory control over that project.

On the first, the developer as decision-taker on the project primarily has regard to his direct costs and benefits in evaluating viability, and finding the option that gives the best value for money. But his options are conditioned by factors external to himself. For example, there are the project stakeholders: the landowner cannot aim for the highest site value without impinging upon the developer's profit margin; the developer cannot aim for the highest development profit without impinging upon the site value; and neither party can aim for the highest occupation value without impinging upon the ultimate consumer. And, in aiming at value for money, the developer will be constrained by the options available to each of the parties (other land by the developers, other developers by the landowner, other developments by the consumers). He will also be influenced by the potential impacts of the project on the environment. If these are ignored in the project planning, it could undermine the viability of the project as well: too much generation of traffic from a shopping centre will clog the adjoining roads and so discourage customers from shopping. In brief, there must be implicit or explicit negotiation, bargaining and compromise, between the many parties directly or indirectly concerned, compromise being necessary if a firm project is to be reached.

This applies equally to the planning authority. It needs to take account of the impacts on the environment that the developer does not propose to internalize (for example loss of trade in a competing shopping centre); the conclusions from the consultation, formal and informal, which lead to the need to reflect various interests of differing weights; and the wish to have "planning gain" from the developer which might lead (Ch. 15 below) to acceptance of elements of the project that otherwise would not be acceptable under the ruling policies. A further complication to this multifaceted situation is that, with the vertical hierarchy in planning authorities, differing weights are often found at local, county and central government level, and within the latter, between differing departments who have different priorities (compare transport, employment, agriculture, etc.). Thus, there needs to be negotiation and bargaining between these different pressures, and trade off between them, if a firm planning decision is to be reached.

So far, we have considered bargaining trade-offs *within* each of the two main actors in the process, the developer/landowner and those concerned with planning regulation. But the bargaining also takes place *between* the two. Left to themselves, the project planners might not offer "planning gain", which involves expenditure off site (if only because it would be resisted by the landowner as eroding his site value, and therefore puts the particular developer in a non-competitive position *vis à vis* others who did not contemplate the planning gain). But faced with an authority that was indicating that the "voluntary" planning gain under an agreement was a precondition of the planning permission, they would "offer" the gain to the detriment of the site value, on the proposition that other developers would also be ready to do so. The trade-off to the landowner is the gain of the permission, albeit at the cost of reduced land value (Working Party on Economics 1995: ch. 11).

It could be, in the developer/planning authority negotiation and bargaining, that the planning authority would give way. For example, if it were to be too firm on ensuring support for a particular policy against the developer, it might lose elements of the development itself that it would rather not forgo. An instance is the grant of offices in a non-office area (against policy) which is able to finance conservation of the same site, which would otherwise be non-viable.

Negotiations and bargaining of the kind described here could be conducted as mere horse-trading. But since the process needs to be formalized (meetings, representations on plans, applications, submissions) it would benefit also if there were some "common denominator" in the confrontations. This common denominator could take its flavour from the facts brought out above (§2.4): that each planning application, and the negotiation and bargaining that precedes or follows it, relates to variations (i.e. options) on the theme of a development project. If one of these options could be regarded as a datum, then those carrying out the negotiations/bargaining could be seen to be postulating another option. What brings these options into a "common measuring rod" is that they can all be described rigorously in terms of their differences in impact on community sectors (see §12.7). It would then be practicable for each of the protagonists to show explicitly the differences in disadvantage (costs) and advantage (benefits) to him and others, whether this be done in objective (measured) terms or subjectively (without measurement). Each would be implicitly or explicitly using the cost–benefit approach, which is typically employed in appraisal of a physical development project (§2.7.1 above).

Given the different methods in the cost–benefit family that would be employed, and given that each of the methods can "nest" within the comprehensive community impact analysis (Table 10.2), it can be seen that the negotiations/bargaining would be eased if the continuing dialogue were summarized and updated in discussion by reference to the comprehensive analysis. This would mean that the discussions could be made around specific factors stemming from a rigorous evaluation rather than in the current style of "beating the air" with a miscellany of planning considerations.

This bargaining and negotiation would be eased if the logic be accepted of imposing some onus on the developer to go beyond the minimum requirement for outline applications, for the following reasons:

- Although the landowner/developer has the right to put forward proposals for utilizing the development rights in the land, they do not own the rights until the planning permit has been granted; these have remained nationalized since 1 April 1948, with any denial of those rights by the State as owner carrying no liability for compensation, except in limited circumstances.
- The government releases these rights to the developer, under the planning system which ". . . is designed to regulate the development and use of land in the public interest". And even while that guideline is obscure, the purport is quite clear: the interests of the public as well as the landowner/developer need to be taken into consideration.

Thus, in each planning application the landowner/developer is asking for the grant of nationalized development rights for which he is not obliged to pay *directly* (although he may "offer" to do so indirectly as "planning gain"), and in the use of which he will be making an impact on the community, either or both beneficial and adverse. From this it follows that there also should be an obligation on the landowner/developer to demonstrate clearly why they should have the privilege of using without payment the nationalized development rights to pursue their own objectives in a manner inconsistent with the public interest. The very least would seem to be an assessment of the adverse impacts the development would impose on the community, and an offer for amelioration. This is on all fours with the new requirements for environmental assessment (Ch. 5), although it is not suggested that the generality of applications should be caught up in anything like the complex procedures in the new Regulations. This suggested shift of onus is on all fours with the DoE guideline flowing from the shift of weight in Section 54A of the 1990 Act (§14.3d):

> An applicant who proposes a development which is clearly in conflict with the development plan would need to produce convincing reasons why the plan should not prevail.

14.9 Application to appeals inquiries

Although the preceding has been presented in relation to bargaining on development control decisions, the approach applies also to appeals. For these the logic is however even stronger for using the approach presented here rather than the traditional. For one thing, in the cases that tend to go to appeal the issues are somewhat sharper than in those that do not. For another, more time and resources are available on each side for analysis and presentation for the decision itself, once it comes to a hearing. For another, the very necessity to make statements of case prior to the hearing in an inquiry permits of the opportunity for each side to agree some fundamentals.

Traditionally the exchange at an inquiry is founded on expert terms of reference from either side, which in the ultimate (even though the experts have full integrity) are conditioned by the tactics of each of the opposing advocates, who are not instructed to find the *best option in the public interest*, but rather to promote the *case of their client* and denigrate *that of the opposition.* Instead there could be a prior analysis of the options of the applicants and the planning authority on the lines of the generic model, presented beforehand (possibly by an independent consultant who would not be involved in the inquiry) and subscribed to by each of the parties subject to dissent on detail. Then the opposing evidence could be developed as variations around the central theme introducing, for the relevant options, differing versions of the effects (Box 10) and of the community sectors and their sectoral objectives (Box 11) and the weighting to be given to these sectors in reaching the overall conclusion (Box 12). In this way the planning inquiry would become more structured and meaningful.

14.10 A framework for judging sustainable development

The concept of sustainable development was, on its introduction by the Brundlandt Commission, presented very simply as "development that meets the needs of the present without compromising the ability of future generations to meet their own needs" (World Commission on Environment and Development 1987). Although the term "development" was not defined, its meaning has been taken in the literature to be much the same as in economics generally.

Sustainable development means that per capita utility or wellbeing is increasing over time. In this it differs from economic growth, which means only that real GNP per capita is increasing over time. But for either the growth or development to be sustainable requires that the increase in GNP is not threatened by "feedback" from either biophysical impacts (pollution, resource problems) or from social impacts (social disruption) (Pearce et al. 1993: para. 1).

Whereas these concepts are reasonably clear, their operation is not. A search for the meaning of the term found some 25 acceptable definitions (Pearce et al. 1989: annexe). In general these rules explore what is needed to make the concept workable in practice, with a view to setting principles, policies and targets whereby to achieve the aim set out in the concept. The efficacy of such rules can then be judged as time proceeds through monitoring the results in order to judge success in application.

Based on this approach the following operational rules have been suggested (ibid.):
(a) the conditions for achieving sustainable development include the requirement that future generations be compensated for damage done by current generations, e.g. by global warming;

(b) the compensation is best secured by leaving the next generation a stock of capital assets no less than the stock we have now (the "constant capital" requirement);

(c) the capital in question is both "man-made" (Km) and natural (Kn), the latter including for example environmental assets.

These rules are difficult in application (Jacobs 1991). A distinction is made between renewable resources (e.g. forests or fish) whose stock can be maintained by appropriate management; and non-renewable resources (minerals) which must be inevitably depleted, and therefore requires substitutes for the maintenance of stock. In addition, stocks of natural capital will be affected by the assimilation of wastes deposited in the soil, air and water. With good management, some (flow wastes) can be assimilated by the natural environment through bio- and geochemical processes of dispersal, decomposition and recomposition, when they disappear or are rendered inert. By contrast, others, such as heavy metals or nuclear waste cannot be absorbed in this way and will adversely affect natural capital (stock wastes). Furthermore, environmental quality in soil, air and water can be affected by the utilization of the Earth's surface, as for example in carbon dioxide which affects the ozone layer and global temperatures.

The aim of sustainable development was adopted by the British Government in 1991 in its first comprehensive white paper on the Environment (DoE et al. 1991):

The Government therefore supports the principle of sustainable development. This means living on the earth's income rather than eroding its capital. It means keeping the consumption of renewable natural resources within the limits of their replenishment. It means handing down to successive generations not only man-made wealth (such as buildings, roads and railways) but also natural wealth, such as clean adequate water supplies, good arable land, a wealth of wildlife and ample forests.

Among the many measures and instruments to be adopted for the advancement of sustainable development, the government will be using the land use planning system (DoE 1992b):

The Government has made clear its intention to work towards ensuring that development and growth are sustainable. It will continue to develop policies consistent with the concept of sustainable development. The planning system, and the preparation of development plans in particular, can contribute to the objectives of ensuring that development and growth are sustainable. The sum total of decisions in the planning field, as elsewhere, should not deny future generations the best of today's environment.

In advancing this aim (DoE 1992: para. 6.7):

One major responsibility is to ensure that development plans are drawn up in such a way as to take environmental considerations comprehensively and consistently into account. In this way environmental development decisions are taken against an overall strategic framework that reflects environmental priorities."

The DoE has published guidance on the way in which environmental considerations can be fed into the development plan as it is prepared (DoE 1993b).

For the purpose of this guide, environmental appraisal of development plans is:
- an explicit, systematic and iterative review of development plan policies and proposals to evaluate their individual and combined impact of the environment.
- an integral part of the plan-making and review process, which allows for the evaluation of alternatives
- based on a quantifiable base line of environmental quality.

This being so, there is the possibility of introducing into development planning the concept of sustainability. The simple means of so doing is to include the criterion of sustainability in development control alongside the great range of considerations that need to be borne in mind when considering whether or not to give the planning permit. The implication is that "non-sustainability" could be then included as a reason for refusing the permit. For this it will be necessary to improve upon our prediction of the consequences both on site and off site for the capital involved in the application, in terms of the needs of future generations, having regard to all the uncertainties that will abound over the long term future under review.

Such environment/sustainable development consideration must necessarily enter into the *balance* of considerations that might be made in leading to the planning permit. In this, however considerable the weight to sustainable development that is offered, it is only one of the considerations that much be taken into account in reconciling priorities in the public interest. Others are economic and social considerations, conservation and the environment.

Given such an understanding of the problem it is then possible to see the generic model as a framework for drawing conclusion on the questions raised. Given the wide array of possible interpretations of the concept of both "sustainability" and "development" there will be the need for a specific definition to be adopted. This could come at various stages of the formal generic model, for example, in the description of the planning and implementation process (Box 1), in the options (Box 3), in the decision base/framework (Box 9). The application of the specification to the project in hand will emerge in relation to Box 8, which describes the predicted change in the system as a result of the of the project, by reference to each of the system elements (Table 7.9). In this way a picture can be obtained of the changes that the project will produce in terms of:
- people (1(a));
- nature (1(b), 2(b));
- man-made assets (1(b), 2(a));
- cultural built heritage (1(b));
- the economy (1(a), 2(c)).

In relation to these elements will be predicted the impact (Box 10) and the evaluation in terms of preferences (Box 11). Then comes the summary as a basis for the evaluation (Boxes 12 and 13) where the evaluation criteria are introduced, in respect of sustainability.

In this way, assessing the effect on the categories of capital in question can be readily integrated into the methods and techniques for predicting *effect assessment*; and the *impact assessment* can be incorporated by judging the repercussions for the various generations who would be impacted. This starts with the "current generation", which will enjoy the benefit for the life of the project under consideration, and then those future generations, which would enjoy them or otherwise following the augmentation or depletion of the capital in question.

Given this analysis for sustainability, it would then be possible to ameliorate or enhance the effects of the proposal with a view to achieving a greater sustainability. Since the approach must of necessity be experimental for some time, a monitoring of the method of analysis use in the development control will contribute to the learning curve over the decades of progress towards achieving sustainability.

14.11 Adaptation of the generic model for development control

As indicated above (§14.7) the approach advocated here for "striking the balance" is based upon the approach of community impact analysis and evaluation. This does not mean that the application in the varying circumstances presented (informal consultation prior to the application, bargaining before or after the application, confrontation at inquiry, etc.) would necessarily involve the generic model as described above. In negotiation and bargaining, for example, it would be sufficient for the various parties to present arguments in terms of impacts, community sectors and net benefit to the various sectors, rather than in terms of planning considerations alone, detached from implications for people. For major and contentious planning applications and planning inquiries, the model could be used more formally. For informal discussion the simplified approach could be employed (see Ch. 8). What matters is that the *approach* be *consistent* on all occasions in order to avoid the confusion that stems from disputation that is unstructured, and does not have the common purpose, of finding the options acceptable to different parties. Such a consistent approach can be achieved by using the generic model as a comprehensive checklist of the steps to be taken in sequence, with the content to be adjusted for simplicity or complexity in particular cases. For example, in the simplest case, all that might be needed is the identification of the options in the project under consideration, always using one option as a datum, in order to identify precisely what the differences are, between the parties with a view to focusing on those differences in terms, for example, of impacts on community sectors.

14.12 Lament for planning decisions[5]

At the opening of this discussion on development control, we made the case that this very important aspect of the planning process, in fact its cutting edge, was quite neglected compared with the concentration of plan-making; and that there has been inadequate guidance from central government, or from the inquiries into planning that they or others had set up. Later we examined how planning control decisions and inspectors' appeal decisions were made in practice, by reference to a comprehensive review of the considerations taken into account in either case. This has shown a somewhat bewildering picture. In both cases many considerations were taken into account stemming from a variety of sources. Although mostly relevant to the development application or appeal, they are nonetheless so variegated that they are difficult to interrelate in a manner that will lead to a rational decision; that is, one that can be defended with reason. Furthermore, their typical use in reports of planning officers to committee, or inspectors within the Ministry, not only lack a structure around which they can be placed in context, but also fail to be specific as to their impact on *people*, which is the *raison d'être* for planning analysis, or show the reasons for their inevitable weighting as between *different groups of people*.

If this be a reasonably accurate and fair analysis of the majority of planning decisions in practice, the only conclusion can be that the vast machinery of planning control, and the countless decisions made under it, are not doing justice to the importance of the subject matter with which they are concerned, namely the enduring social control of our built environment. The significance relates to much more than the form and content of development on a particular piece of land. First, such development, once built, lasts for a considerable time over the life-cycle of the fabric, during which it conditions the activities of people on site and also has repercussions for those off site, in terms of what have been called linkages and effects (Table 7.9). Secondly, it then endures further as the starting point for its renewal at the close of the life-cycle. Thirdly, each such decision is part of a chain of associated decisions, for each subsequent development needs to take into account both what exists and what is in the pipeline. And, finally, the development permitted is a physical expression of the values and mores of the generation, which is enduring and difficult to erase. As has been well said by Winston Churchill, "We shape our buildings and then they shape us".

From this a simple question emerges. Is it not about time that the planning system, with its flexibility and discretion, was capable of producing and operating planning decision rules generally acceptable to its practitioners, politicians and the public, and so avoid continuation of the friction and waste of time and resources that has come from the previous 40 years. We are certainly sufficiently endowed to do so in richness of resources: educated, competent and experienced manpower; organization for professional administration; availability of financial resources to test planning applications and appeals; and the lessons of millions of cases since 1948. Can we not pursue *quality* in planning decision-making as vigorously as we have pursued *speed* in making them (Audit Commission 1992)?

---

5. With deference to Barbara Wootton in her *Lament for economics* (1938).

In brief, if only the effort over the years into reducing delays in the development control machinery had been accompanied by establishing an improved decision-making process! As put by one planning officer, "Is fastest best?" (Thompson 1987). To which could be added, "Is best not the fastest?"

CHAPTER FIFTEEN
Planning gain/obligation

15.1 Origins

Within the procedures of development control under the TCP Acts there grew up since the 1970s a practice which was popularly termed "planning gain" (Jowell 1977). This in essence related to certain "offerings" by a landowner/developer in the negotiations and bargaining, which were pursued prior to or in parallel to the planning application itself. They were pursued in parallel simply because the elements in question did not fall within the matters which can be "properly" dealt with under planning conditions, under the administrative rules set out in the Regulations and Circulars. There were varied reasons: positive undertakings the planning authority finds legally difficult to secure by conditions; proposals offsite not covered by the planning application itself; proposals not strictly related to the use and development of land as such; financial contributions to local government services not related to the development in question. For these reasons the "offerings" were the subject of an agreement, governed by contract and not planning law.

The practice was the subject of considerable controversy from the beginning, as to its nature, legitimacy, role in planning, economic impact (Jowell 1977, Property Advisory Group 1980, Loughlin 1981, 1982, Keogh 1985, Grant & Jowell 1983, Royal Town Planning Institute 1983). To some, planning gain should be abandoned since "we are unable to accept that, as a matter of general practice, planning gain has any place in our system of planning control" (Property Advisory Group 1980). Nonetheless, the Department of the Environment accepted its role in 1983 and attempted to regularize its practice (DoE 1983):

> "Planning gain" is a term that has come to be applied whenever, in connection with a grant or planning permission, a local planning authority seeks to impose on a developer an obligation to carry out works not included in the development for which permission has been sought, or to make some payment or confer some extraneous right or benefit in return for permitting development to take place.
>
> It is . . . distinct from any alterations or modifications which the planning authority may *properly seek* to secure to the development that is the subject of the planning application (para. 2). But the planning gain must be reasonable, depending on the circumstances (para. 5) and tests of such reasonableness are presented (paras 6–8).

The Department sought to confine the scope of the planning gain agreements to that already provided by the scope of conditions on permissions (DoE 1983):

5. If a planning application is considered in this light it may be reasonable, depending on the circumstances, either to impose conditions on the grant of planning permission, or (where the Authorities purpose cannot be achieved by means of a condition) to seek an agreement with the developer which would be associated with any permission granted. But this does not mean that an Authority is entitled to treat the applicant's need for permission as an opportunity to obtain some extraneous benefit or advantage or as an opportunity to exact a payment for the benefit of ratepayers' at large.

Despite this attempt, many local authorities used the opportunities to make a foray into exactions from the landowner/developer. These, by and large, accepted the system as a necessary price (in money or kind) for obtaining the permission, since the price tended to be lower than the enhancement in development values, which were gifted with the permission.

Although questionable in many ways (law, policy, ethics, etc.), the practice has nonetheless shown a robustness and general acceptance, and has flourished (Rosslyn Research Ltd 1990). It has done so because (Lichfield 1989a):

The practice is a common-sense response to the contemporary situation. With the firm abandonment by the current Government of the third post-World War II attempt at collecting betterment (in the Community Land and Development Land Tax Acts) landowners/developers/financial institutions can make fortunes out of a planning permit for using development rights which are still nationalized (the relevant provisions of the Town and Country Planning Act 1947 never having been repealed). Concurrently, under the present Administration, local government has restrictions on its financial resources and freedom to spend. Thus, the tax which planning gain imposes on the development industry, which it is generally prepared to accept to obtain the planning permission, offers a way of assisting local government in the financial trammels in which it finds itself, and comforts the taxpaying public in seeking "social justice".

Despite the DoE Circular 22/83 the controversy continued, in which some wished to go backwards and some forwards. To opponents the practice was iniquitous, being seen as "planning blackmail" and a "planning black-market", since the Section 52 Agreements were not made public; and planning authorities were diverted by the gain from granting decisions on "planning merit". To proponents it was a golden opportunity to use negotiating strength, so that many local authorities have over the years introduced policies in their local plans which have set out their approach to planning gain agreement (Healey et al. 1992). In doing so, some have openly presented to the developers a "shopping list" or "planning gain package" as a basis for bargaining. Some have done so within a wide context of infrastructure, including job training, child-care facilities, housing for local needs, help to community groups, environmental improvements and transport. These practices have been encouraged by the shift in DoE policy to favouring more specific expressions of practice and policy on planning obligations in their plans (DoE 1992b).

15.2 Evolution

After some years of controversial practice, it became necessary for the Department of the Environment to attempt once more to clarify the situation. This they did in a Consultation Paper (DoE 1989b), which "substantially reaffirmed" the guidance in Circular 22/83, which it would supersede. In doing so it introduced certain welcome clarifications, including their intention to abandon the controversial title of "planning gain" and replace it by "planning agreement", for the name had "come to be used very loosely to apply to both normal and legitimate operations of the planning system and also attempts to extract from developers payments in cash or in kind for purposes that are not directly related to the development proposed but are sought as the price of planning permission. The Planning Acts do not envisage that planning powers should be used for such purposes, and in this sense attempts to exact "planning gain" are outside the scope of the planning process".

15.3 The changes

The clarifications just mentioned were introduced by the PCA of 1991, by substituting a new Section 106 in the Town and Country Planning Act 1990, replacing Section 52 of the 1971 Act, which had consolidated previous planning legislation. The new provisions adopted the term "planning obligations" instead of "planning gain" or "planning agreements" and, while introducing what has been termed technical changes, also reflected some of the fundamental criticisms of the former system. The formal changes were:

15.3.1 Statute: Planning and Compensation Act, 1991
As just indicated, the largest and most significant change is the first embodiment of planning gain in statute, thus legalizing, regularizing and institutionalizing the practice which had until then been granted by agreement.

15.3.2 Administrative (DoE 1989b, 1991b)
- *Officially in the planning system*
  While Circular 16/91 (Annexe B1) "substantially reaffirms, with some amendment, the advice given in DoE Circular 22/83 which it supersedes" there are differences in treatment in the two Circulars.
- *The name*
  The reasons for departure from the term "planning gain" are clearly aimed at its

association with its dubious past, which includes (Circular 16/91, B3):
> ... attempts to extract from developers payment in cash or in kind for purposes that are not directly related to the development proposed but are sought as the price of planning permission ... offers from developers to a local authority that are not related to their development proposal ... in this sense "planning gain" is outside the scope of the planning process ... is imprecise and misleading ...

The implication is therefore that the term "planning obligations" will avoid this notoriety, presumably because its existence is now legitimized in statute, with the intention to tighten up and regulate its position. But there is no parallel justification for adopting that particular new name, except that it ". . . reflects the fact that obligations may now be created other than by agreement between parties (that is, by the developer making an undertaking)".

- *Reasonableness*
  There seems to be no departure from the general policy in Circular 22/83 that the alternative facing the authority is: ". . . depending on the circumstances, either to impose conditions on the grant of planning permission, or (where the authority's purpose cannot be achieved by means of the condition) to seek an agreement with the developer which would be associated with any permission granted".

Although the boundary between conditions and agreements is maintained, there are rules on the test of reasonableness of imposing obligations on the developers through agreements (Circular 16/91: Appendix B), which modify and enlarge the rules in Circular 22/83, section 6.

15.3.3 Conclusion
The changes relate to detail, albeit significant detail. All in all, planning obligations have legitimized and institutionalized planning gain and have clarified some important policy details. In this perhaps three stand out:
- Planning obligations can be seen as part and parcel of the development application itself, even though the obligation relates to land other than that included in the initial application.
- The infrastructure which is the prime purpose of the obligation is no longer limited to the physical but can take in social facilities also.
- The gain can also directly relate to the conservation/preservation of the natural environment.

15.4 Planning obligations and development control

15.4.1 The issue
For the reasons given above, while the role of planning gain in development control under Circular 22/83 was somewhat dubious, the role of planning obligations under the 1991 PCA and Circular 16/91 will be better institutionalized, for two reasons:
- the concept and practice is defined more clearly (§15.3 above);

- the planning obligation element is seen as part of the package relating to the planning application itself.

If the second reason be accepted, then another objection to the practice of planning gain, and thereby planning obligations, can be overcome. This is the possibility of influencing an authority to give permission to a developer who has made a bid in terms of the obligation, even where he would infringe planning policies. This practice has been frowned upon by the Department of the Environment (DoE 1991b):

> ... local planning authorities should take care to ensure that the presence or absence of such arrangements or extraneous benefits does not influence their decision on the planning application ...

and

> An unacceptable development should never be permitted because of unrelated benefits offered by the applicant, nor should a development that is otherwise acceptable be refused permission simply because the applicant is unable or unwilling to offer such unrelated benefits.

By applying the tests introduced here (net benefit to people), an "extraneous" or "unrelated" benefit becomes one that will not result in net benefits to the people impacted, compared with one that excluded the whole or part of the planning obligations which are offered as the inducement. In essence, given that adherence to policy will, in the view of the LPA, by definition, convey net benefit to the community, would there be *greater* net benefit by adherence to the developer's original proposal or to the original plus gain/obligation. If the latter exceeded the former, it follows that the net benefit being offered by the developer, even though departing from policy, would be greater than the net benefit conveyed by adherence to the policy itself.

If the practice described above were followed in negotiation and bargaining, in the generality of applications it would help to crystallize discussion around the common denominator, and avoid the tendency for somewhat disjointed debate and argument. In this it is not visualized that all parties would *formally* use the cost–benefit family in its rigour, but rather employ the approach behind the method. It would be sufficient if discussion ensued on common language: what are the comparative disadvantages (costs) and advantages (benefits), to whom and how much; and how could the disadvantages to all affected parties be ameliorated, still leaving net advantage to those who would be paying the cost of the amelioration.

For major applications, which by definition would raise complex issues leading to inquiries, there would be the time and resources available for greater formality in using the cost–benefit family. This would then penetrate to the Inspector's report and the logic for the decisions.

Given this approach, the authority could show that the "presence or absence of such arrangement for extraneous benefits does not influence their decision on the planning application" (DoE 1991b). They could show, by definition, that they were not being influenced just by the contributions, in money or kind, made under the planning agreement; they were not being "money led". This in the CIA will emerge as only one kind of benefit, on one particular sector. By contrast, the authority could show that it *had been* influenced by greater net benefit to the community when departing from policy; in other words that they are pursuing the "use and development of land in the public interest".

15.4.2 Application of the generic model

This approach can now be summarized by showing how it would fit into the generic model:

- *Box 1*: This would clearly be development control.
- *Box 2*: The proposal without the planning obligation.
- *Box 3*: The original proposal plus the modification with the "planning obligation".
- *Box 8*: These would stem from the two options in Box 3.
- *Box 9*: What would be the net benefits to the various sectors of the modified project, including the package of planning obligations, compared with that without; and would the planning obligations contravene the DOE advice on "influencing an authority" to permit an "unacceptable development" (DOE 1991).
- *Box 10* and *Box 11*: The options would be compared by a simplified method.
- *Box 12*: A comparison would be made between the two option, in the light of the questions in Box 9.
- *Box 13*: Consultation, etc. as necessary, leading to the answer:
- *Box 14*: Report and recommendation.
- *Box 15*: Choice and decision leading to planning agreement as necessary.
- *Box 16*: Communicate the decision.

CHAPTER SIXTEEN
The cultural built heritage

16.1 What is the cultural built heritage?[1]

At any moment in time, any society is using its *general heritage* from the past, namely all that it inherits from its forebears. Its varied character can be categorized as follows:

(a) physical stock: natural resources or man-made, the latter being divided into works and buildings attached to the land (immobile) and those that are not (mobile);

(b) activities: in consumption, production, religion, arts, knowledge, folklore and tradition.

By definition, the natural resources are not part of man's *cultural* heritage. That term relates to that part of the man-made general heritage which expresses some indefinable but recognizable element that current society values especially, and wishes to pass on to posterity. Within this the cultural built heritage (CBH) is that part of the man-made (built) immobile physical stock which any particular society has *deemed* to be *cultural*. This could cover a wide array of isolated objects, such as archaeological sites, ancient monuments, individual buildings or groups, streets and ways connecting the groups, places surrounded by buildings, objects such as single standing columns, statues, etc.; or it could extend to whole conservation areas, be they ones that in themselves have heritage value or, having no such value, are nonetheless of importance because they are surrounding or nearby part of the CBH.

Although it is the cultural quality of the built heritage that contemporary society wishes to pass on, it must for this purpose protect the bricks and mortar in which the cultural quality is fused. It is for this reason that the preservation/conservation of the cultural built heritage is in practice identified with the fortunes of the real estate which houses the cultural quality. In consequence, those concerned with the preservation/conservation of this heritage must take full account of the principles and practice of the real estate management or administration with particular reference to (Lichfield 1988: chs 1, 7):

- the life-cycle of the built heritage from its inception to ultimate death;
- the role of obsolescence in this life-cycle, which of necessity threatens the cultural element;
- the logic (primarily economic) behind the decisions of the estate owner to under-

---

1. Based on Lichfield (1988: ch.4)

take renewal of the property to combat obsolescence;
- the logic (primarily economic) behind the decisions to choose redevelopment as opposed to renewal by rehabilitation or adaptation;
- the implications for the objectives and economics of the management decision of the introduction of protection/conservation;
- the way in which the bar upon redevelopment of the CBH creates "conservation as a special case of renewal".

16.2 The statutory framework for conservation[2]

In order that any particular generation may take steps for conservation of the CBH, it is necessary for them to decide the particular items they would like to protect with a view to passing on: they must prepare an *inventory* of possible candidates, with the help of those knowledgeable of the cultural heritage. Such an inventory typically encompasses more items than it is feasible to protect. Accordingly, there must be a selection into a *list* that will receive the protection of law.

This is the beginning of the statutory framework for conservation, which can, broadly speaking, be subdivided in four steps (Lichfield 1988: ch. 5):
1. content of inventory or list:
   - what kind of artefact is to be included
   - what are the criteria for inclusion or exclusion
   - how to grade the conservation quality
2. machinery for protection of list
3. the effects of listing
4. securing permission to alter or demolish the objects on the list.

This framework has been applied in Britain to four different categories of the CBH, as follows (Suddards 1982, NLP et al. 1986):
- *Listed buildings:* the preparation of the list of buildings of architectural and historical interest was begun with the TCPA of 1947 and is now embodied in the PLBC Act 1990. The list is compiled by the Department of National Heritage for the guidance of local planning authorities for use in relation to planning control, and the buildings are graded to show their relative cultural importance. Where a building is listed, it cannot be altered internally or externally without a "listed building consent" by the local planning authority or the Department;
- *Conservation area:* the designation of conservation areas was introduced in the CAA 1967 and subsequently incorporated into the mainstream of planning legislation in the TCPA 1971, and currently in the PLBC 1990, Part II. The requirement is on the local planning authorities, with the power of designation also with the Secretary of State for National Heritage (and English Heritage in London), to prepare schemes of preservation and enhancement of their conservation areas, namely those ". . . areas of special architectural or historic interest, the character

---

2. DoE (1992d, 1994b).

or appearance of which it is desirable to preserve or enhance". Following designation no building in a conservation area shall be demolished without a "conservation area" consent of the local planning authority or the Secretary of State.

- *Ancient monuments:* Ancient Monuments and Archaeological Areas Act 1979, consolidating previous legislation starting in 1882: An "ancient monument" comprises *scheduled monuments* (the Schedule being drawn up by the Secretary of State), and *any other* which the Secretary of State feels is of public interest by reason of the historical, architectural, traditional, artistic or archaeological interest attaching to it; and also any *protected* monument which is a scheduled monument, and any other under the ownership or guardianship of the Secretary of State or local authority. Under the legislation it is an offence to destroy or damage these categories of monuments without specific "scheduled monument consent".
- *Archaeological areas:* an archaeological area is one of archaeological importance ". . . which appears (to either the Secretary of State or the relevant local authority) to merit treatment as such". Because of their nature, the conservation is of a different character. It is not in fact one of preservation but of allowing archaeological investigation in areas as they are uncovered in development. The control makes it an offence to carry out, or cause to be carried out, certain operations in the designated area without first serving an operations notice and allowing a period of 6 weeks to elapse after the service of the notice.

The brief indication just given of the categories and their protection is only the tip of the iceberg of very complex legislation under the different Acts indicated, supported by administrative law. This is provided by the Department of National Heritage (formerly Department of the Environment) and its conservation arm, the Historic Buildings and Monuments Commission or English Heritage; and also the policies of the local planning authority which are embodied in local development plans (for London, see NLP et al. 1986).

It is this battery of statutory and administrative law, supported by case law, that needs to be incorporated in development control for these categories of the heritage.

16.3 The development control issue

The statutory framework for conservation just described has the following effect on development control introduced above (Ch. 14) of the *cultural built heritage* (CBH). In effect it:

- introduces a distinct stream of policies into development plans, at the central and also local government level, which need to be taken on board as part of the development plan;
- introduces another potential array of "other material considerations" deriving from conservation alongside the normal development control applied to the *development* which is the subject of the planning application for the item of the CBH;
- introduces that control for items of the CBH where *change* is proposed, even where the change is not strictly *development* under the TCPA, Section 70.

The protection of both listed buildings and conservation areas is administered by the Department of the Environment (in consultation with the Department of National Heritage) alongside normal development control over the particular item of the built heritage, since the powers are completely incorporated in the Planning Acts (PLBC Act 1990). But as regards ancient monuments and archaeological areas, the control is not in the Planning Acts per se but in the Ancient Monuments and Archaeological Areas Act of 1979, with its own Regulations, and so on. Since these are not linked with the normal development control, they are not of direct concern here.

Although the control of the listed buildings and conservation areas is fused with normal development control, it has its special dimension. This arises not only because of the complexity and dual permits that are required, but also because of the special presumption that emanates from the fact that the conservation policies themselves are incorporated in statute (unlike for example, the Green Belts or Areas of Outstanding Natural Beauty). This gives the conservation policy a particular weight in the balancing that needs to be made on decisions affecting the CBH (Millichap 1995).

16.4 Negotiation, bargaining and mediation

We saw above (§14.4) that, although the planning decision (and the associated listed building and area conservation consents) are the culmination of applications, there is considerable scope for negotiation and bargaining between the various interested actors. This is certainly so in conservation, with the added participation of the conservation interests, as Table 16.1 shows.

Table 16.1 Actors in conservation.

1. Property development	2. Heritage tenure	3. Town planning
Owner occupier	Department of National Heritage	Department of the Environment
Developer (both private and public)	English Heritage Local planning authority Conservation societies	English Heritage Local planning authority Local authority (by-laws) Consultees Public

Three distinct groups (concerned with property development, heritage tenure and town planning) are enabled through the planning system to bring their influence to bear on the case. It is the differences between their values, aims and objectives that give rise to conflicts which need to be resolved. There is added complexity when certain actors are found in more than one of the groups. For example, the local authority as owner or developer of the cultural heritage in groups 1, 2 and 3; the two Departments in groups 2 and 3; the local planning authority, with its conservation policies, in group 2, and town planning policies in group 3; the public in groups 2 and 3.

Whereas conflicts of the kind just described could arise in many different situations (for example, in the original listing, in the offer for sale of a listed building, in the designation of a conservation area), the main thrust of our concern here is the point in time when *change* in the CBH property is being considered in the planning system. It is here that the conflicts need to be resolved and, generally speaking, within a reasonably short space of time.

The mode of resolution is not essentially distinct from that which takes place for the general built heritage under the planning system. Here the two critical steps are the application by the property development group for planning permission and listed building/conservation area consent, and the ultimate decision taken on the application by the town planning and conservation group (local planning authority, English Heritage Conservation, Inspector or Secretary of State on appeal). But the path between these two steps is not a simple one. As in the planning process generally, the path is hacked out of a wide array of considerations by the various parties to the conflict via the process of discussion, negotiation, bargaining and mediation (Ch. 13).

This somewhat bewildering array can be simplified if it be realized that each of the groups is implicitly or explicitly arguing, negotiating and bargaining around one theme, namely, what cost would it be faced with should the change to the objects embodying the cultural heritage be implemented through the planning permission (subject to conditions and Section 106 agreements) or be refused within the planning system. Within this, each of the separate actors (stakeholders) can be thought of as formulating its view on how it would *make* its decision on the change to the property in question, if given the formal opportunity, through a weighing up of the costs and benefits to itself.

16.5 Considerations in the conservation and planning decision

16.5.1 General policy

In making their decisions, the Secretaries of State and the local planning authority must have regard to the government policies on conservation as follows (DoE/DNH 1994b):

1.2 The function of the planning system is to regulate the development and use of land in the public interest. It has to take account of the Government's objective of promoting sustainable economic growth, and make provision for development to meet the economic and social needs of the community . . . The objective of planning processes should be to reconcile the need for economic growth with the need to protect the natural and historic environment.

1.3 The Government has committed itself to the concept of sustainable development – of not sacrificing what future generations will value for the sake of short-term and often illusory gains . . . This commitment has particular relevance to the preservation of the historic environment, which by its nature is irreplaceable. Yet the historic environment of England is all-pervasive and it cannot in practice be preserved. We must ensure that the means are available to identify what is special in the historic environment; to define through the development plan system its capacity for change; and, when proposals for new development come forward, to assess their impact on the historic environment and give it full weight, alongside other considerations.

It is in the weighing up of these considerations that the DoE/DNH gives guidance in PPGN15. The main points are now summarized under a series of headings which cross-references to the PPGN.

16.5.2 Use

3.8 Generally the best way of securing the upkeep of historic buildings and areas is to keep them in active use. For the great majority this must mean economically viable uses if they are to survive, and new, and even continuing, uses will often necessitate some degree of adaptation. The range and acceptability of possible uses must therefore usually be a major consideration when the future of listed buildings or buildings in conservation areas is in question.

3.9 Judging the best use is one of the most important and sensitive assessments that local planning authorities and other bodies involved in conservation have to make. It requires balancing the economic viability of possible uses against the effect of any changes they entail in the special architectural and historic interest of the building or area in question. In principle the aim should be to identify the optimum viable use that is compatible with the fabric, interior, and setting of the historic building. This may not necessarily be the most profitable use if that would entail more destructive alterations than other viable uses. Where a particular compatible use is to be preferred but restoration for that use is unlikely to be economically viable, grant assistance from the authority, English Heritage or other sources may need to be considered.

3.10 The best use will very often be the use for which the building was originally designed, and the continuation or reinstatement of that use should certainly be the first option when the future of a building is considered. But not all original uses will now be viable or even necessarily appropriate: the nature of uses can change over time, so that in some cases the original use may now be less compatible with the building than an alternative . . . Policies for development and listed building controls should recognise the need for flexibility where new uses have to be considered to secure a building's survival.

16.5.3 Demolition

3.16 While it is an objective of Government policy to secure the preservation of historic buildings, there will be very occasionally be cases where demolition is unavoidable.

3.17 There are many outstanding buildings for which it is in practice almost inconceivable that consent for demolition would ever be granted . . . Indeed, the Secretaries of State would not expect consent to be given for the total or substantial demolition of any listed building without clear and convincing evidence that all reasonable efforts have been made to sustain existing uses or find viable new uses, and these efforts have failed; that preservation in some form of charitable or community ownership is not possible or suitable (see paragraph 3.11); or that redevelopment would produce substantial benefits for the community which would decisively outweigh the loss resulting form demolition. The Secretaries of State would not expect consent to demolition to be given simply because redevelopment is economically more attractive to the developer than repair and re-use of a historic building, or because the developer acquired the building at a price that reflected the potential for redevelopment rather than the condition and constraints of the existing historic building.

16.5.4 Listed building consent

Planning permission and listed building consent need to considered separately. On the first, the normal criteria will apply; on the second there are four special criteria alongside general conservation policy (DoE 1994b: §3.5):

(i)   the importance of the building, its intrinsic, architectural and historic interest rarity, in both national and local terms;

(ii) particular physical features of the building . . . which justify its inclusion in the list;

(iii) the building's setting and its contribution to the local scene;

(iv) the extent to which the proposed works would bring substantial benefits for the community in particular by contributing to the economic regeneration of the area or the enhancement of the environment (including other listed buildings).

16.5.5 Flexibility

In conservation areas the control relates essentially to the demolition of unlisted buildings, but not to alterations and extensions (unless they embrace partial demolition) nor to their interior nor to new buildings.

4.14 Section 72 of the Act requires that special attention shall be paid in the exercise of planning functions to the desirability of preserving or enhancing the character or appearance of a conservation area. This requirement extends to all powers under the Planning Acts, not only those which relate directly to historic buildings. The desirability of preserving or enhancing the area should also, in the Secretary of State's view, be a material consideration in the planning authority's handling of development proposals which are outside the conservation area but would affect its setting, or views into or out of the area.

4.19 The Courts have recently confirmed that planning decisions in respect of development proposed to be carried out in a conservation area must give a high priority to the objective of preserving or enhancing the character or appearance of the area. If any proposed development would conflict with that objective, there will be a strong presumption against the grant of planning permission, though in exceptional cases the presumption may be overridden in favour of development which is desirable on the ground of some other public interest.

16.5.6 Conservation area

In conservation areas the control relates essentially to the demolition of unlisted buildings, but not alterations and extensions (unless they embrace partial demolition) nor their interior nor new buildings (DoE 1993c: §4).

Special attention shall be paid in the exercise of planning functions to the desirability of preserving or enhancing the character or appearance of a conservation area. This should extend to the consideration of development proposals that are outside the conservation area but would affect its setting, or views into or out of the area.

Direct loss to the conservation area in question could be outweighed, in the public interest, for example by a local or national need for the development in question or the prospect of wider environmental benefits outweighing the direct loss.

16.5.6 The planning considerations

In considering an application, the Authority can decide to grant listed building or conservation area consent and refuse planning permission, or vice versa. Clearly, the denial on the one will preclude action on the other. But because of the presumption in favour of preservation/conservation, authorities are enjoined by the Department of the Environment to be relaxed in their planning control (land-use allocation, density, plot ratio, daylighting, etc.) and building regulations and fire safety, where this would enable historic buildings to be given a new lease of life (DoE 1994b: §2.18, §3.26). This relaxation is of particular importance where the planning application relates to a change of use, since "new uses for old" buildings may often be the key to their preservation. And although there is a preference in an historic building for the use for which it was designed, and if changed from its original purpose for reversion to that use, it is recognized that in many cases it must be accepted that a continuation of the original use is not a practical proposition and it will often be essential to find appropriate alternative uses.

A particular instance of this interplay between the conservation and the planning stream arises where the new use that is desirable in the interests of conservation would require a planning permission that would be contrary to a policy in the plan, as for example the use for offices in historic buildings in an area which is allocated for residential purposes, on the grounds that the higher-value use is needed for financial viability. In essence, the Authority must trade-off a breach of its policy (which would deny the benefits for the community on which the policy is based) against the objective of conservation. A contrary instance would be where the planning authority favoured the demolition of the listed building despite its historic value, simply because the new use to be provided on the site would advance non-conservation policies in the plan.

From this brief review it is seen that many situations can arise where it would be desirable to give both kinds of consent, in order to enable the development works or change in use to proceed while at the same time keeping the erosion of the cultural heritage to a minimum. This presents a particular case of the *balancing* needed to reach a defensible planning decision. To this we now turn.

16.6 Approach to the balanced decision

In the interplay described above, many options could clearly emerge for the proposed future of buildings which are listed or in a conservation area, with each option reflecting the degree of cultural quality which would be retained after works have been carried out (Feilden 1982: ch. 1). It is the options so generated that would be subject to evaluation in accord with the generic method. To this we now turn, by highlighting those elements in the generic model that have particular relevance in this chapter.

16.7 Use of the generic model, by reference to boxes

- *1* Development control, in the light of the modifications described above (§16.5).
- *2* The development project with its impact on the CBH, or a change in its setting (in physical or activity terms) in that "the setting of a building of special architectural or historic interest is often an essential feature of its character . . ." so that authorities are required ". . . to have special regard to . . . the desirability of preserving the setting of the building" (DoE 1994b: para. 2.14).
- *3* This would follow the possibility of options brought out above.
- *6* Use and also implementation, in respect of the differing degrees of conservation of the cultural heritage.
- *9* Particular questions would be the extent to which the cultural heritage would be eroded or enhanced under the options, compared with its current and projected cultural quality; and the balance between conservation and economic prosperity.
- *10* The special feature here would relate to the differential effects of the options on the cultural quality, which would flow from a *conservation impact assessment*. The effect would be easy to introduce under options of conservation or redevelopment (since in the latter there would be complete loss). But it would be more difficult if the options related entirely to conservation (for example, facadism at the one extreme to reversion to original use at the other). This would bring in the need to compare the *heritage* quality of the buildings that would be impacted by the development, with that which would result under the options. On the first, there is typically no more than the limited information included in the information provided for listing and in the rich architectural histories. A similar description of the expected effects following development could only result in a highly subjective and individual assessment of the difference. Something more precise is needed. Such measurement/valuation has been attempted on similar lines in a

variety of sources which in effect break down the cultural qualities into sub-divisions, each of which would be assessed by judgement, on the proposition that judgement on many sub-elements in common use could be better justified (Lichfield 1988: ch. 10). In addition, some assessment of the characteristics of the effect would be helpful, in particular of the timing, in the sense of the predicted life of the building following works. Clearly, the longer the life, the greater the significance (ibid.: §10.4.6).

- *11*   The particular feature here would be the community sectors who would be impacted by the conservation. The direct conservation interests would clearly be of importance. But the repercussions would go wider, as brought out in Table 16.1, which shows the wide-ranging distribution of the benefits and costs of conservation in an actual case, where the options related to redevelopment or conservation.
- 12   The preference by sector of the options and their ranking would flow directly from Table 11.
- 13   Consultation, etc., as necessary.
- 14   Report and recommendation to the LPA.
- 15   Choice decision by LPA, Inspector or Secretary of State.
- 16   Communicate decision

16.8 Weighting in conservation

There would be a special element in the weighting as between the sectors from the presumption in favour of conservation. In effect this would mean greater weight to the sectors which had beneficial or adverse impacts from the conservation element. For example, in Table 16.2, this would relate to sectors 9, 11, 13, 16, 18 and 20. In brief, these would reflect the *use value* for the conservation of the heritage, divided as between direct beneficiaries (9, 10, 11, 12, 13, 16), indirect beneficiaries (18), non-users who would have the option to use if they so wished (option value), or to reserve for posterity (bequest value), or others (existence value).

Although it is not difficult to accept the principle of weighting from the presumption, it is difficult to place a numerical value on such weights. No guidance is available in the statute, administrative or judicial law that introduces the presumption; there is no reason for the weighting to be uniform between the sectors; the weighting should discriminate between the cultural quality indicated by the grading (I, II*, II and III). In effect, the weight to be given to the presumption can be tested only by considering the opportunity cost of favouring the conservation option against the others (Lichfield 1988: ch. 10). Seeing that the decision-taker would be the local planning authority acting in the public interest, the opportunity cost would be social as well as private.

In demonstration, a method of comparison in relation to a series of options is presented in Figure 16.1. The options are indicated in the legend to the Figure. Decrease or increase in heritage value from the do minimum (DM) is shown in the horizontal axis, on a points scale. The opportunity cost (the net benefit figure by comparison

Table 16.2   Distribution of benefits and costs of conservation.

PRODUCERS AND OPERATORS				CONSUMERS			
Impact of conservation				Impact of conservation			
Sector	Description	Redevelopment	Conservation	Sector	Description	Redevelopment	Conservation
2	3	5	6	7 8	9	11	12
1 Owners of CBH property	Property values			2 Occupiers of property	Occupation values		
3 Owners of property nearby general	Property values			Occupiers of property	Occupation values		
5 Local government on site off site	Costs			6 Local services	Occupier		
7 Local planning authority	Operating costs			8 Local ratepayers	Rate assessments		
9 Local conversation agencies	Capital and operating costs			10 Local residents			
11 Central conservation authority loans grants	Capital and operating costs			12 Local residents			
13 Central government conversation authority	Capital and operating costs						
15 Local economy goods services	Employment			16 Local community residents workforce	Environment Culture Employment		
17 National economy taxation revenue taxation costs imports prices maintenance	Economic flows			18 National citizens taxpayers	Heritage prestige Tax assessment		
				20 Posterity	Opportunities for heritage		

Notes: 1, 3: "owners" include developers, financiers, etc.; 6: local services include shops, hotels, restaurants, etc.; 9: local conservation agencies include the local authority, voluntary bodies, etc.
Source: Lichfield (1988: table 12.1).

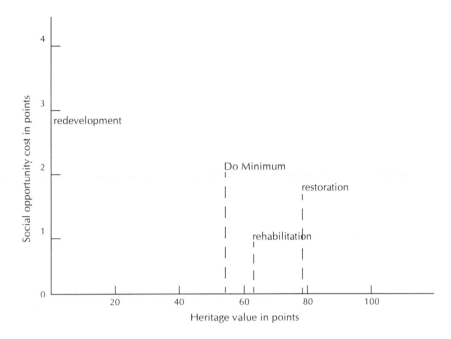

Figure 16.1 Social opportunity costs to community of conservation of the heritage (source: Lichfield 1988: diagram 10.2).

with the most financially profitable alternative of redevelopment) is shown in the vertical axis. Since so many of the costs and benefits are unpriced, we cannot show the social opportunity cost in money terms, but only by some points system, thus giving a ranking. Following in Table 16.3 are the conclusions from Figure 16.1.

Faced with this analysis, the decision-taker could then see what marginal net benefit to the community would need to be given up to achieve the marginal difference in heritage value. He would then be able to offer judgement on the trade-off between the two, on behalf of the community.

Table 16.3 Ranking of conservation options.

	Post-project heritage value in points	Ranking in social opportunity cost
(a) Do minimum	48	3
(b) Rehabilitation	65	1
(c) Restoration	82	2
(d) Redevelopment	0	4

CHAPTER SEVENTEEN

Integrating the environment into development planning

17.1 Environmental protection[1]

Protection of what is today called the environment is long rooted in common law and practice, relating to nuisance and tort. The protection grew in the 19th century in public law, as a reaction to the environmentally disastrous consequences of the first industrial revolution. These early and prescient concerns have been given greater urgency and moment by the growing world awareness since the Second World War of man's impact on the natural resources and environment of planet Earth, and the need to take appropriate measures for their protection: of forests, which can be seen as part of the traditional concern of planning, to the pollution of the ozone layer and the seas, which cannot (Ward & Dubois 1972).

To reflect these concerns, there was a decided increase in the 1970s in what is today called *environmental law*. Legislation was passed in a somewhat piecemeal manner to protect against pollution in air, water and land; against risks from hazardous substances, disposal of waste and nuclear power; and for the protection of natural resources and of ecosystems. This fragmented approach began to be consolidated in major statutes in the 1980s, for example, the WCA 1981, WA 1989 and EP Act 1990. The latter introduced further integration in control, via Her Majesty's Inspectorate of Pollution (HMIP) at the national level and Local Authority Environmental Health Departments at the local level. This energy was stimulated by the activity of the EEC in producing its own community legislation, which was binding on the member countries (Haig 1984).

Considerable stimulus was given to this progress by successive Royal Commissions on Environmental Pollution, founded in 1970 and producing 13 reports since on a variety of matters, and by government in the publication of *This common inheritance,* a bulky White Paper providing a comprehensive programme on Britain's environmental strategy (DoE et al. 1990) and initiating its annual updating (DoE et al. 1991, 1992e, 1994a).

---

1. Ball & Bell (1991).

17.2 Environmental protection and the planning system

There can be no doubt that environmental protection has been an objective of the planning system from its beginning, albeit under the name "amenity". The genesis of the system was the protection against environmental pollution of the nineteenth-century legislation in public health, housing and environmental nuisance (Ashworth 1954). These origins were reflected in the 1909 HTPA, which inaugurated the British planning system, in its objective of "securing proper sanitary conditions, amenity and convenience". This objective was repeated and taken further in the TCPA of 1932, Section 1, that planning schemes included the object "of preserving existing buildings or other objects of architectural, historic or artistic interest and places of natural interest or beauty, and generally protecting existing amenities whether in urban or rural portions of the area". And these objectives were again taken further in the progression to the 1947 Act and beyond, in the protection of open space, areas of natural beauty, green belts, areas of scientific interest, buildings and areas of historic and architectural interest, and so on

Accordingly, *This common inheritance* (1990) included as a major section (Part III) the subjects of land use, countryside and wildlife, towns and cities and the heritage. In essence, the aim was to enumerate the environmental assets under each of these heads and to show how these could be protected and enhanced. This approach echoed earlier contributions (e.g. CDPUK 1983) and has been followed by major contributions about sustainable development (Blowers 1993, DoE et al. 1994a). Environmental protection and planning have since the nineteenth century tended to grow separately. They are converging. There is space here only to highlight the moves to integration between the environment and development planning.

An important step in this direction is provided in DoE guidance on planning and pollution control (DoE 1994d). This guidance indicates several desirable practices in integration. Examples are how the planning system is being used to consider whether the development should take place at all, its impact on amenity, the potential for risk or contamination, the prevention of nuisance, and the impact on the road network of any change in traffic flows. This led to enumeration of appropriate pollution policies in local plans and how environmental considerations can become other material considerations in development control.

17.3 Environmental assessment in Britain under the EA regulations

17.3.1 Origins

The environmental assessments introduced in Chapter 5 burst upon a ready world with their introduction in the US NEPA of 1969, and have since spread (Wathern 1988). They reached Britain with the impact studies made on a voluntary basis between 1972 and 1987 for major projects: about 100 for major oil and gas developments and a further 100–150 on reservoirs, energy, roads, resource based mills, ports, etc. (Clarke 1988). Government issued no official guidelines for the purpose, but commended to local authorities the theory and methodology of a series of research reports commissioned from the University of Aberdeen, whose *Manual for practice* became the basic text for the informal assessments (Clarke et al. 1981).

- Despite the early start, it was not until 1988 that in Britain the legal requirement for assessment was introduced, following an EEC Directive (EEC 1985, DOE 1988a).
- Between July 1988 and September 1993, just over 1,800 environmental assessments had been submitted in the UK (Frost et al., undated).
- By contrast to the pre-1988 *Manual*, guidance of the DOE (DOE 1988a,b, 1989a) was primarily *procedural* and did not set out to offer guidance on methodology and technique, and indeed left the form of the environmental statement to the individual developer. Thus, while the environmental assessments of necessity follow the requirements of the Regulations, and while the Aberdeen *Manual* has influenced the *method*, it has not been adopted as a standard. From a review of about 100 assessments made under the Regulations it was found that, on the whole, the quality of the statements was very varied, patchy and poor, and left much to be desired (Coles 1990). Accordingly, no attempt is made here to generalize on a standard method for environmental assessments under the DOE Regulations.[2]

17.3.2 Procedures for EA

Projects to be assessed under the Regulations

Following the EEC Regulations, the British Regulations and Circulars asked for assessments to be made only on what are regarded as likely to produce significant environmental effects. Those that come within the DOE planning system either require an assessment in all cases, since they relate to major projects where environmental implications seem bound to occur (Schedule 1), or are "likely to have significant effects on the environment by virtue of factors such as their nature, size or location" (Schedule 2). This formula is not at all easy to interpret and is subject to guidance, which indicates that the need is likely to arise (DOE 1989a: appendix 2):

- for major projects which are of more than local importance;
- occasionally for projects on a smaller scale which are proposed for particularly sensitive or vulnerable locations;
- in a few cases, for projects with unusually complex and potentially adverse environmental effects, where expert and detailed analysis of those effects would be desirable and would be relevant to the issue of principle as to whether or not a development should be permitted.

In addition, the government can introduce a further category of projects to be assessed (DOE 1989: pt III). Environmental assessments are also required for appropriate projects that are outside the DOE planning system but have their own competent authority and procedure (trunk roads and motorways, power stations and overhead lines, etc.), arise in Scotland or Northern Ireland, or are approved by Private Act of Parliament (DOE 1989: pt III).

---

2. This situation will change on the finalization of a consultation draft on preparing environmental statements (DoE 1994c).

Integration of environmental effects in the planning system

There are great variations in the nature of the planning system in the various countries of the Community: in the government departments responsible, the making of plans, the degree to which they are legally binding, the amount of discretion in development control, etc. (Davies et al. 1989). Thus, it could be no part of the Community Directives to indicate how environmental assessments *should be* related to the planning system itself in a particular country, and whether and how the two processes be integrated.

The British Regulations and DoE Circular, clearly aim at integration (DoE 1988a: para. 12):

> The local planning authority or the Secretary of State or an Inspector shall not grant planning permission pursuant to an application to which this regulation applies, unless they have first taken the environmental information into consideration.
>
> The *environmental information:*
>
> . . . is the information contained in the environmental statement (i.e. the information gathered in conjunction with the planning application) and any comment made by the Statutory Consultees and representations from members of the public, as well as to other material considerations.
>
> And the *environmental assessment:*
>
> . . . refers in this Circular to the whole process required to reach the decision, that is the collection of information on the environmental effects of the project, the consideration of that information which must be carried out by the local planning authority and the final judgement resulting in development consent or refusal.

As a preliminary to demonstrating below how to integrate the environmental and planning considerations, it is useful to recognize that the scope of the impacts to be included from either source could differ in their nature, as is now indicated.

Environmental considerations   It is well recognized that the *scope* of effects/impacts flowing from physical development can be very variegated, in several ways:

- wide-ranging: covering natural resources, natural environment, man-made environment, social, economic, fiscal, cultural, urban, family.
- size: from the trivial to the cataclysmic, those which create extreme hazard to life through catastrophe (flood, nuclear, etc.)
- characteristics: direct (on site), indirect (off site); real or technological, pecuniary or transfer; first-order or second-order (induced or secondary)
- duration: short, medium or long term; permanent or temporary; reversible or irreversible/irretrievable
- timing: before construction through blight, during construction, on commissioning (operation), at the end of the operational life in decommissioning.

Given this variegation, it is apparent that not all environmental assessments can

hope, or be expected, to cover the whole range. Therefore, there is need for some pre-selection of the *scope*. This can follow various criteria, such as:
- aspects of the environment: the decision-taker seeking the assessment could be concerned with only certain of the aspects
- skills available: the availability of relevant skills comprising environmental natural scientists (ecology, soil, atmospheric pollution, hydraulic, noise, etc.), or environmental social scientists (economics, sociology, anthropology, culture, etc.)
- significance: whether or not to be concerned with *all* impacts of the kind selected, or to have a cut off at those which are deemed to be significant.

Taking each in turn as described under the 1988 Regulations and Circular:
- The scope of the impacts to be assessed are presented as a *minimum*, with an emphasis on the natural sciences. Possible additions are indicated.
- There is no limitation of scope by reference to availability of potential skills. This is not to say these all exist. The experience and know-how in environmental assessment rests primarily with the natural environment scientists who have best responded to the challenge. The social scientists have lagged behind. This is understandable in relation to sociology/anthropology, for their practice has not developed in this direction. It is less understandable for the economists who have a tradition in economic impact analysis, even though this has flourished in the very different world of project appraisal and assessment, using cost–benefit analysis rather than environmental impact.
- The scope is further circumscribed by asking that the assessment be carried out only for the impacts that are "likely to be significant". This is elusively delineated in the DoE Circular, although not in the EEC Directive.

Planning considerations   A quite different picture emerges when the scope is looked at from the viewpoint of the *planning system*, within which the environmental assessment is to be integrated. Taking again the criteria in turn:

ENVIRONMENT   Although the question is rarely brought out into the open, it would be generally accepted that in the theory and principles of urban & regional planning (with its synonyms of land use, physical, environmental planning) the concern is with the *whole* of the human environment. This might seem to be a tall order for a field of planning which is primarily relating to the "use and development of land". But it is not strange if it be accepted that this implicitly must include activities of people on that land, in all their multifarious interpenetrating lives (economic, social, cultural, spiritual, etc.).

This wide-ranging concern for the *whole* of the human environment is in practice scaled down in the *administration* of the planning system, where responsibilities for different aspects are divided between local or central government departments. There is ever present an uneasy line where on matters dealt with within the planning system there are also legislative powers outside the Planning Acts. This is exemplified by the disinclination to deal with social housing in development plans, the conducting of public inquiries for major transportation schemes under the aegis of the Department of Transport, and in the separate procedures for environmental assessment of projects outside the planning system (e.g. power stations, afforestation, ports & harbours).

SKILLS   In the theory and practice of planning, which relates to the total human environment, the contribution is to be sought from a very wide range of skills. But the experience and know-how of the professions primarily concerned with administering the planning system, as reflected by the Royal Town Planning Institute, is narrower. The weight until the 1950s was with the development skills that originally founded the Institute (architects, surveyors, engineers and lawyers). This limitation was recognized in the seminal Schuster Report of 1950 (Committee on Qualifications of Planners 1950), which brought out that many more skills were required.

Since the Schuster Report there has been vast growth in the opportunities for contributions to planning of both the skills visualized by Schuster and those needed for environmental assessment. But the doors to the Royal Town Planning Institute are primarily for the general and not the specialist planner (RTPI 1990). The consequence is that, with the growth of environmental concern, there have already been initiatives to open up new professional organizations to cater for the new skills.[3]

SIGNIFICANCE   Whereas the term "significant" has considerable importance in the environmental assessment field, it hardly appears in the planning field per se. Its counterpart would appear to be the highly entrenched concept of "material considerations", which is embodied, outside the plan itself, in the critical guidelines to local planning authorities in deciding whether or not to give planning permission, in the TCPA 1990, Section 70.

Such considerations have been dealt with in Chapter 13 as "planning considerations". But instead of the word "significance", the law relates to "material". That is nowhere defined but has been considered at length by the courts. From a recent authoritative review of judgements, the following conclusions were drawn (Layfield 1990):

1. To be accepted as a material consideration, a statement of policy or practice, an argument or submission, an averral of facts or circumstances, must
(a) comprise or contain a factor that should be taken into account in reaching the decision sought, and
(b) must be a planning consideration, that is one that is related to the character, use or development of land, and
(c) the planning consideration must fairly and reasonably relate to the application and the application (or appeal) site and its surroundings.

Thus, having decided whether the consideration is or is not *material* there is then the question *how material*? This would not appear to be susceptible to any scientific probing but:

---

3. For example, *Directory of environmental scientists*, Institute of Environmental Assessment, United Kingdom Environmental Law Association.

3. Whenever a consideration is established as a material consideration regard must be paid to it, but the weight to be given to the consideration in the process of forming a decision is a matter for the decision-maker, the local planning authority, Inspector, Minister or the Court, as the case may be.

They in turn must be influenced by what the law or policy conveys in terms of "paramount or special consideration", such as the desirability of preserving or enhancing a conservation area; the "very special circumstances" permitting development in the Green Belt; interference with an Area of Special Scientific Interest (SSI) or an Area of Outstanding Natural Beauty (ANOB).

Balancing the environmental and planning considerations   This sets the scene for *balancing* the "provisions of the development plan so far as material to the application" and the other "material considerations" in the everyday practice in development control. This balancing will extend to the environmental considerations to be integrated with the planning decision. The intent is clear, but what is not so clear is how the balancing is to be carried out. It was not made clear for environmental assessment in one place where it might have been expected: in the Department's commissioned Research Studies mentioned above, which culminated in the Manual for Practice (Clarke et al. 1981). The assessment there was on the "description and identification of impacts" and was "intended to assist planners and others in making a balanced assessment of proposals within the existing framework of planning control and other statutes" (Clarke et al. 1981: viii). But how to introduce such balance in the assessment does not come through.

This situation would be eased if the balancing of planning considerations was more robust in planning analysis generally, for then the environmental considerations could be more readily coped with. But this is not so, as we saw above (§12.8). Our proposals for improvement were there given (§12.9) and exemplified by application of the generic method (§12.10). We now proceed to demonstrate this balancing by two studies integrating the environmental information in the planning decision.

17.4 Planning/environmental assessment and the generic method

17.4.1 Project planning and environmental assessment

We saw above (§12.4) how the landowners/developers project planning is carried out in relation to the local planning authority's development control. We now look at the relation between the project planning and the environmental assessment required by the local planning authority.

From the landowner's/developer's viewpoint, the application to the planning authority, and appeal to the Minister on an adverse decision, is but an incident, albeit critically important, in the project planning process itself. From this it follows that, since the *procedural* consideration of the environmental assessment and statement are linked with the planning application, the consideration of the environmental information is implicit within the whole of the project planning process, for relevant projects under the Regulations. It comes into the supply-side analysis of site characteristics, into the demand-side assessment (environmental implications will affect both prospective on site consumers and offsite neighbours), the evolution of options and the impacts from them, the evaluation of the options, the amelioration of the undesirable impacts, and so on. In brief, an informal environmental assessment is part of the project planning process (Lichfield 1988).

The relevant professions are in the main already to be found in the development planning of projects where, in addition to the development skills (planners, architects, engineers, surveyors, lawyers), there are also skills conversant with ecology, landscape, retailing, economics, sociology, and so on. To these can be added specialist environmental social and natural scientists: respectively for social, economic and planning impact; and noise, atmospheric pollution, water pollution, and so on. In brief, project design and appraisal leading to planning applications continues in much the same way, with the difference that the impacts to be assessed in scope are wider and taken into greater depth. In consequence, the team may need to be enlarged, and a team leader/co-ordinator is needed more than ever for the complex technical operation (typically the generalist planner or a project manager skilled in project design, etc.). When the process enters the appeal stage, this team functions alongside the solicitor and barrister who will be more directly concerned with the management, strategy and tactics of the inquiry proceedings themselves.

The project design process is not carried out in isolation in the development team's office. In the need to prepare a scheme acceptable to the landowner/developer and to those who control the planning permission, there is the need in the design process as well as in the formal stages to consult or take note of many different interests beyond those of the developer/landowner and the planning authority. These could include the various governmental departments and other statutory bodies who need to be formally consulted, those groups and individuals who are informally consulted, those who need to be considered because of objections or support, neighbours (owners and occupiers) within or adjacent to the proposed development, and lobbyists from any of those interests. This results in a highly complex decision-taking process characterized by negotiation/bargaining (see Ch. 13).

17.4.2 The Prospect Park study, using the generic method with adaptation (Lichfield 1992c, Lichfield & Lichfield 1992)

Context

The study was made to support a planning application and environmental assessment for a major office development for the new headquarters of British Airways (BA). It was directed by Dalia Lichfield as planner/analyst in the planning and environmental team that prepared the application for planning permission, and then presented evidence at the inquiry into the application. In this, in addition to the evaluation, the evidence referred also to planning policies and other material considerations bearing on the planning decision. We now present the study by reference to the generic method.

A Planning

Box 1   The typical plan-making and implementation process will be just as for the
project in the generic model. Here a particular issue is whether or not the development planning process and the more recent environmental planning process are integrated in the one team, or whether they are being pursued separately. The substance
of the following analysis will not be affected, but there will be implications for the
management of the study process.

B Projects in the system

Box 2   The case relates to an actual proposal by British Airways for its new headquarters building at Prospect Park, Hillingdon District, near Heathrow. The building
(designed by Niels Torp) was located on a site of 250 acres in the green belt in Hillingdon Borough (Fig. 17.1). Currently it is mostly unused, having been worked for

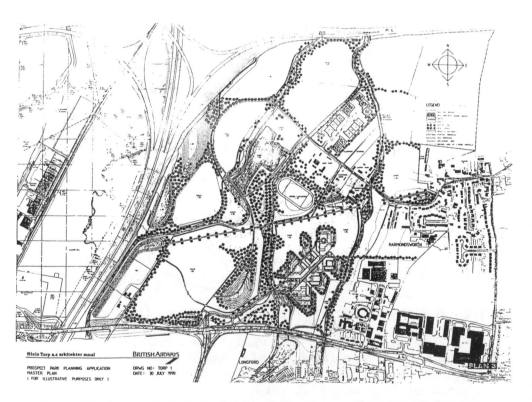

Figure 17.1   Prospect Park: application site as existing (*source:* Lichfield & Lichfield 1992:
176).

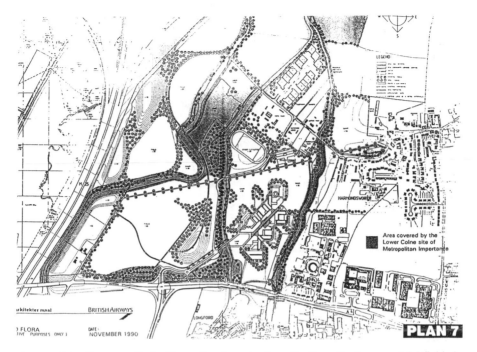

Figure 17.2 Prospect Park: illustrative layout of application site (*source:* Lichfield & Lichfield 1992).

gravel and then for refuse landfill, resulting in soil contamination and dereliction.

The proposed development (Fig. 17.2) comprises the BA headquarters (covering some 8 ha) with the remainder as a country park as part of the Colne Valley Park. Within this would be built a community and leisure centre designed to minimize the visual impact of the buildings on surrounding residents and on the Park, with mitigation of potential effects on traffic, water regime and air pollution.

Box 3   The options were the proposal (Fig. 17.2), which included the mitigation and enhancement of the potential effects through the design process; or refusal, implying continuation of the current conditions, which would doubtless deteriorate.

Boxes B4–5   Although in the generic model no need was seen to describe the pre-project system (indeed it was thought undesirable to do so because of the complexity), in environmental assessment there is the tendency to audit more systematically the pre-project system following baseline studies. There is no reason why this should not be done in the generic model; indeed the changes that are predicted could be grafted on to form a description of the post-project system.

C Options specification of plan variables

Box 6   The variables were: implementation/construction; use upon completion, sub-divided into office buildings, leisure centre, park, water features and roads/parking; operation and management, subdivided similarly.

Box 7   The project options by plan variable were fully set out in the submission with the application.

D System change

Box 8   Changes in the urban & regional system through project option were not identified in tabular form but via plans, fully described in the submission with the planning application.

E Decision space/framework

Box 9   The context was simply to provide an accompanying submission to the planning application, including the environmental statement, backed up with full information which described the proposal and culminated in the planning case for approval.

F Effect assessment

Box 10   In this Box there is a departure from the traditional environmental effect assessment, since there is also the need to include effects which do not clearly come under the Environmental Assessment Regulations but certainly should be included in relation to planning considerations. The Prospect Park study included recognizable "planning effects" in addition to the traditional "environmental effects", from traffic, landscape, recreation and employment.

The "scoping" of the effects "likely to be significant" were made by the team of environmental scientists. The selection is shown by initials in the relevant boxes related to plan variables in Figure 17.3, in each of the "during construction", "post completion" and "operation/management" phases, the initials referring to the natural or social environmental scientists concerned with the preparation of the effect assessments. These followed the traditional methods and techniques for such assessments. Since these were not uniform in character, it was necessary to produce a summary of the effects in a standard format. Figure 17.4 presents the information in accordance with the standard steps in environmental assessment (Fig. 5.2) adapted to the particular case. We now proceed to describe the method by reference to the numbered steps in the left-hand column.

(1) PROPOSALS   The proposals that are the subject of the assessment are described. They are divided into the three aspects that could each require separate assessment, namely:

Figure 17.3 Planning and environmental scoping: preliminary schedule of potential y significant effects.

Elements likely to be affected

PROPOSALS	Flora & fauna	Soil	Water SURFACE	Water GROUND	Air	Noise	Climate	Landscape VISUAL	Material assets	Cultural heritage CONSERVATION	Safety	Traffic on roads	Recreation	Identity of settlement	Employment
(a) Physical development after completion															
1. Office buildings				HAL				TORP NLP			BER			NLP	
2. Leisure centre				HAL				TORP NLP			BER			NLP	
3. Park and water features	BER			HAL				TORP NLP		NLP				NLP	
4. Roads and parking	BER							TORP NLP			BER	WSA			
(b) Operations & management															
1. Buildings						HAL					BER	WSA	BA NLP	BA NLP	NLP
2. Leisure centre						HAL BA					BER	WSA	BE NLP		NLP
3. Parks and water features	BER							BA NLP			BER	WSA	BA NLP		NLP
4. Roads and parking												WSA			NLP
(c) During construction	HAL BER	HAL BER	HAL	HAL BER	HAL BER			TORP NLP							

Notes: BA: British Airways plc; BER: Berridge Environmental Laboratories Ltd.; HAL: Sir William Halcrow & Partners; NLP: Nathaniel Lichfield & Partners; TORP: Niels Torop & Partners; WSA: W. S. Atkins & Partners.
Source: Lichfield & Lichfield 1992.

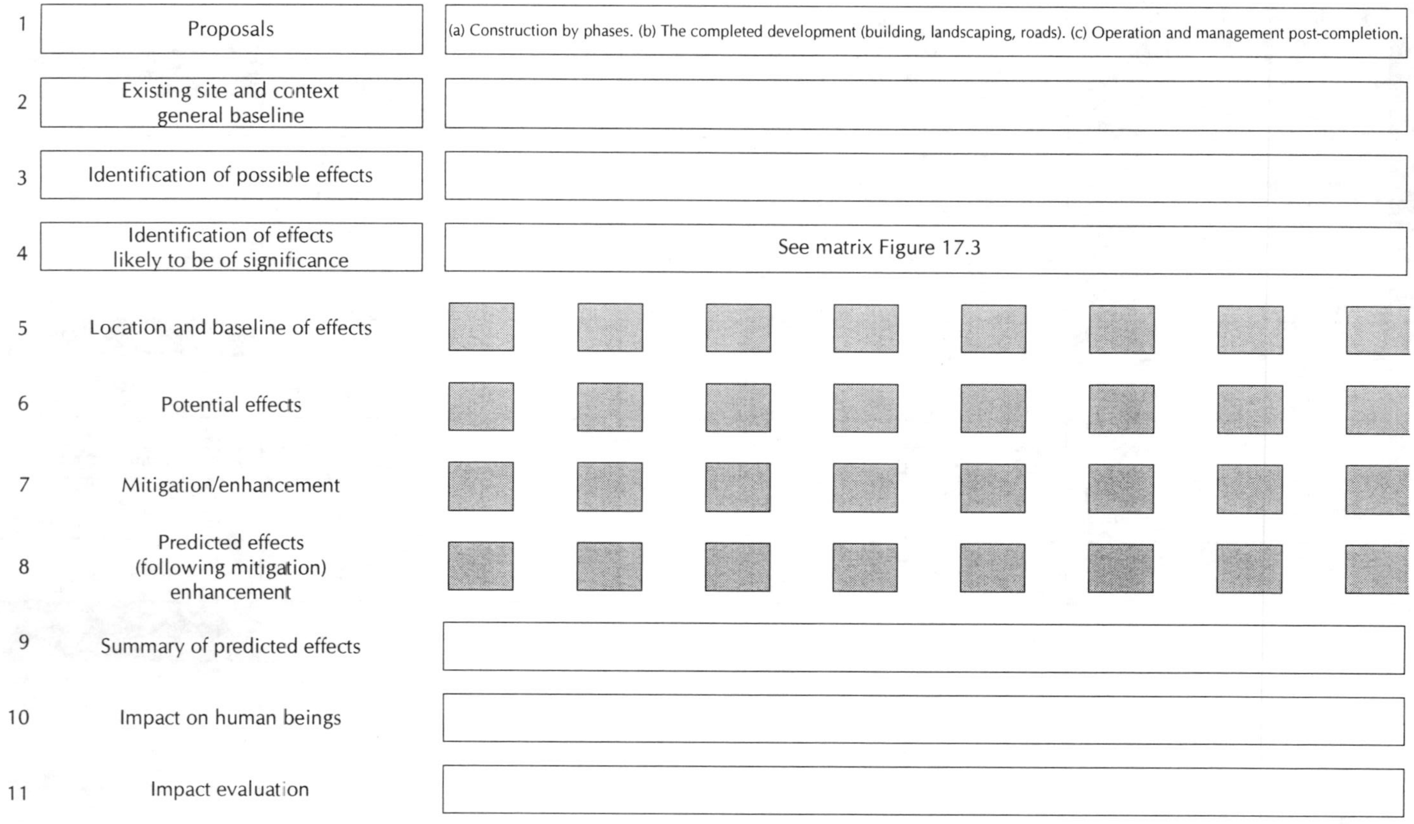

	(a) Construction by phases. (b) The completed development (building, landscaping, roads). (c) Operation and management post-completion.
1 Proposals	
2 Existing site and context general baseline	
3 Identification of possible effects	
4 Identification of effects likely to be of significance	See matrix Figure 17.3
5 Location and baseline of effects	
6 Potential effects	
7 Mitigation/enhancement	
8 Predicted effects (following mitigation) enhancement	
9 Summary of predicted effects	
10 Impact on human beings	
11 Impact evaluation	

Figure 17.4 Environmental and planning assessment of effects: framework for evaluation (*Source:* Lichfield & Lichfield 1992).

- the construction process to implement the development, by phases.
- the physical development (e.g. the site, infrastructure, buildings, spaces and their landscaping, roads, etc.) to which are related the socioeconomic activities which would be carried out in that development
- the operation and management of the development post-completion

(2) GENERAL BASELINE FOR SITE AND ITS CONTEXT   This describes the condition of the development on the site as a whole, and its context off site, as a starting point for the analysis.

(3) SCOPING OF EFFECTS   The traditional technique for identification of *effects*, using the term as defined above (§5.2.1), is the *impact* matrix (§5.3a). However, this is felt to be inadequate, since it fails to reflect the whole gamut of the proposals indicated in (1) above. Instead, the quite different approach is adopted of identifying the effects in relation to predicted changes in the urban & regional system from the project, based upon Figure 1.3 above. The approach is shown in Table 17.1, whereby the effect on the system elements is shown in the columns, produced via the development process in the rows.

The latter presents a possible wide array of actions which could produce change in the baseline situation which is either current at the time of the exploration, or what would occur "without the project" in the passage of time, namely blight, displacement, retention, construction, operation and management on completion and decommissioning. As with matrices generally, a cross would be inserted in any of the cells where it is visualized that an action in the various phases in the rows is likely to have a potential effect on any element of the system in the columns.

IDENTIFICATION OF SIGNIFICANT EFFECTS   Having identified the totality of the possible effects, it is then necessary to reduce their number to those that *should* be assessed since they are "likely to be significant". This would employ the techniques of scoping outlined above. In the Prospect Park case considerable knowledge had been built up by the design team (which included the environmental scientists) when the identification of significant effects became necessary. They accordingly were able to carry out the scoping process via discussion, which resulted in an enumeration of the "effects likely to be of significance" shown in Figure 17.4. The process was helped by breaking down the proposals into the three constituents shown in Figure 17.3, which were the plan variables of Box 6 above. The detailed environmental assessment was pursued for each significant effect by the consultant specified in the relevant box in Figure 17.3. This precluded the need to use the matrix Table 17.1. If it had been employed, the identification could then be recorded in the matrix by adding a distinguishing mark (e.g. a circle or line) in the relevant cell.

(5) SPECIFIC LOCATION OF SIGNIFICANT EFFECT AND ITS BASELINE   Having identified the significant effects to be explored, it is then necessary to identify the location for each, and then its baseline, to a greater degree of detail than was possible in the general baseline referred to in (2) above.

Table 17.1 Effect matrix: predicted effect on the urban and regional systems through plan variables.

Plan variables			System elements						
			1. People	2. Natural resources	3. Utilities	4. Transport	5. Telecomms	6. Urban fabric	7. Open space
No.	Development process	Current without	With project						
1.	Datum			1 2 3	1 2 3 4 5 6 7 8	1 2 3 4 5	1 2 3 4 5 6	1 2 3 4 5	1 2 3 4 5 6	1 2 3 4 5 6
2.	Blight									
3.	Displaced									
4.	Retained									
5.	Construction									
6.	Completion									
7.	Management/ operation									
8.	Decommission									

Notes: 1. System elements could be subdivided for the particular case (e.g. for natural resources), as could plan variables (e.g. temporary retained).
2. For more detail on the content of the columns, see Table 1.1 above.

(6) PREDICTION OF POTENTIAL EFFECTS   In essence this establishes the difference for each effect picked out for assessment, between the specific baseline (5) and the situation predicted as a result of any of the proposals (1). This difference was explored by traditional "impact assessment". Each assessment was summarized in terms of the effects, and their characteristics, in relation to the following headings which are typical in the literature:

- *Kind*   Natural resources, natural environment, material assets, cultural heritage, social, economic and financial.
- *Type*   Direct and indirect.
- *Timing*   When, in probability, the effect will be felt, and whether certain or uncertain.
- *Reversibility*   Can the effect be reversed following experience.
- *Significance*   Judgement based upon the preceding.

(7) MITIGATION AND ENHANCEMENT   From the preceding assessment, the view was formed as to whether the effect is acceptable as predicted, acceptable in the sense of an appropriate norm for the time at which the impact will be experienced. If it is not, then the question arises: can the adverse effect be mitigated/ameliorated, or the beneficial effect be enhanced. If so, the means of doing so will be part of the development planning and design of the project.

(8) PREDICTED EFFECTS FOLLOWING MITIGATION AND ENHANCEMENT   As a result of the preceding step, a series of predicted effects will emerge, which will have been mitigated/ameliorated or enhanced to the degree thought desirable, by the planning team and promoter, in the light of the views of those whom they consult (the local planning authority, consultees, the public at large, etc.). The effects would be assessed by reference to their size and characteristics, as in 6 above. This is the *effect assessment*.

(9) SUMMARY OF PREDICTED EFFECTS, FOLLOWING MITIGATION/ENHANCEMENT   It is this summary of predicted *effects* which are included in Box 10.3 and supported by the report on Environmental and Planning Assessment.

G Impact evaluation

Box 11   This summary of predicted *effects* (following mitigation and enhancement) is the starting point for Box 11. It will be noted that the range of effects departs from the traditional effect assessment of the Regulations and DoE Circulars in two respects:

- the effects are not limited to the compulsory list of the Regulations (Schedule 3.1) but go beyond to include others (as authorized by Schedule 3.2) which are pertinent. The addition will stem not simply from those effects seen as *environmental* but also from others which are more familiar in *planning* assessments;
- but whereas the mandatory effects to be considered under the Regulations include "human beings", these are omitted from Figure 17.4. The reason is not

that they are not pertinent or important. On the contrary, the reason is that they are the centrepiece of the *impacts* as opposed to the *effects*, and so justify the separate analysis shown in Box 11, via CIA.

This analysis is demonstrated in Tables 17.2–17.6, which give a simplification of the generic model for sub-Boxes 10.1–10.2 and 11.1–11.3. It is Table 17.6 that draws conclusions on the evaluation of the *impact* on human beings as between Options A and B, which are derived from the *effects* in Table 17.2. Tables 17.3 and 17.4 are the bridge between Tables 17.2 and 17.5. We now explain the tables in greater detail.[4]

A first step was to use the summary of effects in line 9 of Figure 17.4 as a basis for proceeding. How this was done is shown in Table 17.2, which presents for each of the elements the essentials relating to the preceding lines 1–8 of Figure 17.4, namely the predicted effects relating to the options (i.e. of carrying out the BA Scheme (A) or refusing permission which would result in do minimum (B)). Table 17.3 shows the impacts from the effect of these options and from the assumed sectoral objectives, the difference in net benefit and assumed preference.

Whereas the technique of preparing Table 17.3 from Table 17.2 requires that the *particular* community sectors are attached to particular *effects*, the planning assessment requires that the effects are attached to a *particular* sector; this is more readily used by planners charged with making an impact assessment in relation to *people*, and elected representatives making decisions on behalf of a *community*. This grouping of different impacts by sector is shown in Table 17.4 by enumerating a community sector by a number that refers to that sector in the notes to Table 17.5. Then follows aggregation in Table 17.5. Given the varied impacts on a community sector, it follows that any community sector could have different preferences for the options in respect of each impact (column 6). Then it is necessary to deduce the preference for each sector (column 7).

The "sectoral objectives" of those impacted are postulated in column 4 of Table 17.4. They were derived from common sense and also from interpretation of the social survey commissioned by BA (Solon Consultants 1990). By consistently comparing the impact of (B) on the sectors via their sectoral objectives with that of (A), it was possible to judge the difference in impacts between the options. Column 5 shows by a negative sign (–) where the impact of B would be *less* beneficial than A; conversely by a positive sign (+) where the impact of B would be *more* beneficial; and whereby an equal sign (=) where there would be no difference, that is, neutrality. This comparison thus assesses the preference (shown in column 6) which each sector may have for option A or B, on account of its perception of the predicted impacts which that sector is likely to experience.

Box 12   From Table 17.5 it is possible to summarize the evaluation of the impacts by community sector, as shown in Table 17.6.

---

4. All Tables 17.2–17.6 relate to the post-completion stage, except 17.4 which, for illustration, is in the construction stage. Only the first page of the tables is shown here.

Table 17.2 Prediction of potential and mitigated effects of selected environmental elements, post-completion.

Elements	Baseline	Potential effect	Mitigating/enhancing measures	Predicted effect following Option (A)	Predicted effect following (Option (B)
1	2	3	4	5	6
Air	No specific smell or dust problems.	No significant effects envisaged.			
Soil	Poor-quality soil. Ground contains large quantities of waste, largely lined with small quantities of toxic or other hazardous materials. Varying levels of contamination on site.	Soil contamination.	Capping of M25 bunds and landfill sites with impermeable materials. New layers of topsoil where necessary.	Containment of waste and prevention of soil contamination.	Site will continue to have poor-quality soil with risk of exposure of waste.
Water (surface)	Site crossed by the Wraysbury River, the River Colne and the Duke of Northumberland River. River Colne is a primary drainage river. Two existing lakes serve no specific drainage function.	Stormwater discharge causing pollution and overload of flow.	For office building and leisure centre, stormwater run-off will enter River Colne via an oil/petrol interceptor. Attenuation of flow will be provided. Effective capping of parkland areas and creation of new clay-lined water courses to prevent contamination.	Improvement in flow and quality of surface-water drainage. Reduced potential for pollution of water features. Improvement in surface-water ecology.	Continuation of generally poor drainage and poor capping of existing waste material, leading to an increasing risk of pollution of water courses and lakes.

Notes: (A) and (B) are two optional schemes: permission and refusal, respectively.

Table 17.3 Evaluation of impacts on selected environmental aspects, with preference by community sectors, post-completion.

Elements	Impact (A)	Impact (B)	Assumed sectoral objectives	Difference in net benefit (B–A)	Assumed preference
1	2	3	4	5	6
Air	No significant impacts envisaged			=	0
Soil	Improvement in overall soil quality, allowing park to be developed, permitting better recreational access and enjoyment.	No change.	Improvement in soil and parkland amenity.	–	A
Water	Increase in opportunities to enjoy angling and bird watching and for other recreational users.	Infill of lake, leading to reduction in angling and bird-watching activities.	Increased open-air recreation.	–	A
	Minimization of water contamination.	No change.	Improvement in water quality.	–	A

Table 17.4 Prospect Park: impacts of system elements in the natural environment on community sectors during construction, with preference.

Sectors experiencing effects	Impact (A)	Impact (B)	Sectoral objectives	Difference in net benefit (B–A)	Preference
1	2	3	4	5	6
(Soil) 2a, 2b, 2c	Minimization of potential soil contamination effects on local ecology.	No change.	Improvement of overall soil quality.	O	O
(Air) 2a, 2b, 2c 8a, 8d	Prevailing winds may lead to occasional low concentrations of odour drifting eastwards.	No change.	Minimization of LFG emissions.	+	B
(Water) 2a, 2b, 2c	Potential minor pollution of watercourses affecting limited areas.	No change.	Minimization of water/ leachate contamination.	+	B
(Flora & fauna) 2c	Short-term reduction in size of SMI before parkland.	No change.	Minimization of ecological damage.	+	B

Notes: The numbers under the effects (e.g. soil) relate to the checklist of community sectors.
"+" gives greater benefit to Option B, with "–" the reverse and "O" or = being neutral.

Table 17.5 Prospect Park: evaluation of impacts on community sectors, post-completion.

Community sectors		Impact by sector		Difference in net benefit (B–A)	Preference	
		A	B		By impact	By sector
1	2	3	4	5	6	7
1	BA as operator	Reduction in productive time lost.	Continuation of productive time lost.	–	B	B
		Improved prospects for staff recruitment, retention and productivity.	Continued impediments to optimum productivity.	–	B	
		Much greater efficiency in the corporate operation of BA.	Continued impediments to optimum efficiency.	–	B	
		Greater productivity per worker.	Continued impediments to optimum productivity.	–	B	

Notes: The full list of community sectors in Tables 17.6–17.10 is as follows:

Onsite landowners/producers/operators
1 British Airways as an operator
3 Local authority on site
5 National Rivers Authority

Offsite
7 Department of Transport
9 National
 (a) Economy
 (b) National perception

Onsite consumers
2 Current users
 (a) Residents
 (b) Workers
 (c) Recreation
 (d) Inter-village access
4 New users
 (a) Workers/visitors
 (b) Recreation
 (c) Inter-village access

Offsite
6 Road users
8 Adjoining occupiers
 (a) West Harmondsworth and school
 (b) Rest of Harmondsworth
 (c) Longford
 (d) Government offices
 (e) Industrial estate
 (f) Other British Airways offices

Table 17.6 Prospect Park: summary of evaluation of impacts on community sectors, post-completion.

Community sectors		Preference by sector
Item	Description	
Landowners/producers/operators		
1	British Airways as an operator	A
5	National Rivers Authority	A
7	Department of Transport	A
9	National economy	A
	National perception	A
Consumers		
2	Current users	
	Residents	A
	Recreation	A
	Inter-village access	A
4	New users	
	Workers/visitors	A
	Recreation	A
	Inter-village access	A

Box 13   The evaluation consultation was not in this case in relation to the total evaluation display in the Tables, but rather through social survey conducted with onsite residents, in particular at Harmondsworth (Solon Consultants 1990). This was taken into account in finalizing the effect and impact assessments.

Box 14   Table 17.6 thus provides a summary of the community impact *evaluation* of the two options, which integrates environmental assessment in the planning decision in terms of net benefit to the community. In this case study, the report and recommendations were contained in the Outline Application Support Document which included the conclusions from the community impact evaluation.

Box 15/16: choice and decision communication   This role falls to the Inspector's report and Secretary of State's decision letter.

Conclusion from CIE

1. Environmental effects

(iv) On completion : On balance, the effect of implementing Option A would be the creation of a higher amenity environment for both work and recreation, with greater containment and control over the potentially hazardous infill and better leisure facilities. This would be a considerable improvement upon the existing conditions which remain if no development takes place.

(v) During construction: On balance, Option A may have greater adverse short-term effects on noise and visual intrusion. That apart, there should be no significant differences in the environmental effects of the two options during construction.

Conclusion on environmental effects: compared with Option B (refusal) the environmental effects following completion would be beneficial, even though there would be greater adverse effects during construction.

2. Impacts on human beings

(i) On completion: Here the summary of preferences is quite clear. All the sectors would prefer the proposals (Option A) with four groups of adjoining occupiers being non-certain (8 a, b, c and d), the non-certainty arising because of benefits to these sectors being offset through the adverse consequences for the views into open fields. The question then arises: if these groups of adjoining occupiers were to resolve the non-certainty by favouring Option (B), would this be sufficient to overturn the predominance of preferences for (A) in the other sectors? The likelihood here is very small indeed. The preference for (A) includes thousands of people amongst local residents and BA staff, as well as public and private bodies, whereas the preference for (B) comes from local residents in terms of their opposition to the threat to the Green Belt. However, this threat would not in fact be serious since the proposed development will not prejudice the role that the site currently plays in the Green Belt, but would strengthen its long-term standing as Green Belt (through designation of 250 acres of park and also some private open space). In addition there would be real benefit in the improved use, appearance and enjoyment of the park area in the project.

Thus, the net effects of development in the Green Belt would be the loss of view into open fields, which will affect only a small number of adjoining residents. It is unlikely that their preferences would outweigh the preferences of the much larger numbers beneficially affected.

(ii) During construction: Certain of the sectors would favour the development option (A) whereas a greater number would favour the scenario on refusal (B). In order to strike a balance it is necessary to explore the sectors further:

- Landowners/producers/operators: The local authority on site and the National Rivers Authority (sectors 3 and 5) would prefer a refusal (B). As against this, the Department of Transport and the National Economy would prefer permission (A). The outcome is non-certain.
- Consumers: Here the preference for refusal would run throughout all sectors (who would be likely to experience more adverse than beneficial effects) except for vehicular travellers on A4/M4/M25, who would enjoy the increased road capacity leading to reduced congestion at the junction with the project.
- Conclusion during construction: During construction the net outcome will be adverse for most of the sectors if permission is granted for the option A, with only a few offsetting preferences, mostly related to the reduction in congestion on the road intersection with the project and the jobs that are generated

in construction. On balance the preference would still be for option B. This is inevitable, since adverse construction damage is an essential concomitant of large-scale construction works which must obviously take place where permission is granted. But it would be short lived in relation to the life of the completed project.

(c) Overall conclusion on impacts on human beings: The adverse impacts during construction act against option A, as might be expected. However, the long-term impacts from option A after completion would decidedly create net benefits for the bulk of the community, with only an element of non-certainty for adjoining occupiers, which is not significant in number. These are mostly related to amenity. Their magnitude should be gauged by comparison with the poor environment that would continue on the application site under Option (B). Since the adverse impacts during construction in respect of option A are short term by definition, and the beneficial impacts of option A are long term, the conclusion is fairly straightforward: the disbenefits during construction cannot outweigh option A, with its decided long-term benefits on operation and management.

In brief, there would be net benefits from implementing BA's proposed development rather than have a continuation of current conditions which would flow from refusal.

Conclusion from case study

If environmental assessment be incorporated into development planning, by the appropriate interdisciplinary team, then the generic model can be used to cover the whole of the project assessment, including environmental aspects. If the environmental assessment is pursued independently, then there is no difficulty in incorporating its output into the generic model, although the study management could be made more complex. In such a case, the environmental assessment would tend, if conventional practice is followed, to culminate in typical scientific statements which have the following defects for decision-making in planning; they are not directly addressed to the people affected; it is not practicable to draw conclusions other than in relation to the scientific outputs; by the same token it is difficult to incorporate the inclusion into the planning decision framework. We thus have a situation where environmental assessment can help planning assessment through its more rigorous approach to the framework for decision-taking; but of itself it cannot readily provide the decisions simply because of the limitation in terms of impacts on people and having no evaluation approach which is built up on the family of cost–benefit methods.

17.4.3 The Manchester Airport study: using the generic method with further adaptation (DNLA 1994: appendix 1)

Context
The case study comprises evidence presented by the writer at the public inquiry 1994 into the application by Manchester Airport to build a second runway. It related spe-

cifically to one of the eight issues raised by the Secretary of State on "the extent to which the benefits from the proposed second runway outweigh its environmental and other effects". To address the issue the evidence was built around an integrated planning and environmental assessment adapted from the generic model. The adaptation in this study generally followed that of Prospect Park (§17.4.2) but with greater simplification: Prospect Park Tables 17.2–17.6 are condensed into Manchester Airport Tables 17.7–17.9.

A Planning

Box 1   The application was for building of a second runway and related operational works, within the framework of the Airport Development Strategy. The evidence was presented on behalf of the objectors in Manchester Airport's Joint Action Group (MAJAG). Representing a third party, the witness could not work as part of the Airport's planning team as in Prospect Park but needed to rely upon the Airport's documentation for the public inquiry.

B Project in the system

Box 2   The alignment of the second runway was chosen after a planning and environmental assessment of four options, as described in the Airport's Statement of Case.

Box 3   The options were implementation of the second runway, R2, which reflected the mitigation and enhancement of the potential effects through the design process; or R1, refusal, implying continuation of the current conditions, which would impose a limit on the capacity of the Airport

Boxes 4–5   The post-project system was described in the Statement of Case in relation to the regional airport system.

C Specification of plan variables and project options

Box 6   The variables were: construction; use upon completion of the project as a whole, that is on airport and off airport.

Box 7   The specification of project options by plan variables were set out in the Airport Statement of Case

D System change

Box 8   Apart from economic development that the economic impacts would generate (jobs and income) there would be the large scale associated urban development which would be a corollary of the airport expansion.

E Decision/space framework

Box 9 The context for the evaluation was presentation of proof of evidence for MAJAG.

F Effect assessment

Box 10 Unlike Prospect Park, where the planner/analyst worked with the planning/ environmental team, as witness for a third party objector he was unable to do so. Instead, the effect assessment needed to be gleaned from the Manchester Airport documentation, in particular the Statement of Case, Airport's Environmental Assessment and the numerous proofs of evidence at the Inquiry, most in favour and some against the second runway. All this material was insufficiently integrated to enable a clear summary to be produced of predicted effects following mitigation/enhancement, based on Figure 17.4. It was thus not possible to complete the appropriate Table for the purpose, shown in pro forma Table 17.7.

Table 17.7 Manchester Airport second runway: prediction of potential and mitigated effects and impacts. Options A and B: construction and post-completion variables.

Construction								
1	2	3	4	5	6	7	8	9
Element	Baseline	Potential effects	Mitigation/ enhancing	Predicted effects		Sector experiencing	Impacts on sectors	
				A	B		A	B
Land use								
Landscape								
Conser- vation								
Archae- ology, etc.								
Post-completion								
1	2	3	4	5	6	7	8	9
Element	Baseline	Potential effects	Mitigation/ enhancing	Predicted effects		Sector experiencing	Impacts on sectors	
				A	B		A	B
Land use								
Landscape								
Conser- vation								
Archae- ology, etc.								

G Impact evaluation

Box 11 In the absence of this summary of effects it was thus necessary to move to the situation visualised in the simplified method described above (§9.3.3), i.e. directly to the impact evaluation. This was produced around the framework of pro forma Table 17.8, corresponding to Table A of the earlier case studies. The community sectors are pragmatically enumerated (column 1) and then the *impacts* visualized from the documentation for each of the alternatives (columns 2–3). For each impact is postulated a sectoral objective, leading to the preferences for the options (columns 4–5). The Table is supported by an Annexe giving the reasoning behind its content and the related sectoral objectives.

Table 17.8 Manchester Airport second runway: prediction of impacts on community sectors and preferences. Option A and B: construction and post-completion variables.

Construction					
1	2	3	4	5	
Sectors by category	Impacts of A	Impacts of B	Sectoral objectives	Preferences	
				A	B
Producers					
on site off site					
Consumers					
on site off site					
Completion					
1	2	3	4	5	
Sectors by category	Impacts of A	Impacts of B	Sectoral objectives	Preferences	
				A	B
Producers					
on site off site					
Consumers					
on site off site					

Source: DNLA (1994).

Box 12 Pro forma Table 17.9 summarizes Table 17.8 by repeating the sectors and their preferences as a basis for drawing conclusions.

Box 13 Evaluation, consultation, negotiation and bargaining were hardly practicable in an Inquiry situation, except in so far as it took place in the adversarial atmosphere changes.

Table 17.9 Manchester Airport second runway: summary of preferences of community sectors. Option A and B: construction and post-completion combined.

Sectors by category	Preferences	Preferences
	A	B

Box 14   The evaluation report and recommendations were in effect the proof of evidence for the conclusions and opinion. Refusal for permission of the second runway was urged on the following conclusions drawn from the Integrated Planning and Environmental Assessment. From them it will be seen that while the evaluation itself did not reach a clear-cut conclusion on the Secretary of State's Issue 8, owing to the circumstances of a third party witness, it did help to clarify complex arguments on the two basic options, and thus serve as a decision support function, as follows:

- I have displayed on the one hand the economic impacts, which are largely benefits, and on the other hand the secondary and environmental impacts, which are largely disbenefits. In simple terms, this display shows the economic benefits which would be widespread over the Region, as against the widespread and varied disbenefits to the residential and working communities.
- From this display it has still not been possible to give a numerical answer in terms of totality of costs and benefits to the Secretary of State's Issue Number 8, namely as to *"The extent to which the benefits from the proposed second runway outweigh its environmental and other effects"*. But nonetheless, from the display and its back-up in textual analysis, the following conclusions can be drawn which are highly relevant to the Inspector's recommendation.
- Based on a comparison of the options R2 and R1 my judgement in relation to Issue 8 is: the *marginal loss* in economic benefits to the North West as a whole by not building R2 would be outweighed by the *marginal benefits* in avoiding the adverse environmental and other effects which would be experienced if R2 were abandoned.
- To be sure this is my judgement. In it I might add that I am, in accordance with contemporary thinking, urging for this planning decision that a considerable *weight* be given to environmental issues as against economic growth *per se*. My weight on the environment is more than that of the Airport. To them the economic benefits outweigh all else.

Boxes 15–16   The choice decision and communication role falls to the Inspector's Report and Secretary of State's decision letter.

17.5 Strategic environmental assessment and strategic plan evaluation

17.5.1 Context

Having presented above (§17.4.2–3) the integration of environmental assessment and community impact evaluation for *projects*, we now explore such integration at the strategic level, *for policies, plans and programmes*. The approach outlined here is built up from the preceding as follows:

- EA for projects leading to EA for policy, plans and programmes, called strategic environmental assessment (SEA) (see Ch. 5).
- CBA for projects (see Ch. 6) leading to CIE for projects (see Ch. 7) incorporating SEA.

From these it is necessary to show how the preceding can be advanced to integrate the SEA of Chapter 5 with the plan evaluation of Chapter 11, building up on the integrated planning and environment model in §17.4 above.

17.5.2 The DoE approach

The approach to this strategic integration is well articulated in the Department of the Environment's Planning Policy Guidance on the topic (DoE 1992b):

> Most policies and proposals in all types of plan will have environmental implications, which should be appraised as part of the plan preparation process. Such an environmental appraisal is the process of identifying, quantifying, weighing up and reporting on the environmental and other costs and benefits of the measures which are proposed. All the implications of the options should be analyzed, including financial, social and environmental effects. A systematic appraisal ensures that the objectives of a policy are clearly laid out, and the trade-offs between options identified and assessed. Those who later interpret, implement and build on the policy will then have a clear record showing how the decision was made. In the case of development plans, this should be set out in the explanatory memorandum or reasoned justification. But the requirement to "have regard" does not require a full environmental impact statement of the sort needed for projects likely to have serious environmental effects.

Following these initial guidelines for *appraisal* of environmental policy (DoE 1991a), the DoE issued guidance on *integrating* environmental policies in development plans (DoE 1993a). The guidance in essence relates to providing the very preliminary impacts to plan-making: the SEA for environmental considerations, to take its place alongside the introduction of other plans with consideration of traffic, housing, conservation, and so on. As such it goes beyond the pioneer approach to the strategic environmental assessment in the preparation of the revised Lancashire Structure Plan, which used an "impact matrix" to predict the implications of 164 individual policy statements for some twelve environmental components (receptors) that might be affected, with an indication of the comparative orders of magnitude (Pinfield 1992). This is a useful starting point for the generation of alternatives.

17.5.3 Towards integration of environmental and land-use policies

But although the DoE guidance refers to evaluation alternatives, the guidance does not show how the environmental evaluation can be made of *integrated* environmental and planning assessments. Some pointers follow.

As introduced above (§5.5) it is hardly practicable to be specific about SEA methodology, since it can be applied to such a variety of situations, involving both the level of the plan, and also whether the SEA is to be applied to policies, plans, programmes or projects. The same is true for integrated environmental and land-use planning. But some pointers can be given by reference to the indication above (Ch. 11) that any plan contains within the one document a variety of subplans which make up policies, proposals, projects, programmes of projects and strategies. For this reason, in the strategic integration of environmental and land-use planning, it is necessary within each plan to distinguish these particular components, in order that the generic evaluation methodology can be applied to each in its own way.

For example, the end state of policies or strategies which are to be evaluated can be crystallized only by the development of scenarios that attempt to describe what would occur in the future if the policy or strategy were implemented. Projects would not be treated in isolation as in the generic model, but rather all interrelated projects in the plan would be seen as "programmes of projects" and analyzed as such. This would overcome, for example, the objection mounted against the evaluation of particular road proposals that are part of a wider network, whose cumulative impacts should be considered at the strategic level. Account could then be taken, for example, of synergistic impacts (the total impacts exceeding the sum of individual impacts); threshold/saturation impacts (where the environment may be resilient up to a certain level and then becomes rapidly degraded; Therivel et al. 1992).

This approach can be tackled by the adaptation of the generic model for the integration of environmental assessment and planning evaluation in projects. In essence, once the options have been generated (Box 3), it will be necessary first to generate the effects assessment (Box F) and then to evaluate the impact of those effects (Box G). But, clearly, the analysis under these headings will not be as precise and quantitative as in project evaluation, bearing in mind that even then there is considerable qualitative judgement. Indeed, the analysis of the effects must be general, indicative and non-site specific and the evaluation between options must necessarily be strictly ordinal (Verheem 1992). But however qualitative the data, analysis and judgement, the aim would be as comprehensive an evaluation of options as possible, using as a basis the generic model above, with a view to choice in the process of plan-making.

17.6 Mutual interaction of environmental and planning assessment

17.6.1 The impact of environmental assessment on planning assessment

The use of environmental assessments in Britain, in the 200–250 studies in the 15 or so years prior to their formal introduction by Regulation, has certainly made some impact on technical/professional aspects of planning assessment. It seems likely that the impact will be more widespread following introduction of the Regulations: not only are there likely to be more assessments made over a comparable period, but because of the Regulations, an informed assessment will be pursued throughout many projects which do not in the end need formal actual assessment.

That there will be such an impact is to be expected, for it will follow the general experience of the planning process. In this, additions to the disciplines traditionally involved in planning have made a contribution from their specialist vantage point to the planning process, in theory, principle and application on the job, for example in relation to transport, retailing, economics, conservation. This enrichment has been absorbed by the central and local government planning authorities and practitioners, resulting in the planning process being carried out at a higher level in the evolutionary spiral than before. It was always welcome, even though it rocked the boat on occasions. And it is being welcomed again in environmental assessment.

However, the need for the environmental statement brings out an irony in relation to the development planning process and planning applications. Under DoE Regulations, Circulars and advice, the environmental assessment tends to be somewhat formidable. It *must* contain an array of specified information, and in addition it *may* include further information in response to any request by the planning authority (DoE 1989: appendix 3). This leads to the wide scope of the matters that need to be considered for *possible inclusion* (DoE 1989: appendix 4), it being borne in mind that this does not in itself define the *scope* of the EA, which is to be confined to the *main and significant* effects.

By contrast, nowhere in the long history of regulations and circulars relating to development control is there *required* a "planning statement", even where the project is of some significance. Often, in practice, it is the minimum only which tends to be provided, although in many cases a fuller statement is offered. Thus, compared with the typical planning statement for applications, the EA could certainly be the tail wagging the planning dog.

Another irony in the comparison is that, despite the depth of the content in the environmental assessment, the scope of the matters covered is *narrower* than those that could reasonably be incorporated in the planning statement under current practice. This stems from the fact that the specified minimum environmental information clearly appears to put its weight on the *natural* and *man-made* environment, with only a passing reference to *human beings*. By contrast the accepted intent of any planning analysis or statement, and the evidence that could be presented at a public inquiry, go well beyond this limitation. It would certainly embrace the social and economic impacts from the development relating to human activity in the natural and built environment.

This limitation of impacts in EA clearly stems from the content of statements prescribed in the EEC Directive (DoE 1989: appendix 7). But this hardly explains the restrictive content provided for in the Regulations, since the Directive presents the *minimum* that must be provided for in the assessment and not the *maximum*. This limitation is striking when Schedules 1–3 of the Regulations (DoE 1989: appendix 3) are compared with the content of environmental impact assessments and statements in the foundation statute, the US NEPA of 1969. Section 102 there authorizes and directs all agencies of the Federal Government in preparing environmental impact statements (EIS) to:

> Utilize a systematic and interdisciplinary approach which will insure the integrated use of the natural and social sciences and the environmental design arts in planning and decision-making which may have an impact on man's environment.

This implies that the Act calls for environmental statements to relate to the *human environment as a whole*, that is *all* impacts affecting human beings. To be sure, in the early years most such statements were prepared narrowly, referring only to the "natural environment". As regards economic impact, the field was already developed and readily adaptable, under the different title of cost–benefit analysis (see Ch. 6). Although this was not so for social impacts, these were rapidly dealt with by the emergence of the methodology of social impact assessment (SIA), where the focus is on the demographic, social and economic impacts (Ch. 5).

From this it follows that the "ironies" must rapidly turn into "anomalies", unless planning practice is amended. First, the development planning process must embrace the environmental planning process. Secondly, planning statements must be as deep as environmental statements. Thirdly, since planning analysis and assessments do now embrace the socioeconomic, there must be pressure for the environmental assessments in Britain to do likewise where they are used in the planning context. Fourthly, in the planning application process, as much concern is needed as in environmental assessment for the presentation and evaluation of options. All this is perhaps implicit in the following (DoE 1989: pt I, para. 1):

> Authorities already obtain from developers such information as they consider necessary to determine a planning application, including information about environmental effects; and those effects are among the material considerations which a planning authority must already take into account when considering a planning application. What is new about EA is the emphasis on systematic analysis, using the best available techniques and sources of information, and on the presentation of information in a form which provides a focus for public scrutiny of a project and enables the importance of the predicted effects, and the scope for modifying or mitigating them, to be properly evaluated by the planning authority before a decision is given.

For the planning assessment to be comprehensive, it will be necessary for it to move forwards on two counts: to embrace within it the rigours of environmental assessment, as evolved by the social and natural environmental scientists; and to reach a higher level than previously in making planning assessments (appraisals, evaluation) which do as much justice to the planning situation as do the environmental assessments for the environment. Taking a leaf from environmental assessment, the planning assessment must contain *information, assessment* and *statement*. This will require some uplift in the conventional methodology of planning decision-making and assessment.

17.6.2 The impact of planning assessment on environmental assessment

So far we have seen that the methodology of environmental assessment can contribute towards the improvement of planning assessment. But this is not to suggest that environmental assessment can be used for development planning decisions, for the reason that it has certain built-in weaknesses for this purpose. Some are:

- it adds nothing new in concept to the development planning approach, albeit it has strengthened the project planning process by formalizing environmental analysis within it
- the output tends to relate to the scientific dimensions of the effects, with the rather weak recognition of the significance of the repercussions on human beings, which is the kernel of planning
- this weakness is aggravated by the relative simplicity of the *evaluation analysis* in environmental assessment, which amounts to a testing of the *significance* of the *effects*, rather than a comparison of their costs and benefits of their impacts on people, as in planning
- the considerations relating to environmental assessment are not readily in practice integrated into the planning assessment, more particularly in the weighing up and balancing of those considerations against planning policy in order to reach a planning decision.

From this it follows that planning assessment can make a considerable impact on environmental assessment, not so much in methods and techniques but in the integration of the output of the assessment into the everyday world of development planning and development control decisions. This requires that the environmental impact be seen as material consideration in relation to the planning decision, perhaps because it is part of the plan policies, and even if not, because it is clearly another material consideration. Furthermore, its "significance" is seen not simply in terms of degree of environmental effect but also in the weight given to it in *balancing* the plan and other material considerations in the planning decision.

CHAPTER EIGHTEEN

Green belts

18.1 Historical introduction: the sanctity of green belts

Green belts, to restrict a town's growth and avoid consequential evils, are a long held dream (Blom 1986: ch. 1). They were promoted by Ebenezer Howard, proposed by Raymond Unwin for London in 1935 ". . . to provide a reserve supply of public open spaces and of recreational areas and to establish a green belt or girdle or open space . . .", and they figured in the post-war Abercrombie Plans for Greater London, Glasgow and Edinburgh.

Before the Second World War, green belts had to be provided by agreement with landowners or through land acquisition. It was with the powers available under the Town and Country Planning Act 1947, to restrict development without local authority purchase or liability (MHLG 1955) that the contemporary huge green belt areas became feasible, and were inaugurated throughout the country via structure plans, now extending to several million acres (Cullingworth 1994: ch. 5).

The purposes of green belts are now recognized to (DoE 1988: paras 4, 5):

- check the unrestricted sprawl of large built up areas
- safeguard the surrounding countryside from further encroachment
- prevent neighbouring towns from merging into one another
- preserve the special character of historic towns
- assist in urban regeneration
- provide access to open countryside for the urban population.

18.2 Current provisions for the protection of green belts

Despite their sanctity, green belts seemed to come under threat, as did other planning safeguards, in the early years of the Thatcher Administration (Thornley 1990). A spate of allowed appeals implied a weakening of resolve for green belt protection. However, following opposition (Elson 1986: ch. 10) two Circulars continued the protection (DoE 1984a,b) ". . . unless the circumstances of the case are such that there is an exceptional need to make land available for housing". This stand was maintained by successive Secretaries of State for the remainder of the Thatcher years and into the early 1990s.

18.3 Presumptions in green belt policy

So strong has been the policy of protecting the green belts against incompatible developments that there has been built up a series of *presumptions* towards that end, namely indications by the Department of the Environment of the significant weight it attaches to its protection in the making of decisions. These presumptions derive from different sources, as the following shows.

18.3.1 Presumption in favour of development plan

As brought out above (Ch. 12) the introduction of Section 54A of the 1990 TCPA changes the relative weight of the plan against other material considerations (DoE 1992a: para. 25): "In effect, this introduces a presumption in favour of development proposals which are in accordance with a development plan". Thus, if the plan indicates a green belt, there is a presumption against development within it.

18.3.2 Exceptions to presumption in favour of development[1]

Exceptions to presumption in favour of development relate to presumption against inappropriate development in a green belt (DoE 1988: para. 12). "The general policies controlling development in the countryside apply with equal force in green belts, but there is, in addition, a general presumption against inappropriate development within them". What is inappropriate can be derived from the definition of what is appropriate (DoE 1988: para. 13): "Inside a green belt, approval shall not be given, except in very special circumstances for the construction of new buildings or for the change of use of the existing buildings for purposes other than agriculture and forestry, outdoor sport, cemeteries, institutions standing in extensive grounds, or other uses appropriate to a rural area".

18.3.3 Onus for giving reason

There is a long-standing general presumption in favour of allowing applications for development, having regard to all material considerations, unless that development would cause demonstrable harm to interests of acknowledged importance. This has carried with it the reinforcement that, in seeking to refuse, the authority has clearly to demonstrate the reasons: "Except in the case of inappropriate development in the green belt the developer is not required to prove the case for development he proposes to carry out; if the planning authority consider it necessary to refuse permission, the onus is on them to demonstrate clearly why the development cannot be permitted".

As this shows, the exception to this long-standing approach relates to "inappropriate development in the green belt" which again reinforces the presumption in favour of the green belt. This is further reinforced by the new obligation on applicants, which is introduced in relation to the new provisions on presumptions in favour of the development plan as follows (DoE 1992a: para. 25): "An applicant who proposes a development which is clearly in conflict with the development plan would need to produce convincing reasons to demonstrate why the plan should not prevail".

---

1. The reference here is to PPG2 on green belts.

18.3.3 Conclusions

This review of presumptions in green belt policy would appear on the surface to amount to an impenetrable ring fence around the boundaries of the green belt, through the mesh of which "inappropriate development" would find it difficult to penetrate. Accordingly, except for the hesitancy of the 1980s, DoE and Local Authority policy has been robust in its defence of green belts, as appeal after appeal shows. But despite this, permissions have been given around the proviso of "exceptional circumstances", in what would otherwise be considered as "inappropriate development". Summarizing the above presumptions, there is an array of guidelines that need to be brought to bear on particular cases:

- in green belts there is no presumption in favour of development generally and also a presumption against inappropriate development. This is strengthened by the presumption in favour of green belts shown in development plans; in this the ruling criterion is one or other of "exceptional", "special", or "very special" circumstances;
- in the case of inappropriate development in the green belt, the developer has the onus to demonstrate the case for his proposal.

18.4 Planning considerations

A reading of Inspectors' or Ministerial decisions on green belts, and judicial review on green belt appeals, shows the difficulties in reconciling these presumptions, and also in identifying the particular "exceptional, special or very special circumstances" which will justify going against the presumption. The result is that in these decisions there is a subtle argumentation, attempting to piece these guidelines together as they apply to a particular case. It is not our purpose here to attempt to pursue the lines of this argumentation, but to consider whether there is a role for CIA in elucidating the circumstances.

On this it is revealing to consider the views of Mr Justice Auld, in one such case (Tesco Stores Ltd 1990), when in doing so he cited an earlier case (Phersson 1990). After reviewing the guidelines introduced above, he went on to suggest that: "Paragraph 13 of PPG2 and its reference to very 'special circumstances' undoubtedly includes circumstances of planning benefit or advantage arising from a proposed development" and then pointed out that the other two judges involved, Stuart-Smith and Staughton, L, JJ: ". . . do not talk of a balancing of harm and advantage in this way, but of a balancing harm by reason of inappropriateness against the developer's claim of 'very special circumstances' justifying the grant of permission".

In essence, although the presumptions against development in the green belt would seem on the surface to be so solid, there are clearly openings to show the "exceptional, special or very good circumstances" which would justify the grant of permission. In essence, the judicial guidance seems to convey that these "circumstances" can be construed as providing planning benefits or advantages following from a permission, which can be compared with damage which would be caused by the inappropriateness to the green belt of the permitted development. It opens up the possibilities of balancing/weighing harm (cost) and advantage (benefit). The question is thus raised: how can the generic model assist in the balancing/weighing as a basis for presenting evidence, on appeal or judicial review. To this we now turn.

18.5 The generic model and green belt development[2]

1   Whether or not to grant permission for development in the green belt.

2   Project for which the application is made.

3   The datum option is to pursue the strong presumption against development which is inappropriate. If this be the datum, then other options are developments which, while being inappropriate, may nonetheless be justifiable in the "exceptional, special or very special circumstances". The options either could be on their own or associated with a package of related proposals for other areas of the green belt, aiming to improve them, as in the Prospect Park case (Ch. 17). It is these options that are progressed through the succeeding steps of the generic method until:

10   The effect assessment would be handled as demonstrated in the Prospect Park Study (Ch. 17).

11   It is the impacts on the various community sectors, linked with the effect assessment, which presents the basis for balancing of impacts, in the words of Stuart-Smith and Staughton, L. JJ:
   • harm (cost) and advantage (benefit);
   • harm (by reason of inappropriateness) against benefit drawing from "very special circumstances".

12   The conclusions would emerge from the balancing and weighing of the differences between the impacts.

---

2. This model, adapted into the Planning and Environment Framework, was used in the Prospect Park study (§17.4.2).

CHAPTER NINETEEN

Deregulation

19.1 Context

As brought out above (§1.6), the planning system is aimed at regulating market processes in the public interest. With this in mind, government has introduced a formidable system of regulation: in statute, regulations, statutory and administrative guidance. Accordingly, government has periodically found it necessary to hack away at the thicket, for example in the Macmillan "bonfire of controls" in the 1950s (Cullingworth 1994). The Thatcher Government made a major attempt from 1985 when pursuing its market-orientated policies towards widespread deregulation of enterprise, under the general title *Lifting the burden,* (Minister without Portfolio 1985) by both removing established controls, and critical scrutiny of proposals for new controls.[1]

19.2 The approach to lifting the burden

The initial framework for *Lifting the burden* was provided in an imaginative Report of civil servants from various Departments (DTI 1985). This policy had some welcome objectives. In essence (Minister without Portfolio 1985):
- legislative and administrative regulation has grown enormously over this century. While introduced from the best of motives, does it not now constitute a drain on our national resources (in central and local government regulatory manpower), in being too complex and confusing for efficient business. Thus, to the extent it goes further than necessary it will lower profits in firms, raise prices, stifle competition and deter new firms from entering the market or prevent others from expanding;
- there is a need therefore to examine rigorously the cumulative weight of legislation to ensure "that its benefits outweigh its costs (we want less-better-regulations)";
- in this "we must maintain our quality of life. But we have to strike the right balance".

The deregulation measures that the government introduced covered many aspects of our socioeconomic life, categorized by the government department concerned:

---

1. The Major Government has made another attempt, which is not reviewed here.

environment, customs and excise, taxation, health, social security, employment, trade and industry, transport, agriculture, fisheries and food, home departments (Minister without Portfolio 1985). Of particular concern here are the past and proposed measures on the environment, articulated under the two forces that need to be balanced, planning and enterprise; and also the closely related field of transport which is discussed within "the wider role of government" (Minister without Portfolio 1985: chs 3, 5; in later White Papers this was referred to as "Government and Business"). While the environmental deregulation was aimed, *inter alia*, at planning measures, the value of the planning system was stressed (Minister without Portfolio 1985: para. 3.1–3):

- The Town and Country Planning system has not changed in its essentials since it was established in 1947. "In many ways it has served the country well and the Government has no intention of abolishing it".
- ". . . the Government's policy is to simplify the system and improve its efficiency" to ". . . speed the planning process and to facilitate much needed development which helps create jobs – in construction, commerce and industry and in small firms".
- ". . . the Government is equally concerned to protect and enhance the environment in town and country, to preserve our heritage of historic buildings and rural landscape, to conserve good agricultural land and to contain the green belt. This too requires a planning system that works efficiently and effectively, and strikes the right balance between the needs of development and the interests of conservation."

Because "the Government want to stem the flow of regulations", following the recommendation in *Burdens on business* (DTI 1985) they set up a system for assessing and monitoring *new* regulations to ensure that they ". . . are not introduced unless the need for them has been clearly demonstrated". For this they devised a three-pronged approach (DTI 1985: 13; Minister without Portfolio 1985: para. 8.1):

1. "structural analysis of each proposal, to be prepared and published by the initiating agency concerned, including a systematic assessment of its impact on business enterprise;
2. "critical scrutiny of the proposal, in particular on the assessment, by *a small task force in Central Government* with real teeth; and
3. "regular *overviews* by the task force of proposals in the pipeline and the scope for eliminating, simplifying, or rationalizing the existing requirements systems".

On (i) the *structural* analysis would be in the form of *impact* analysis, that is (DoT 1985: appendix 9): ". . . a means of systematically identifying and appraising *all* the main costs and benefits implied by adoption of the proposal, and of giving appropriate weight to its implications for business and for jobs".

It would normally include brief statements of the essential purpose of the proposal; the problem the proposal is intended to tackle; any alternative to the particular line of action proposed; and the *costs and benefits* of the proposal, quantified where possible. The analysis of the impact of the proposal on *business* ". . . should be a key element in the total assessment, which would bring out any implications for investment and operating costs, employment and competition . . ." This would require an assessment of cost compliance (CCA). A preliminary CCA was based on a 17 point checklist on costs and benefits: setting out the purpose of the Regulation, the impact of the Regulation on business, and its wider impact (Secretary of State for Employment et al. 1986: para. 3.10). Where the preliminary CCA needs further probing there is a detailed CCA, which (Secretary of State for Employment et al. 1986: ch. 3.12): ". . . is a full-scale analysis of the costs and benefits to business, government and the economy as a whole, using more quantitative evidence".

The machinery for making the analysis and assessment was set up through a Concordat between regulatory or sponsoring departments, in consultation with an Enterprise and Deregulation Unit of the Department of Trade and Industry (Secretary of State for Employment et al. 1986: annexe 3):

> The EDU act in some sense as a proxy for the voice of business within Whitehall and considers Departments proposals from the viewpoint of business. In this the Unit can also call upon the business experience of the Advisory Panel on Deregulation . . . a group of 9 businessmen who advise on implications and regulatory proposals for the business world.

In terms of environmental planning there was one important weakness. The responsibility for the "systematic assessment of its impact on business, government and the economy as a whole" was naturally the "initiating agency concerned", which in most cases would be the sponsoring department for the sector involved. The intention would appear to be that the department would screen, and the task force would review, the "impact of the new regulations" with a view to avoiding adverse consequences on business, industry and the economy. But there seemed to be inadequate scope for consideration of the ramifications, outside the economy itself, on the environment, countryside, urban areas, etc.

Thus, our question is: have the proposed and prospective deregulation measures in environment and transport been in the wider community and public interest, as opposed only to that with which the White Paper is legitimately concerned, the interests of business and industry and thereby the economy. More precisely, is the deregulation achieving the objective stated in *Lifting the burden* of striking the *right balance* in maintaining quality of life (Minister without Portfolio 1985: para. 8.7).

19.3 The removal of established controls

19.3.1 The controls under review

Early in its programme the government listed proposed changes in the legislative and administrative framework to which they would be turning their attention over the ensuing years (Minister without Portfolio 1985: ch. 2). For our purpose these fell into two main categories:

- Affecting the context of development as a whole, and thereby the role and function of development control:
  - Environment: Enterprise Zones, Amendment of GDO, Simplified Planning Zones, Use Classes Order, General Development Order, simplified planning procedures, pre-inquiry stages at major public inquiries, small firms in relation to working from home, building regulations, environmental protection, outdoor advertisements, by-passing the planning system through private bill procedure, Listed Buildings.
  - Transport: taxis and car hire, long distance express coaches, domestic air routes, air services to Continent, local bus services, merchant shipping, International road haulage permit and heavy goods vehicles.
- Affecting directly the practice of development control:
  - Environment: presumption in favour of development, use of conditions in planning permission (Circular 1/85), relaxation of onerous conditions on environmental nuisance in favour of statutory nuisance controls, relaxation of restraint on competition in retailing and other commercial development.
  - Transport: removal of DOT involvement in development control on trunk roads.

19.3.2 The procedures for removal

By contrast to the procedures for scrutiny of new proposals, those for removing established ones were far less evaluative. To be sure, each of these measures was preceded by consultation on White Papers, Green Papers, etc.; indeed that relating to Use Classes was exhaustive. But there was the general impression that the dirigiste conviction politics, that surprisingly goes with deregulation and liberalization, has not reacted sympathetically to the formidable comments, criticism and warnings on the introduction of these measures (e.g. Lloyd 1987).

The effect of these deregulation measures is to remove from the planning system cases that would otherwise have been tackled under normal development control. This does not mean however that the effects/impacts of such deregulation are not of concern in the planning system. Five well known examples will illustrate that they are:

- In enterprise zones, there is something seriously wrong where the huge urban complex in Canary Wharf, Docklands, could go ahead without adequate consideration of the transportation assessment for commuting to work in Docklands and on the financial services industry in the City of London.
- Results on the ground in Docklands outside the Enterprise Zone, where market-led planning leaves much to be desired for the environment, leads to apprehension about the abandonment of conventional development control in simplified planning zones.
- The 1987 Use Classes Order (DoE 1987) was justifiably based on a long overdue review of the 40 year old Order. But the fact that the freedom to change uses (e.g. in the B1 class) was made *retrospective* has caused serious problems for planning authorities. Their relevant constraint policies in plans, approved by the Secretary of State, have been thrown into disarray, and their efforts to stop the damage have certainly not helped to "simplify the planning system".

- Moving to transport, the abandonment under the Transport Act 1983 of the public (Passenger Transport Executive) monopoly on local bus services, and admitting competition, has doubtless led to reduction in fares. But there is a tendency to close down non-paying services, since the new commercially orientated entrants to bus services are not taking up those routes which in themselves were not financially remunerative, and previously were cross-subsidized by the remaining remunerative services out of the public transport pot. In narrow efficiency terms this "Beeching" exercise is justified by the marketeers. But what seems to have been overlooked is the possible repercussions on the consequential everyday activities of the city or region served by the buses. In Greater Manchester, for example, this deregulation undermined the basic objectives of the Structure Plan in terms of ability and accessibility of the population concerned (Lichfield 1987b).
- Again in the transport field, another consequence of deregulation lies in increasing the burden on the roads of heavy goods vehicles, within Britain and from the Continent. The result is the Continental transporters attempting to negotiate the lesser roads which clearly cannot cope with them, with congestion for other traffic, atmospheric pollution and danger to pedestrians.

19.4 Decommissioning of development

The blanket deregulation of the kind described has the effect of leaving without control certain activities in the built environment which had previously been subject to regulation. In terms of those activities the market will then decide. This abandonment to the market without control arises also in situations of decommissioning development. The consequence is that enterprises which could be the economic mainstay of a town or region (industrial, shipbuilding, coalmines, steelworks, etc.) have been closed down without any machinery for considering, ameliorating and coping with repercussions. At a time of disinvestment in major industry of the kind experienced in the past decade or so, this has had serious consequences for the basic resources in the locality which is discarded, such as its labour power, associated dependants and community infrastructure.

Thus, coalmines affecting whole communities have been closed simply following financially based argumentation by the Coal Board on the one hand, and its resistance by the National Union of Mineworkers who argued in favour of jobs, however "uneconomic", on the other. And yet, when a new coalmine is to be opened, such as Belvoir, considerable resources are spent, and correctly, in ensuring that the prospective planning and environmental considerations are fully reckoned with.

19.5 Application of the generic model

In the preceding it is seen that three different situations have arisen, each of which could benefit from the application of the generic model:
  (a) Scrutiny of new regulations by a rigorous evaluation method, even though it does not cover a broad enough canvas.
  (b) Removing established regulations (i.e. deregulation), following general consultation without however recognizable evaluation analysis.
  (c) Allowing the market to decommission development, with perhaps serious consequences, without any provision for public inquiry or scrutiny.

Accordingly, in the scope of this book the question arises: how can community impact evaluation assist in these cases?

For (a) the case hardly needs making. The evaluation methodology which was set up by the government for the scrutiny of new deregulation proposals would appear to have rigor of a similar order to that of CIE in terms of impact and cost–benefit analysis. But, as indicated, the scope of the analysis is restricted in terms of the very objectives of the government in pursuing deregulation, and its appreciation of the planning system and regard for the environment (Minister without Portfolio 1985: para. 3.1–3).

As regards (b), the very system set up for scrutinizing new regulation proposals in (a) in itself makes a compelling case for a similar system for the removal of established regulations. Such regulations have been introduced in the past to perform some social purpose which must have been fully ventilated at the time. Business and government have thereby been operating accordingly and have set in train the pertinent policies and programmes, by definition to the benefit of the public. Thus, the repercussions in terms of impacts, and their costs and benefits, is called for as much as in (a). From this it flows that community impact analysis would be a relevant and useful method. It would be particularly important where the deregulation is introduced at the macro or national level (as in the change in the 1987 Use Classes Order or privatization of the bus services or deregulation of heavy goods vehicles). Clearly, the beneficial or adverse impacts cannot be uniform over the country and thereby could give rise to serious consequences in particular localities.

The situation is different under (c), where the market introduces the change rather than the government through regulation or deregulation of the market. Clearly, planning permission cannot be required for closing a mine. But how, in such a situation, can "economic de-development and community benefit" be reconciled, if the opportunity is not taken to place the issues on the public agenda. In procedural terms, the de-development could certainly be at least ventilated at local level, and the planning and environment machine be adapted to bear informally in making its contribution to the amelioration of the adverse impacts. This necessity has been brought to the fore in the realization that coalmines have been closed down without even provision being made to avoid pollution to water supplies. There has been no formal impact assessment for the decommissioning, resulting in responsibility allocated for the mitigation of pollution.

CHAPTER TWENTY

CIE study management

20.1 Approach

The preceding chapters delved into community impact analysis/evaluation as an academic/professional exercise. Here we are contemplating the carrying out of that exercise in relation to specific commissions, for example as staff of a local planning authority, or development agency (public or private), or consultants to them, or expert witness at a public inquiry. How should they set up the study from inception? In this we are assuming that the academic/professional steps, procedures and techniques will be carried out in the appropriate manner, simplified as necessary, and we are searching for the context of doing so in relation to the planner's/analyst's environment. In this we will be referring as appropriate to the generic method as a skeleton for the management process, but only in that connection.

20.2 Role of evaluation in the planning process[1]

20.2.1 Kind of evaluation

Since this essay into study management process relates to *evaluation* in the restricted sense used in this book, and not to general "tests", the latter are excluded and would need to be considered separately. The management of the evaluation will vary according to whether it is to be carried out as a discrete operation (that is on material supplied by the planning team at particular points in the planning process) or conducted by the evaluation analysts throughout the planning process as integrated members of the team. Whether a discrete or integrated contribution, the evaluation analysis needs to be adapted to whether or not the planning process is linear or cyclical, that is, respectively, proceeds sequentially step by step or which goes steadily from lesser to greater depth. In the cyclical approach, which of two directions should be followed to find the best of the preselected options: to use the analysis of the preselected options to generate further options, which combine the best features of the initial studies and rejects the worst; or to use evaluation analysis as a learning process with a view to generating an entirely new set of options. Instead of the evaluation analysis being faced with a series of firm options for final choice, there could be

---

1. Lichfield et al. (1975).

many options, of which there would be "coarse screening" to reduce this number to a manageable shortlist for later detailed evaluation.[1] In this approach the evaluation methodology would be essentially the same at the two levels, but would be simplified for the first. But in this it is essential that the same principles be pursued in order that there be consistency. If not, there is the possibility that the shortlist may not contain the best of the options analyzed at the coarse stage.

20.2.2 Role of the evaluation analyst in the planning team

When setting up the team it will be necessary to designate that member who would be the evaluation analyst. He could clearly double this capacity with another, preferably, although not necessarily, grounded in economics. The evaluation process will also be influenced by whether or not the team is working in collaboration with one concerned with planning at the higher level (a district looking to a county council) or at a lower level (a county council looking to a district). In either case the evaluation will be influenced by respectively the local or strategic contribution.

The evaluation criteria that will aid choice should be defined at the outset in agreement with the planning team, so that they can be incorporated by the team into their design criteria, with a view to their aiming at options which stand a good chance of being accepted according to the agreed criteria (Lichfield et al. 1975).

20.3 Setting up terms of reference for evaluation analysts

The terms of reference for the evaluation analyst would clearly be part of, and integrated with, the terms of reference for the planning team as a whole. Within this the analyst would need to address himself to the relevant issues and questions (Box 9), and so be able to deliver conclusions and recommendations which are pertinent (Box 14). For this purpose it is necessary to probe the answers to a series of questions such as the following, which were derived largely from experience on actual consultant commissions carried out between 1966 and 1994 (see Ch. 8 and the Appendix).

20.3.1 Who is the commissioning client?

The case studies showed a considerable variety: government departments, regional agencies, local authorities, landowners, development companies, new town development corporations, industrialists, civic groups, airport authority, public transport executive, business association, conservation agency.

20.3.2 What is the planning for?

Then there is the purpose of the plan-making. Is it being offered by private individuals or a group who consider that their town or region needs a plan and wish to put their

---

1. This was done in the selection of the four final sites for the Roskill Commission evaluation; the list was reduced by stages from 78 initial possible locations (Commission on the Third London Airport 1969). In the West Midlands Regional Study, 1971, the original 107 options were reduced by stages to the final 4 considered in depth in the evaluation (NLP 1971).

money where their mouth is, and employ consultants for the purpose (Naples: 1988). Is it a landowner/developer who, in order to publicize his proposals, seeks to place them within a wider planning framework in relation to a planning application (Virginia Water 1988) or an appeal (Brentford 1984) or a plan inquiry (Brigg 1987). Is it a master plan for a new town prepared by consultants at the request of government at the initial stage of setting up a corporation to plan and build the town (Peterborough 1969). Is it a plan prepared by the local planning authority as part of the statutory planning process (Cambridge 1966 or Barnstaple 1981). Is it for options in improving an existing transport system. (Greater Manchester 1983–4 or Stevenage 1971), for choice in pedestrian crossings (1980), or in an urban shopping policy (Dublin 1980).

20.3.3 What kind of plan?
- Strategy, policy, plan, project, programme or projects.
- Scale of plan (regional to local).
- Content of plan (a new town, renewal, road building).
- Aspects of a particular kind of content (private motor-car, public transport, pedestrian crossing).

20.3.4 What variant is needed from the generic model?
Although the generic method of evaluation is common to all types, the individual analysis requires adjustment according to the circumstances. For example, the scale of plan analysis must be carried out with the level of generality which is consistent with the data available and the nature of the proposal: regional plans are concerned more with broad locations and strategies than with detailed layout and data of a more aggregated nature than for local or subregional studies.

20.3.5 What are the time/money resources available for the study?
These will condition the depth into which it is possible to go in the study itself, and will thereby influence the study programme. For example, one month was available for Ipswich (1970) and three years for Manchester (1983–4).

20.3.6 Is the decision-taker the same as the commissioning client?
Generally speaking it is, but not always. For Naples (1988) the decision-taker on the privately commissioned study invited the consultants to put themselves in the role of regional government, and draw conclusions accordingly.

20.3.7 Who are the stakeholders?
Alongside the commissioning client and decision-taker there will also be agencies who as "stakeholders" will have a say in implementing the choice of the plan following evaluation (Box K/15).[1] These need to be identified early to ensure their participation and collaboration in the study with a view to bringing out their particular interest in the evaluation.

---

1. For example, London Transport as bus company in the Stevenage study, 1969.

20.3.8 Who are the consultees?

Since in any planning process there needs to be a large array of consultations it is valuable from the outset to prepare a list of those who must be approached and those who may be. This will provide a schedule of potential inputs into the planning process and also dates that need to be adhered to for the purpose.

20.3.9 How are the public to be involved?

Since in any planning process there will need to be consultation with the public, and procedures for their participation, it is useful at the outset to identify the location of this public and also the names of relevant representative organizations. This again will influence the work programme and provide target dates for the approach.

20.4 Best use of available resources

As in any planning study, evaluation requires the time of personnel, which need to be financed from the budget made available for the evaluation. For this purpose the budget would need to be divided between the evaluation analyst per se and the remainder of the planning team called upon to give support by way of input, data, and so on. Thus, the role of the analyst in the planning team is critical for budgeting purposes. Having regard to the complexity of the evaluation process, and the importance of pointing to the "best" option, it is apparent that for any particular study there could be wide variation between the minimum and the desirable. As all planning studies are not as generously funded as they might be, it follows that the financial budget will act as a constraint upon the scope of the evaluation study itself.

For this reason, although this topic has been considered here at the end of the study management process, it should be agreed early on in order to move smoothly through an efficient study process. For this purpose, either a particular figure might be mentioned on the initial approach to the evaluation analyst, or he would be invited to carry out the reconnaissance, following which a firm budget can be suggested on the basis of the circumstances discovered.

The time and budget available for the planning study will clearly affect the quality of the evaluation process, by constraining the numbers of people involved in the process, the quality and quantity of research and information, and the number of possibilities which can be examined. Even where these constraints are tight it may be possible to consider many possibilities initially; but the options are likely to be specified at a very coarse level, and the evaluation is likely to be based on few crude parameters. There is a danger in this type of analysis that evaluation could be misleading in not being based on comprehensive analysis of all effects, impacts, sectoral objectives and data. The result may lead to rejecting a good option which might have shown up well in the comprehensive final evaluation.

Where a small time/resource budget is likely to limit the initial number of possibilities the evaluation can handle, it is useful to generate the initial list as widely as possible. This can be achieved by grouping the options into variants, and for each major variant to generate few suboptions that represent the range of possibilities for that subset. Depending on the number of options generated and the time available, there can be short-listing stages to reduce the numbers of possibilities to those that make up the final shortlist. For example, iterations after the shortlist may well be dispensed with if they involve searching for an alternative that is only marginally better. Thus, although the shortlist may not contain the truly optimum solution, the one that would be preferred if all the desired information were available, it should contain at least one alternative which is close to being satisfactory.

20.5 Best use of data

At the beginning of the evaluation analysis it is not readily known what data are available and will be required. But it is nonetheless important that the planning team be informed of the topic as soon as practicable: they can bear it in mind in the accumulation of data in the planning process, to avoid needing to go back over ground which has already been covered; they can be alerted as to which data are available and which require *ad hoc* collection for the purpose of the evaluation; they are made aware of the nature of the data required and the form in which it is to be supplied.

In this, the evaluation analyst needs to distinguish between the data that are or are not readily available, in the sense of already being in existence or not requiring significant resources to collect. The reasons are twofold: the readily available data would facilitate rapid progress with the evaluation, which would not be held up by survey, etc.; and it might be that data which are not readily available will not ultimately be needed, simply because they relate to matters not found to be significant for conclusions in the evaluation itself, and thereby could have led to a waste of resources in collection. A clear instance here relates to matters in which the options show little significant difference from the readily available data, and thereby do not need further data for clarification.

By the same token, gaps in the readily available data, or mistrust of readily available data that are not particularly reliable and therefore should not be used, should not in themselves hold up the progress of the analysis. Even if no firm data exist, useful judgements and conclusions can be drawn in what can be termed Cycle 1 of the evaluation. This could lead to Cycle 2, which arises where gaps in data give rise to uncertainty in the conclusions, which could be remedied by collecting further data. Proceeding in this way will ensure that any requirements for additional data thrown up by the analysis are clearly essential for the purpose, and thereby justifies the time and money for collection as an aid to clearer conclusions in Cycle 2.

20.6 Link to the generic method

Conclusions on the best use of available resources and data will help to determine whether the generic model for evaluation is to be used in full, or be adapted or simplified. These conclusions need to be formally expressed by reference to the generic model itself. Here only particular points are stressed.

F Box 14: What kind of report should be provided?   The culmination is a report which will present the evaluation itself and answer the questions which have been raised in Box 9 and possibly some which have not. Accordingly, it is valuable from the outset to consider the ultimate reporting package into which the evaluation will fit, for this will influence both the amount, scope and kind of material prepared for inclusion. For example, to whom is the report going to be addressed, which will decide its format: fellow academics/professionals, the planning team, Ministers, elected representatives, the public at large? Since one document cannot accommodate all the addressees just mentioned, it will be helpful to have an executive summary for Ministers and Members; a version which is suitable for the public as an environmental statement or planning statement, and technical appendices for the academic/professionals in order to support the robustness of the analyses and recommendations.

J Box 14: How communicate the decision?   Here all the various requirements in communication need to be considered (see Ch. 7). This will in itself also influence the kind of report which has just been discussed: for example, as a draft for consultation with consultees, the public and other interested bodies or a finalized version for presentation to decision-takers. If the report is visualized simply as a step towards generation of further options, either as an amalgam of those evaluated or as a base for new, then the nature of the report and its emphasis will be tentative and contributory to further work. Should the report relate to the coarse screening stage, as a basis for fine sieving, clearly it will avoid being as conclusive as if there were no further fine sieving stage. Should the report have been prepared as a basis for evidence at a public inquiry, it will need to be designed with an eye not only to its technical content but to its potential weakness under adversarial cross-examination procedures. It will also aim to be polished in its presentation, clear-cut in its views, and confident in its recommendations and opinions. There is no need to go as far as this if the report is visualized for a particular stage in a linear evaluation process, with the intention to open up an array of options for debate and to invite further suggestions.

The implications for the evaluation exercise of these approaches is clear. If there is to be no further generation after the short-listing stage, evaluation prior to short-listing must be thorough enough to ensure that the final list embraces the "best" solution, and that the major distinguishing features between alternatives are clearly identified. On the other hand, if the feasibility of generating a preferred option after the short-listing stage is left open, then the evaluation effort can be concentrated on appraisal of short-listed alternatives and subsequent options, with proportionately less time and effort being devoted to data collection and preparation.

20.7 Evaluation study programme

The conclusions from the preceding should be formulated in the evaluation study network, which fits into the overall planning study network. This will bring out the essential tasks to be performed, by whom (inside and outside the evaluation team itself), at what period, by which deadline, and so on. It would also show the points at which inputs will be expected from outside the evaluation team, and the outputs they would produce for the planning team itself.

APPENDIX

Case studies in planning balance sheet and community impact analysis

The following items are not included in the bibliography, which follows.

Authors are indicated by: NL Nathaniel Lichfield; NLA Nathaniel Lichfield & Associates; NLP Nathaniel Lichfield & Partners; DNLA: Dalia and Nathaniel Lichfield Associates.

NL   1959   Value for money in town planning. *Chartered Surveyor* **91**(9), 495–500.

NL   1962   *Cost–benefit analysis in urban redevelopment: Research Report 20.* Real Estate Research Programme, Institute of Business and Economic Research, University of California,. Berkeley.

NL   1963   with D. H. Crompton. Cost–benefit analysis and accessibility and environment in Newbury. Appendix 2 of *Traffic in towns* [the "Buchanan Report"]. London: HMSO.

NL   1965   Spatial externalities in urban expenditures: a case study of Swanley. *The public economy or urban communities: proceedings of 2nd conference on urban public expenditures, 1964,* Julius Margolis, (ed.), 204–49. Washington: Resources for the Future Inc.

NL   1966   *Cost–benefit analysis in town planning: a case study of Cambridge.* Cambridge: Cambridge County Council.

NL   1966   Cost–benefit analysis in town planning: a case study – Swanley. *Urban Studies* **3**(3), 215–49.

NLA   1966   *Worthing: cost–benefit analysis of alternative road proposals* (unpublished).

NLA   1967   Alternative strategies, in *Report and advisory plan for the Limerick Region*, vol. 2, part XII. Dublin: Stationery Office.

NLA   1967   With E. Mitchell & Sons and Arthur Ling & Associates. *Nottingham: cost–benefit analysis of alternative sets of road proposals affecting plans for redevelopment,* vol. 1 (unpublished).

NL   1968   with Honor Chapman. Cost–benefit analysis and road proposals for a shopping centre: a case study: Edgware. *Journal of Transport Economics and Policy* **2**(3), 280–320.

NLA   1968   with Economic Associates Limited. Comparisons of alternative schemes. Ch. VI in *St Kitts airport: an evaluation* (unpublished).

NLP   1969   *Cost–benefit study of the implications of an airport/seaport at Maplin Sands, report no. 5.* Submitted as evidence to the Roskill Commission by the Thames Estuary Development Company Ltd.

NL   1969   Cost–benefit analysis in urban expansion: a case study: Peterborough. *Regional Studies* **3**(2), 123–55.

NLA   1970   Cost–benefit study, *Commission on the Third London Airport: evidence submitted at Stage III to the Roskill Commission* by Bedfordshire, Cambridgeshire and the Isle of Ely, Essex and Hertfordshire County Councils, ch. 14 (unpublished).

NLA   1970   *Supplementary proof of evidence on weighting of sectors and objectives,* evidence submitted at Stage V to the Roskill Commission on behalf of Bedfordshire et al.

NL   1970   with Honor Chapman. Cost–benefit analysis in urban expansion: a case study: Ipswich. *Urban Studies* **7**(2), 163–88.

NL   1971   with Whitbread Michael Evaluation of the hypotheses, in *Israel's new towns: a development strategy,* vol. 1, ch. 11. Israel: Ministry of Housing.

NLA   1971   *Comparative planning analysis with and without Severndale: proof of evidence submitted to Local Planning Inquiry on the proposed Severndale Shopping Centre, Cribbs Causeway, Bristol* (unpublished).

NL   1971   with Honor Chapman. The urban transport problem and modal choice. *Journal of Transport Economics and Policy* **5**(3), 1–20.

NLP   1971   Planning blight in social cost–benefit analysis: the North Cross route in Camden. *Support document S27/222,* submitted in evidence to the GLDP Public Inquiry (stage 2) by the London Borough of Camden.

NLA   1972   Cost–benefit analysis, planning balance sheet, evaluation of alternatives. Technical Appendix 5 in *A developing strategy for the West Midlands: report of the West Midlands Regional Study.* Birmingham: West Midlands Study.

NLA   1973   *Comparative planning analysis with and without the Stonebridge Shopping Centre.* Proof of evidence, submitted to Local Planning Inquiry into the Proposed Stonebridge Shopping Centre, Bickenhill, Warwickshire (unpublished).

NLP   1973   *Essex County Hall: cost–benefit analysis of alternative locations* (unpublished).

NLA   1974   with A. Proudlove. A planning balance sheet analysis of six alternative routes. In *Conservation and traffic: a case study of York,* Appendix B. York: Rowntree Trust.

NLP   1974   with James Simmie. *Carnaby Newburgh Street: social cost–benefit analysis of alternative schemes* (unpublished).

NLP   1977   with W. S. Atkins & Partners. Costs and benefits of the programme from a national viewpoint. In *Morocco: economic opportunities associated with the steel complex* (unpublished).

NLP   1977   Evaluation of alternatives in the regional planning process. *Proceedings of seminar sponsored by Department of Regional Planning, the Regional Municipality of Halton, Burlington, Ontario.*

NLP   1978   *Planning balance sheet analysis of alternative oil-related industrial strategies in Scotland* (unpublished).

NLP   1980   *Towards a shopping policy for the Dublin sub-region.* Dublin: Dublin City Centre Business Association.

NLP   1980   *East Midlands Airport Runway Extension: east/west: assessment and evaluation of environmental and economic impacts* (unpublished).

NLP   1980   *East Midlands Airport.* Proof of evidence submitted to public enquiry into proposed runway extension (unpublished).

NLP   1980   with Department of Transport, Imperial College London, *Pedestrian crossing: application of social cost–benefit analysis* (unpublished).

NLP 1980 *Regional shopping centre, Burlington, New Jersey: community impact analysis.* (unpublished).

NLP 1981 *Barnstaple Town Centre Study: evaluation of the strategic elements of the options.* Barnstaple: North Devon District Council.

NLP 1983 *Achieving value for money in public Transport: Report No. 2 Trafford Park Case Study,* Vol. 1, 2 & 3. Manchester: Greater Manchester Transport.

NLP 1984 *Achieving value for money in public transport: community impact evaluation in the GMPTE Second Three Year Public Transport Plan: Report No. 6.* Manchester: Greater Manchester Transport.

NLP 1984 *Achieving value for money in public transport: the Park Hospital Manchester case study.* (unpublished).

NLP 1984 Brentford/Quay/Village. *Proof of evidence,* submitted to public inquiry against the refusal of planning permission by the London Borough of Hounslow (unpublished).

NLP 1985 Preliminary evaluation of alternative sites. *Proposed stadium, recreation and shopping complex at Peartree Hill, Oxford* (unpublished).

NL 1986 with Joseph Schweid. Sha'arey Tsedek. In *Economics in urban conservation,* N. Lichfield (ed.), ch. 15. Cambridge: Cambridge University Press 1988.

NL 1986 with Joseph Schweid. Nahalat Shiv'a. In *Conservation of the built heritage,* part III. Jerusalem: The Jerusalem Institute for Israel Studies [in Hebrew].

NLP 1986 *Hurn Airport development.* Proof of evidence submitted to public enquiry against refusal of permission (unpublished).

NL 1987 Community impact evaluation: a study in Greater Manchester Public Transport. In *Transport subsidy policy journals,* S. Glaister (ed.). London: Policy Journal.

NLP 1987 Brigg Inner Ring Road. *Proof of evidence for local plan inquiry.*

NLP 1988 De metodo di valuatzione di impato communitaro. In *Rigenerazione dei Centro Storico: Caso Napoli,* R. Di Stefano & H. Siola (eds), 589–602. Naples: Studi Centro; and Milan Edizione dei Sole.

NLP 1988 Royal Holloway Sanatorium Virginia Water: evaluation of office/residential and residential options. In *Economics in urban conservation,* N. Lichfield (ed.), ch. 16. Cambridge: Cambridge University Press.

NL 1988 The use of community impact evaluation in planning assessment: East Midlands Airport Expansion. *Proceedings of international colloquium on assessment methods in urban and regional planning: theory and case studies, Capri/Naples, April 1988* (unpublished).

NLP 1992 The integration of environmental assessment into development planning: part 2: Prospect Park *Project Appraisal* 7(3), 175–85.

NLP 1993 with ICOMOS. International Economics Committee. Renewal of Chinatown: Westminster. In *Cost benefit analysis for the cultural built heritage. Stage I report: principles and practice,* N. Lichfield et al. (eds). Paris: ICOMOS.

NLP 1993 with ICOMOS. International Economics Committee. *Cost benefit analysis in the cultural built heritage: stage II report: four case studies.* Rome: Ministry of Cultural Goods.

DNLA 1994 Torbay Ring Road: evaluation of the Valley and Plateau Routes. Proof of evidence submitted to local plan enquiry (unpublished).

DNLA 1995 Manchester Airport Second Runway: proof of evidence submitted to local planning enquiry (unpublished).

DNLA 1996 Aosta, Italy: evaluation of three alternative road routes (unpublished).

Categorization of case studies

The following is based upon the prime features of the case studies mentioned above, so bringing out the diversity of the application of CIE.

(1) Plan as a whole

	Israel	1971
	West Midlands	1971
	Limerick	1967

(2) Project

	San Francisco	1962
	Cambridge	1966
	Worthing	1966
	Nottingham	1967
	Edgware	1968
	St Kitts	1968
	Hurn	1968
	Third London Airport	1969
	Essex	1973
	Carnaby/Newburgh St	1974
	Morocco	1977
	Barnstaple	1981
	Royal Holloway Sanatorium	1988
	Prospect Park	1992

(3) Scale of plan

National	Israel	1971
Regional	Limerick	1967
	West Midlands	1971
Town	Ipswich	1970
	Peterborough	1969
Central area	Barnstaple	1981
	Edgware	1968
	Naples	1988
	Westminster	1993

(4) Kind of plan

Strategy	Israel	1971
	Scotland	1978
Policy	Dublin	1980
Plan	Limerick	1967
	Edgware	1968
	Ipswich	1970
	Peterborough	19..
	Barnstaple	1981

(5) Content

New Towns	Ipswich	1970
	Peterborough	1969
	Israel	1971
	Chinatown, Westminster	1990
Old towns	Swanley	1965
	Cambridge	1966
Regions	West Midlands	1971
	Limerick	1967

Roads	Worthing	1966
	Nottingham	1967
	Edgware	1968
	York	1974
Urban regeneration	San Francisco	1962
	Chinatown, Westminster	1993
Conservation	Barnstaple	1981
	Jerusalem	1986
	Jerusalem	1986
	Virginia Water	1988
	Naples	1988
	Naples	1993
Retailing	Cribbs Causeway	1971
	Stonebridge	1973
	Carnaby/Newburgh Street	1974
	Dublin	1980
	Burlington N.J.	1980
Leisure complex	Oxford	1985
Business park complex	Hurn Airport	1986
Green belt	Prospect Park	1992
Housing	Brentford Quay	1984
Industrial plant	Morocco	1977
Blight	Camden	1971

(6) Different aspects of transport modes

Airports	St Kitts	1968
	Maplin	1969
	Third London Airport	1970
	East Midlands	1980
	Manchester	1995
Motor vehicles	York	1974
	Brigg	1987
	Torbay	1994
	Aosta	1996
Public transport	Manchester	1983/4/7
Public vs private transport	Stevenage	1969/1971
Pedestrian crossing	General	1980
Accessibility and environment	Newbury	1963

(7) Time and resources

Within one month	Ipswich	1970
In depth over three years	Manchester	1984

Bibliography

Abelson, P. 1979. *Cost–benefit analysis and environmental problems*. Farnborough, England: Saxon House.

Ackoff, R. L. 1962. *Scientific method: optimising applied research decisions*. New York: John Wiley.

Ackoff, R. L. 1963. Towards quantitative valuation of urban services. In *Public expenditure decisions in the urban community*, H. G. Schaller (ed.), 91–117. Baltimore: Johns Hopkins University Press.

ACTRA (Advisory Committee of Trunk Road Assessment) 1977. *Report of the Advisory Committee on Trunk Road Assessment* ["Leitch Report"]. London: HMSO.

Alexander. E. R. 1986. *Approaches to planning: introducing current planning theories, concepts and issues*. London: Gordon & Breach.

Alexander, I. 1974. *City redevelopment: an evaluation of alternative approaches*. Oxford: Pergamon.

—1978. The planning balance sheet: an appraisal. In *Australian project evaluation: selected readings*, J. C. Manchester & G. A. Webb (eds), 46–68. Sydney: Australia and New Zealand Book Company.

Allison, L. 1975. *Environmental planning: a political philosophy analysis*. London: George Allen & Unwin.

Alterman, R. & M. Hill 1978. Implementation of urban land use plans. *Journal of the American Institute of Planners* **44**(3), 274–85.

Amos, F. J. C. 1972. Structure plans & corporate planning. *Journal of Planning & Environment Law*, 400–404.

Archibugi, F. 1993. *Introduction to planology*. Rome: Centro di Studi e Piani Economici.

Arrow, K. J. 1967. Values and collective decision-making. In *Philosophy, politics and society*, P. Laslett & W. G. Runciman (eds), 215–32. Oxford: Basil Blackwell.

Ashby, W. R. 1956. *An introduction to cybernetics*. London: Chapman & Hall.

Ashworth, W. 1954. *Genesis of modern British town planning*. London: Routledge & Kegan Paul.

Atkins, R. 1984. A comparative analysis of the utility of EIA methods. In Clark et al. (1984: 241–52).

Attfield, R. & K. Bell (eds) 1989. *Values, conflict and the environment: report of the Environmental Ethics Working Party*. Ian Ramsey Centre, St Cross College, Oxford and the Centre for Applied Ethics, University of Wales College of Cardiff.

Audit Commission, 1992. *Building in quality: a study of development control*. London: HMSO.

Baier, K. 1969. What is value? An analysis of the concept. In *Values and the future*, K. E. M. Baier & N. Rescher (eds), 33–68. New York: Free Press.

Ball, S. & S. Bell 1991. *Environmental law: the law and policy relating to the protection of the environment*. London: Blackstone Press.

Barde, J. P. & D. W. Pearce 1991. *Valuing the environment*. London: Earthscan.

Barlow see Royal Commission 1940.

Barrett, S. & C. Fudge (eds) 1981. *Policy and action: essays on the implementation of public policy*. London: Methuen.

Batey, P. W. H. & M. J. Breheny 1978. Systematic methods in British planning practice. *Town Planning Review* **49**(3), 257–318.

Bator, F. 1958. The anatomy of market failure. *Quarterly Journal of Economics* **72**, 33–46.

Batty, M. 1976. *Urban models: algorithms, calibration, prediction.* Cambridge: Cambridge University Press.

Baum, W. C. 1982. *The project cycle.* Washington DC: World Bank.

Baum, W. C. & M. T. Tolbert 1985. *Investing in development: lessons on World Bank experience.* Oxford: Oxford University Press.

Baumol, W. J. & W. E. Oakes 1979. *Economics, environmental policy and the quality of life.* Englewood Cliffs, NJ: Prentice Hall.

Beanlands, G. 1988. Scoping methods and baseline studies in EIA. In Wathern (1988: 33–46).

Beer, S. 1966. *Decision and control: the meaning of operational research and management cybernetics.* London: John Wiley.

Beesley, M. E. & C. D. Foster 1965. The Victoria Line: social benefit and finances. *Journal of the Royal Statistical Society* **128**(1), 67–88.

Beesley, M. E. & P. Gist 1983. *Cost–benefit analysis and London's transport policy.* Oxford: Pergamon.

Ben Shahar H, A. Mazor, D. Pines 1969. Town planning and welfare maximisation: a methodological approach. *Regional Studies* **3**(2),105–113.

Bentham, J. 1948 (1789). *An introduction to the principles of moral legislation.* New York: Hafner Press.

Bergson, A. 1938. A reformulation of certain aspects of welfare economics. *Quarterly Journal of Economics* **52**(February), 310–34.

Bingham, G. 1986. *Resolving environmental disputes: a decade of experience.* Washington DC: The Conservation Foundation.

Birch, D. 1977. *The community analysis model.* Mimeograph.

Bisset, R. 1988. Development in EIA methods. In Wathern (1988: 47–61).

Bisset, R. & P. Tomlinson 1988. Monitoring and auditing of impacts. In Wathern (1988: 117–128).

Blowers, A. (ed.) 1993. *Planning for a sustainable environment.* London: Earthscan.

Boulding, K. 1952. Welfare economics. In *A survey of contemporary economics*, B. F. Haley (ed.), 1–38. Illinois: Holmwood.

Branch, M. 1966. *Planning aspects and applications.* New York: John Wiley.

Branch, M. C. & I. M. Robinson 1968. Goals and objectives in civil comprehensive planning. *Town Planning Review* **38**(4), 261–74.

Braybrooke, D. & C. E. Lindblom 1963. *A strategy of decision.* London: Collier–Macmillan.

Breheny, M. J. 1984. *Urban policy impact analysis.* London: Economic and Social Research Council.

Breheny, M. & A. Hooper (eds) 1985. *Rationality in planning: critical essays on the role of rationality in urban and regional planning.* London: Pion.

Bridger, G. 1986. Rapid project appraisal. *Project Appraisal* **1**(4), 243–8.

Bridger, G. A. & J. T. Winpenny 1983. *Planning development for projects: a practical guide to the choice and appraisal of public sector investment.* London: HMSO.

Broniewski, S. & S. Jastrzebski 1970. Optimization method. In *Analytical techniques in the urban and regional planning process: threshold analysis, optimization method.* Planning Research Unit, University of Edinburgh.

Bruton, M. J. 1983. *Bargaining and the development control process.* Papers in Planning Research 60, Department of Town Planning, University of Wales, Cardiff. [A summary is given in Bruton & Nicholson (1987).].

Bruton, M. J. (ed.) 1984. *The spirit and purpose of planning.* London: Hutchinson.

Bruton, M. & D. Nicholson 1987. *Local planning in practice.* London: Hutchinson.

Buchanan, C. & Partners 1966. *South Hampshire Study.* London: HMSO.

Buchanan, J. M. & R. D. Tomlinson (eds) 1972. *Theory of public choice: potential application of economics.* Ann Arbor: University of Michigan Press.

Buchanan, J. M. [as contributor] et al. 1978. *The economics of politics.* London: Institute of Economic Affairs.

Burchell, R. W. & D. Listokin 1975. *The environmental impact handbook.* Centre for Urban Policy Research, Rutgers University.

Cadman, D. & L. Austin-Crowe 1993. *Property development.* London: Spon.

Cambridgeshire County Council et al. 1969. *dence submitted at stage III to the Commission on the Third London Airport*. [Unpublished; part of Roskill Commission papers].

Campbell, A. & P. Converse 1972. *The human measuring of social change*. New York: Russell Sage Foundation.

Carley, M. 1980. *Rational techniques in policy analysis*. London: Heinemann.

Carley, M. J. & E. S. Bustelo 1984. *Social impact assessment and monitoring: a guide to the literature*. Boulder, Colorado: Westview Press.

Caso, F. G. (ed.) 1977. *Readings in evaluation research*. New York: Russell Sage Foundation.

Catlow, J & C. G. Thirwell 1976. *Environmental impact analysis* [Research Report 11]. London: DOE.

CDPUK 1983. *Earth survival. the conservation and development programme for the UK: a response to world conservation*. London: Routledge & Kegan Paul.

Chadwick, G. 1978. *A systems view of planning: towards a theory of the urban and regional planning process*, 2nd edn. Oxford: Pergamon.

Chatzimikis, F. 1983. *Community development block grants*. Washington DC: Department of Housing and Urban Development.

Checkofway, B. (ed.) 1986. *Strategic perspective on planning practice*. Lexington: Lexington Books.

Cheung, S. N. S. 1978. *The myth of social cost: a critique of welfare economics and the implications of public policy*. London: Institute of Economic Affairs.

Churchman, C. W. & P. Ratoosh (eds) 1959. *Measurement: definitions and theories*. New York: John Wiley.

Ciriacy-Wantrup, S. V. 1952. *Research conservation: economics and policies*. Berkeley: University of California Press.

Clark, B. D. 1988. Environmental impact assessment: on the eve of legal implementation. *The Planner* **74**(2), 18–22.

Clark, B. D., K. Chapman, R. Bisset, P. Wathern 1976. *Assessment of major investment applications: a manual* [Research Report 13]. London DOE.

— 1978. *Environmental impact assessment in the USA: a critical review* [Research Report 26]. London: DOE.

Clark, B. D., R. Bisset, P. Wathern 1980. *Environmental impact assessment: a bibliography with abstracts*. London: Mansell.

Clark, B. D., K. Chapman, R. Bisset, P. Wathern, M. Barrett 1981. *A manual for the assessment of industrial development*. London: HMSO.

Clark, B. D., R. Bisset, P. Tomlinson (eds) 1984. *Perspectives on environmental impact assessment*. Dordrecht: Reidel.

Coase, R. 1960. The problems of social cost. *Journal of Law and Economics* **3**(October), 1–44.

Cochrane, J. L. & M. Zeleny (eds) 1973. *Multiple criteria decision making*. Columbia: University of South Carolina Press.

Coddington, A. 1968. *Theories of the bargaining process*. London: George Allen & Unwin.

Cohen, E. & A. Ben Arie 1993. Hard choices: a strategic analysis of value incommensurability. *Human Studies* **16**.

Cole, K. J. Cameron, C. Edwards 1983. *Why economists disagree: the political economy of economics*. Harlow: Longman.

Collins, B. J. 1950. *Development plans simply explained*. London: HMSO.

Commission for the Third London Airport 1969. *Papers and proceedings* ["Roskill Commission"]. London: HMSO.

— 1971. *Report* ["Roskill Report"]. London: HMSO.

Commission of the European Communities 1985. Council Directive of 27 June 1985 on the assessment of the effects of certain public and private projects on the environment. *Official Journal* **L175**, 40–48.

— 1992. *Towards sustainability: a European community programme of policy and action in relation to the environment and sustainable development: 5th environment action plan*. Brussels: EEC.

Committee on the Qualifications of Planners 1950. *Report* ["Schuster Report"; Cmnd 8059]. London: HMSO.

Committee of Inquiry 1986. Town and Country Planning. London: Nuffield Foundation.

Council of Environmental Quality (CEQ). *Annual reports on environmental quality*. Washington, DC: Government Printing Office.
Cracknell, B. E. (ed.) 1984. *The evaluation of aid projects and programmes*. London: Overseas Development Administration.
Cullingworth, J. B. 1975. *Environmental planning 1939–69*, vol. 1: *reconstruction and land use planning 1939–47*. London: HMSO.
Cullingworth, J. B. & V. Naden 1994. *Town and country planning in Britain*, 11th edn. London: Routledge.

Daley, H. (ed.) 1980. *Economics, ecology and ethics: essays towards a steady state economy*. San Francisco: W. H. Freeman.
Darlow, W. C. (ed.) 1982. *Valuation and development appraisal*. London: Estates Gazette.
— (ed.) 1988. *Valuation and investment appraisal*. London: Estates Gazette.
Dasgupta, A. K. & D. W. Pearce 1972. *Cost–benefit analysis in theory and practice*. London: Macmillan.
Dasgupta, P. S., A. K. Sen, A. Marglin 1972. *Guidelines for project evaluation*. New York: United Nations.
Davies, H. W. E., D. Edwards, A. J. Hooper, J. V. Punter 1989. *Planning control in western Europe*. London: HMSO.
Davies, H. W. E., D. Edwards, A. R. Rowley 1986a. *The relationship between development plans, development control and appeals*. Working Paper 10, Department of Land Management and Development, University of Reading.
— 1986b. The relationship between development planning, development control and appeals. *The Planner* 72(10), 11–15. [This is a summary of Davies et al. 1986a.]
Davies, H. W. E. & H. Sia 1994. *The impact of the European Community on land use planning in the United Kingdom* [2 vols]. London: Royal Town Planning Institute.
De Jongh, P. 1988. Uncertainty in EIA. In Wathern (1988: 62–84).
Delafons, J. 1969. *Land use controls in the United States*. Cambridge: MIT Press.
Denman, D. R. 1984. *Markets under the sea*. London: Institute of Economic Affairs.
Dobben, K. 1990. *The evolution of an evaluation method: from PBSA to CIE*. Werkstukken 125, Department of Planning and Geography, University of Amsterdam.
Dobry, G. 1974. *Review of the development control system: interim report*. London: HMSO.
— 1975. *Review of the development control system: final report* [Cmd 7133]. London: HMSO.
DOE 1955. *Green belts* [Circular 42/55]. London: HMSO.
— 1970. *Development plans: a manual on form and content*. London: HMSO.
— 1972. *Structure plan notes*. Unpublished.
— 1983a. *Planning gain* [Circular 22/83]. London: HMSO.
— 1983b. *Land for housing* [Draft Circular]. London: HMSO.
— 1984a. *Green belts* [Circular 14/84]. London: HMSO.
— 1984b. *Land for housing* [Circular 15/84]. London: HMSO.
— 1984c. *Memorandum on structure and local plans* [Circular 22/84]. London: HMSO.
— 1985. *The use of conditions in planning permission* [Circular 1/85]. London: HMSO.
— 1986a. *An evaluation of industrial commercial improvement areas*. London: HMSO.
— 1986b. *Assessment of the employment effects of economic development projects funded under the urban programme*. London: HMSO.
— 1986c *Evaluation of environmental projects funded under the urban programme*. London: HMSO.
— 1987. *Evaluation of derelict land grant schemes*. London: HMSO.
— 1988a. *Green belts* [Planning Policy Guidance Note 2]. London: HMSO.
— 1988b. *Environmental assessment* [Circular 15/88]. London: HMSO.
— 1988c. *Integrated pollution control: a consultation paper*. London: DOE.
— 1988d. *An evaluation of the urban development grant programme*. London: HMSO.
— 1988e. *Environmental assessment* [Circular 15/88]. London: HMSO.
— 1989a. *Environmental assessment: guide to the procedure*. London: HMSO.
— 1989b. *Planning agreements* [Draft Guidance para. 3]. London: (DOE).
— 1991a. *Policy appraisal and the environment*. London: HMSO.
— 1991b. *Planning obligations* [Circular 16/91]. London: HMSO.

—1992a. *General policy and principles* [Planning Policy Guidance Note 1]. London: HMSO.

—1992b. *Development plans and regional planning guidance* [Planning Policy Guidance Note 12]. London: HMSO.

—1992c. *Responsibilities for conservation and casework* [Circular 20/92]. London: HMSO.

—1992d. *This common inheritance: second year report* [Cmd 2069]. London: HMSO.

—1993a. *Making markets work for the environment*. London: HMSO.

—1993b. *Environmental appraisal of development plans: a good practice guide*. London: HMSO.

—1993c. *Integrated pollution control: a practice guide*. London: HMSO.

—1993d. *Evaluation of Urban Development Grant, Urban Regeneration Grant and City Grant*. London: HMSO.

—1994a. *Guide on preparing environmental statements for planning projects* [draft]. London: The Department.

—1994b. *Planning and pollution control* [Planning Policy Guidance Note 23]. London: HMSO.

—1995. *Green belts* [Planning Policy Guidance Note 2 (revised)]. London: HMSO.

DOE and DNH 1994. *Planning and the historic environment* [Planning Policy Guidance Note 15]. London: HMSO.

DOT 1986. *The Government's response to the SACTRA report on urban road appraisal*. London: HMSO.

—1992. *The Government's response to the SACTRA report on assessing the environmental impact of road schemes*. London: HMSO.

DOT et al. 1993. *Design manual for road and bridges*, vol. 11: *Environmental appraisal*. London: HMSO.

Dorfman, R. (ed.) 1965. *Measuring benefits in government investments*. Washington DC: Brookings Institution.

Downs, A. 1957. *An economic theory of democracy*. New York: Harper.

Downs, A. 1962. The public interest: its meaning in a democracy. *Social Research* **29**.

Dror, Y. 1971. *Design for policy science*. New York: Elsevier.

DTI 1985. *Burdens on business: report of a scrutiny of administrative and legislative requirements*. London: HMSO.

—1988. *DTI – The department for enterprise*. London: HMSO.

—1989. *British business*. London: Department of Trade and Industry.

Dupuit, J. 1844. On the measurement of utility of public works. *International Economic Papers* **2** [translated from the French].

Eckstein, O. 1958. *Water resource development: the economics of project evaluation*. Cambridge, Mass.: Harvard University Press.

Edel, M. 1980. "People" versus "places" in urban impact analysis. In *The urban impacts of federal policy*, N. J. Glickman (ed.), 175–91. Baltimore: Johns Hopkins University Press.

Edwards, W. & J. R. Newman 1982. *Multi-attribute evaluation*. London: Sage.

Edwards, W. & A. Tversky (eds) 1967. *Decision-making: selected readings*. Harmondsworth: Penguin.

Eilon, S. 1971. *Management control*. London: Macmillan.

Ekins, P. (ed.) 1986. *The living economy: a new economics in the making*. London: Routledge & Kegan Paul.

Elson, M. 1986. *Green belts: conflict mediation in the urban fringe*. London: Heinemann.

Encel, S., P. K. Marstrand, W. Page (eds) 1975. *The art of anticipation: values and methods in forecasting*. London: Martin Robertson.

Esher, Lord & Lord Llewelyn-Davies 1968. The architect in 1988. *RIBA Journal* **75**(10), 448–55; **75**(11), 491–4.

Expert Committee on Compensation and Betterment 1942. ["Uthwatt Report"]. London: HMSO.

Fabrick, M. N. & J. J. O'Rourke 1982. *Environmental planning for design and construction*. New York: John Wiley.

Faludi, A. 1973. *Planning theory*. Oxford: Pergamon.

—(ed.) 1977. *Recent developments in planning methodology: report on the colloquium*. The Hague: Ministry of Housing and Physical Planning.

— 1986. *Critical rationalism and planning methodology*. London: Pion.

— 1987. *Decision-centred view of environmental planning*. Oxford: Pergamon.

Faludi, A. & H. Voogd 1985. *Evaluation of complex policy problems*. Delft: Delftsche Uitgers Maatschappij.

Feilden, B. M. 1982. *Conservation of historic buildings*. London: Butterworth.

Finsterbusch, K. 1985. State of the art in social impact assessment. *Environment and Behaviour* **17**.

Finsterbusch, K., L. G. Llewellyn, C. P. Wolf 1983. *Social impact assessment methods*. Beverley Hills: Sage.

Fisher, A. C. 1981. *Resource and environmental economics*. Cambridge: Cambridge University Press.

Fitzsimmons, S., S Lorrie, P. Woolf 1977. *Social assessment manual: a guide to the preparation of the social well being account for planning water resource projects*. Boulder, Colorado: Westview.

Fleming, M. 1952. A cardinal concept of welfare. *Quarterly Journal of Economics* **66**(3), 366–84.

Foster, C. D. & M. E. Beesley 1987. Estimating the social benefits of constructing an underground railway in London. *Journal of Royal Statistical Society*.

Foster, C. D. 1970. Social welfare functions in CBA. In *Operations research in the social sciences*, J. R. Lawrence (ed.), 305–318. London: Tavistock.

Friedmann, J. 1973. The public interest and community participation: towards a reconstruction of public philosophy. *Journal of the American Institute of Planners* **39**(1), 2–12.

Friedmann, J. 1987. *Planning in public domain: from knowledge to action*. Princeton, NJ: Princeton University Press.

Friedman, M. 1953. The methodology of positive economics. In *Essays in positive economics*, M. Friedman, 3–23. Chicago: University of Chicago Press.

Friedrich, C. (ed.) 1962. *The public interest*. New York: Atherton Press.

Friend, J. E. & W. M. Jessop 1977. *Local government and strategic choice: an operational research approach to the process of public planning*. Oxford: Pergamon.

Friend, J. & A. Hickling 1987. *Planning under pressure: the strategic choice approach*. Oxford: Pergamon.

Frost, R. et al. n.d. *Directory of environmental impact statements July 1988 – September 1993*. School of Planning, Oxford Brookes University.

Gatenby, I. & C. Williams 1992. Section 54A: the legal and practical implications. *Journal of Planning and Environment Law* **12**(2), 110–20.

Geddes, M. 1988. Social audits and social accounting in the UK: a review. *Regional Studies* **22**(1), 60–63.

Gillie, F. B. & P. L. Hughes 1950. *Some principles of land planning*. Liverpool: Liverpool University Press.

Glanford Borough Council 1987. *Brigg local plan*. Glanford: The Council.

Glasson, J., R. Therivel, A. Chadwick 1994. *Introduction to environmental impact assessment*. London: UCL Press.

Glickman, N. J. 1980a. Methodological issues and prospects for urban impact analysis. In *The urban impacts of federal policies*, N. Glickman (ed.), 3–32. Baltimore: Johns Hopkins University Press.

— 1980b. *The urban impacts of federal policies*. Baltimore: Johns Hopkins University Press.

— (ed.) 1980c. Urban impact analysis, *Built Environment* **6**(2), 85–91.

Gold, R. L. 1978 Linking social with other impact assessments. In Jain & Hutchings (1978: 105–116).

Grant, E. L. & G. W. Ireson 1960. *Principles of engineering economy*. New York: Ronald Press.

Grant, M. 1982. *Urban planning law*. London: Sweet & Maxwell.

— 1986. *Urban planning law supplement*. London: Sweet & Maxwell.

— (ed.) 1983. Guidelines for planning gain. *Journal of Property and Environment Law*, 427–31.

GMPTE 1984. *Public transport plan 1985/6–1987/8*. Manchester: Greater Manchester Passenger Transport Executive.

Gregory, R. L. (ed.) 1987. *The Oxford companion to the mind*. Oxford: Oxford University Press.

Haig, N. 1984. *EEC environmental policy and Britain*. London: Environmental Data Services.

Halevy, E. 1972. *The growth of philosophic radicalism*. London: Faber & Faber.

Hall, P. 1981. *Great planning disasters*. Harmondsworth: Penguin.

— 1988. The coming revival in town and country planning. *RSA Journal* **86**.

Hall, P., H. Gracey, R. Drewett, R. Thomas 1973. *The containment of urban England* [2 vols]. London: George Allen & Unwin.

Hardin, G. 1984. The tragedy of the commons. In Markyanda & Richardson (1992: 60–70).

Harrison, A. J. & P. J. Mackie 1973. *The comparability of cost–benefit and financial rates of return*. London: HMSO.

Harrison, A. J. 1977. *Economics and land use planning*. London: Croom Helm.

Harrison, M. L. & R. Mordey (eds) 1987. *Planning control: philosophies, prospects and practice*. London: Croom Helm.

Harte, G. 1986. Social accounting in the local economy. *Local Economy* **1**(1), 45–56.

Haughton, G. 1987. Constructing a social audit – putting the regional multiplier into practice. *Town Planning Review* **58**(3), 255–66.

Haughton, G. 1988. Impact analysis – the social audit approach. *Project Appraisal* **3**(1), 21–5.

Hayek, F. A. 1944. *The road to serfdom*. London: George Routledge.

— 1978. *New studies in philosophy, politics, economics and the history of ideas*. London: Routledge & Kegan Paul.

Healey, P., G. McDougall, M. J. Thomas 1982. *Planning theory: prospects for the 1980s*. Oxford: Pergamon.

Healey, P., M. Purdue, F. Ennis 1992. *Gains from planning: dealing with the impacts of development*. York: Joseph Rowntree Foundation.

Heap, D. 1973. *Town and country planning or how to control land development*. Chichester: Barry Rose.

Hetman, F. 1977. *Society and the assessment of technology*. Paris: Organisation for Economic Co-operation and Development.

Hendon, W. S. 1975. *Economics for urban social planning*. Salt Lake: University of Utah Press.

Herbert, D. T. & J. W. Raine 1976. Defining communities within urban areas: an analysis of alternative approaches. *Town planning review* **47**(4), 325–38.

Hill, M. 1966. *A method for evaluating alternative plans: the goals achievement matrix applied to transportation plans*. PhD thesis, Graduate School of Arts & Sciences, University of Pennsylvania.

— 1968. A goals-achievement matrix in evaluating alternative plans. *Journal of the American Institute of Planners* **34**(1), 19–28. Also in Shefer & Voogd (1990: 3–29).

— 1973. *Planning for multiple objectives*. Monograph Series 5, Regional Science Research Institute, Philadelphia.

1985. Decision making, context and strategies in evaluation. In *Evaluation of complex policy problems*, A. Faludi & H. Voogd (eds), 9–34. Delft: Delftsche Vitgevers Maatschappij.

Hill, M. & Y. Tsamir 1972. Multi-dimensional evaluation of regional plans serving multiple objectives. *Papers of Regional Science Association* **29**.

HM Treasury 1991. *Economic appraisal in central government: a technical guide for government departments*. London: HM Treasury.

Holmes, J. C. 1972. An ordinal method of evaluation. *Urban Studies* **9**(2), 179–91.

House of Commons 1988. *Town and country planning (assessment of environmental effects) regulations* [SI 1199]. London: HMSO.

— 1976–7. *Planning procedures* [eighth report from the Expenditure Committee]. London: HMSO.

— 1977–8. *Planning procedures* [eleventh report from the Expenditure Committee]. London: HMSO.

House of Commons Environment Committee 1986. *Planning appeals, called in and major public inquiries*, vol. 1: *Minutes of evidence*; vol. 2: *Appendices*. London: HMSO.

Hughes, J. T. & J. Koslowski 1968. Threshold analysis: economic tool for urban and regional planning. *Urban studies* **5**(2), 132–43.

Humberside County Council 1986. *Brigg local plan: inner relief road: technical report 152*. Beverley: The County Council.

Hyde, F. 1947. Utilitarian town planning 1825–1845. *Town Planning Review* **19**(3/4), 153–9.

Inter-Agency River Basin Committee, Sub-Committee on Costs and Benefits, 1950. *Proposed practices for economic analysis of river based projects*. Washington DC: Government Printing Office.

Ireland, F. 1973. Best practicable means: interpretation. In *Annotated reader in environmental planning and management*, T. O'Riordan & K. Turner (1973: 446–52). Oxford: Pergamon.

Jacobs, M. 1991. *The green economy*. London: Pluto Press.
Jain, R. K. & B. L. Hutchings (eds) 1978. *Environmental impact analysis: emerging issues in planning*. Chicago: University of Illinois Press.
Janssen, R. 1990. A support system for environmental decisions. In Shefer & Voogd (1990: 159–173).
Janssen, R., P. Nijkamp, H. Voogd 1984. Environmental policy analysis: which method for which problem. *Revue d'Economie Regionale et Urbaine* 5.
Jenkins, W. I. 1978. The case of non-decisions. In *Decision-making: approaches and analyses: a reader*, A. G. McGrew & M. J. Wilson (eds), 318–26. Manchester: Manchester University Press.
Jowell, J. 1975. Development control. *Political Quarterly* 46, 63–83.
— 1977a. The limits of law in urban planning. *Current Legal Problems* 30, 63–83.
— 1977b. Bargaining in development control. *Journal of Planning and Environment Law*, 414–33.
JURUE 1977. The significance of contrary determination in development control. In *Issues in development control*, Working Paper 1, Joint Unit for Research in the Urban Environment, University of Aston.

Kaldor, N. 1939. Welfare comparison of economics and independent comparison of utility. *Economic Journal* 49, 549–52.
Kaplan, A. 1964. *The conduct of an inquiry: methodology for behavioural science*. San Francisco: Chandler.
Kapp, W. K. 1950. *The social costs of private enterprise*. Cambridge: Harvard University Press.
— 1963. *The social costs of business enterprise*. Bombay: Asia Publishing House.
Keeble, L. B. 1969. *Principles and practice of town and county planning*. London: Estates Gazette.
Keeney, R. L. & H. Raitta 1976. *Decisions with multiple objectives: preferences and value trade-offs*. New York: John Wiley.
Kendall, M. G. (ed.) 1971. *Cost–benefit analysis: symposium*. London: English Universities Press.
Keogh, G. 1985. The economics of planning gain. In *Land policy: problems and alternatives*, S. M. Barrett & P. Healey (eds), 203–228. Aldershot: Gower.
Khakee, A. 1991. Scenario construction for urban planning. *Omega International Journal of Management Science* 19, 409–469.
Kitching, L. C. 1963. Conurbation into city region: how should London grow? *Journal of the Town Planning Institute* 49(9), 316–20.
— 1969. Regional planning considerations. In Cambridgeshire County Council (1969: ch. 2).
Kluckholm, C. 1962. *Culture and behaviour*. New York: Free Press.
Knox, P. L. 1989. *The design professions and the built environment*. London: Croom Helm.
Kozlowski, J. 1968. Threshold theory and the sub-regional plan. *Town Planning Review* 39(2), 99–116.
Kozlowski, J. & J. T. Hughes 1967. Urban threshold theory and analysis. *Journal of the Town Planning Institute* 53(2), 55–60.
Kreditor, A. 1967. The provisional plan. In *Industrial development and the development plan*. Dublin: An Foras Forbartha.
Krutilla, J. B. & O. Eckstein 1958. *Multipurpose river development: studies in applied economic analysis*. Baltimore: Johns Hopkins University Press.

Lancaster, K. 1971. *Consumers demand: a new approach*. New York: Columbia University Press.
Lapides, D. N. (ed.) 1971. *Encyclopaedia of environmental science*. New York: McGraw-Hill.
Layard, R. 1970. *Cost–benefit analysis*. Harmondsworth: Penguin.
Layfield, F. 1990. Material considerations in planning: the law. In *Material considerations in town and country planning decisions. Journal of Planning and Environment Law* [Occasional Paper 17].
Lee, N. 1983. Environmental impact assessment: a review. *Applied Geography* 3, 5–27.
Lee, N. & F. Walsh 1992. Strategic environmental assessment: an overview. In *Special issue on strategic environmental assessment*, N. Lee (ed.). *Project Appraisal* 7(3), 126–36.
Lee, N. & C. Wood 1978. EIA – a European perspective. *Built Environment* 4(2), 101–110.
Leopold, L. B., F. E. Clarke, B. B. Hanshaw, J. R. Balsley 1971. *A procedure for evaluating environmental*

impact. US Geological Survey Circular 645, US Department of the Interior, Washington DC.

Levitt, R. & J. J. Kirlin (eds) 1985. *Managing developments through public/private negotiations.* Washington DC: Urban Land Institute and The American Bar Association.

Lewis, A. 1955. *The theory of economic growth.* London: George Allen & Unwin.

Lichfield, D. 1987. *Brigg local plan: the inner ring road: proof of evidence for local plan inquiry.* Unpublished. [See also Nash (1993) for an account of the income distribution aspects of this study.]

— 1989. Planning gain: in search of a concept. *Town and Country Planning* **58**(6), 178–80.

— 1992. Making the assessment link. *Planning* (975), 4–5.

— 1994. Assessing project impacts as though people mattered. *Planning* (1058), 20–29.

Lichfield, N. 1956. *Economics of planned development.* London: Estates Gazette.

 1959. Value for money in town planning. *Chartered Surveyor* **91**(9), 495–500.

 1960. Cost benefit analysis in city planning. *Journal of the American Institute of Planners* **26**(4), 273–9.

— 1964a. Cost–benefit analysis in plan evaluation. *Town Planning Review* **35**(2), 159–169.

— 1964b. Implementing Buchanan: problem of social choice. *Traffic Engineering and Control* **6**(8), 498–500.

— 1964c. Compensation and betterment. In *Report of the Town and Country Planning School.* London: Royal Town Planning Institute.

— 1966. Cost–benefit analysis in town planning: possibilities and problems. In *Operational research in the social sciences*, J. R. Lawrence (ed.), 337–45. London: Tavistock.

— 1967. Scope of the regional plan. *Regional studies* **1**, 11–16.

— 1968a. Goals in planning. In *Report of Town and Country Planning Summer School.* London: Royal Town Planning Institute.

— 1968b. Economics in town planning: a basis for decision making. *Town Planning Review* **39**(1), 5–20.

— 1970. Evaluation methodology of urban and regional plans: a review. *Regional Studies* **4**(2), 151–65.

— 1971. Cost–benefit analysis in planning: a critique of the Roskill Commission. *Regional Studies* **5**, 157–83.

— 1977. Methods of plan and project evaluation. In Faludi (1977: 137–160).

 1978. National strategy for the new-town program in Israel. In *International urban growth policies: new town contribution*, G. Golany (ed.), 163–80. Chichester: Wiley–Interscience.

 1980. Land policy: seeking the right balance in Government intervention an overview. *Urban Law and Policy* **3**, 193–203.

— (ed.) 1983. *The urban impacts of national policies: proceedings of OECD expert meetings.* Paris: OECD.

— 1985. From impact assessment to impact evaluation. In Faludi & Voogd (1985: 51–66).

— 1987a. *Achieving value for money in conservation of the cultural built heritage.* ICOMOS Information 2, ICOMOS, Paris.

— 1987b. Community impact evaluation: a case study in Greater Manchester public transport. In *Transport subsidy*, S. Glaister (ed.), 84–99. London: Policy Journals.

— 1988a. Il metodo di valutazione di impatto communitario: parte terza. In *Rigenerazione die centri storico: il Caso Napoli*, R. di Stefano & H. Siola (eds), 589–602. Milan: Edizione del Sole.

— 1988b. The cost–benefit approach in plan evaluation [in Italian]. In *Metodi di valuazione nella pianificazioni urbana e territoriate: terria e cosi di studio*, A. Barbanente (ed.). Bari: Istituto perla Residenzia e le Infrastrutture Sociali.

— 1988c. *Economics in urban conservation.* Cambridge: Cambridge University Press.

— 1988d. Environmental impact assessment in project appraisal in Britain. *Project Appraisal* **3**(3), 133–41.

— 1989. From planning gain to community benefit. *Journal of Planning and Environment Law*, 68–78.

— 1990a. Plan evaluation methodology: comprehending the conclusions. In Shefer & Voogd (1990: 79–97).

— 1990b. Dialogue in development planning: the changing dimension. In *Current issues in planning*, S. Trench & O. Taner (eds), 187–97. Aldershot: Gower.

— 1990c. Rapporto sull' analisi costi–benefici per il patrimonio culturale costruito. *Restauro* XIX(111–12), 4–134.

— (ed.) 1992. Transportation evaluation. *Project Appraisal* **7**(4), 195–205

— 1992a. What is the balance sheet approach? *Town and Country Planning* **61**(5), 156.

— 1992b. The integration of environmental assessment into development planning, part 1: some principles. *Project Appraisal* 7(2), 58–66.

— 1992c. From planning obligations to community benefit. *Journal of Planning and Environment Law* 1103–18.

— 1993. La conservazione dell' ambiente construito e lo sviluppo: verso un valore culturale totale. In *Estimo ed economia ambientali: le nuove fiontiere nel compo dela valuatzione: studi in onore di Carlo Forte*, L. F. Girard (ed.), 84–112. Milano: Franco Angeli.

— 1994. Community impact evaluation. *Planning Theory Milano* 12, 55–79.

— 1995. Linking *ex ante* and *ex post* in evaluation. In *Proceedings of Conference of the uk Evaluation Society 1995* (forthcoming).

Lichfield, N. & J. Margolis 1963. Benefit–cost analysis as a tool in urban government decision making. In *Public expenditure decisions in the urban community*, H. G. Schaller (ed.), 118–46. Washington DC: Resources for the Future.

Lichfield, N. & H. Chapman 1970. Cost–benefit analysis in urban expansion: a case study, Ipswich. *Urban Studies* 7(2), 153–88.

Lichfield, N., P. Kettle, M. Whitbread 1975. *Evaluation in the planning process*. Oxford: Pergamon.

Lichfield, N. & U. Marinov 1977. Land use planning and environmental protection: convergence or divergence. *Environment and Planning A* 9(8), 985–1002.

Lichfield, N. & H. Darin-Drabkin 1980. *Land policy in planning*. London: George Allen & Unwin.

Lichfield, N. & J. Schweid 1986. *Conservation in the built heritage* [in Hebrew]. Jerusalem: Jerusalem Institute of Israel Studies.

Lichfield, N., W. Hendon, P. Nijkamp, A. Realfonzo, P. Rostirolla 1990a. *Cost–benefit analysis in the conservation of the cultural built heritage: theory and principles* ["Lichfield Report"]. Rome: Ministero dei Beni Culturali.

— 1990b. Rapporto sull' analisi costi–benefici per il patrimonio culturale costruito, I. *Restauro* XIX(111–2), 5–21. [For version in English, see Lichfield et al. 1993].

— 1992. Rapporto sull' analisi costi–benefici per il patrimonio culturale costruito, II. *Restauro* XXI(122), 4–138. [An English version can be found in Lichfield et al. 1993.]

Lichfield, N. & D. Lichfield 1992. The integration of environmental assessment and development planning, part 2: Prospect Park, Hillingdon. *Project Appraisal* 7(3), 175–85.

Lichfield, N., W. Hendon, P. Nijkamp, A. Realfonso, A. D. Rostirolla 1993a. *Conservation economics: cost–benefit analysis in the cultural built heritage: principles and practice* ["Lichfield Report"]. Paris: ICOMOS. [This is another version of N. Lichfield et al. (1990a) without the content relating specifically to Italian practice.]

— 1993b. *Cost–benefit analysis in the conservation of the cultural built heritage: four case studies* ["Lichfield Report"]. Rome: Ministero dei Beni Culturali.

Lipsey, R. 1963. *An introduction to positive economics*. London: Weidenfeld & Nicholson.

Little, I. M. D. & J. A. Mirrlees 1969. *Manual of industrial project analysis in developing countries*, vol. 2: *social cost–benefit analysis*. Paris: OECD.

Littlechild, S. C. 1986. *The fallacy of the mixed economy: an "Austrian" critique of recent economic thinking and policy*. London: Institute of Economic Affairs.

Lloyd, M. G. 1987. Simplified planning zones: the privatisation of land use controls in the UK? *Land Use Policy*.

Loughlin, M. 1980. The scope and importance of material considerations. *Urban Law and Policy* 3, 171–92.

— 1981. Planning gain: law, policy and practice. *Oxford Journal of Legal Studies* 1, 61–97.

— 1982. Planning gain: another viewpoint. *Journal of Planning and Environment Law*, 352–8.

Luce, R. D. & H. Raiffa 1957. *Games and decisions: introduction and critical survey*. London: Chapman & Hall.

Lyden, F. J. & E. G. Miller (eds) 1968. *Planning, programming, budgeting: a systems approach to management*. Chicago: Markham.

Mabogunje, A. L. 1980. *Development policies: a spatial perspective*. London: Hutchinson.

Mace R. L. 1961. *Municipal cost–revenue research in the United States: a critical survey of researched and measured municipal costs and revenues in relation to land use and areas: 1933–1960*. University of North Carolina: Institute of Government.

Mack. R. 1971. *Planning in uncertainty: decision making in business and government administration*. New York: John Wiley.

MacKenzie, W. J. M. 1967. *Politics and social science*: Harmondsworth: Penguin.

— 1975. Choice and decision. In McGrew & Wilson (1982: 16–18).

McAllister, D. M. 1980. *Evaluation in environmental planning: assessing environmental, social, economic and political trade-offs*. Cambridge, Mass.: MIT Press.

McAuslan, P. 1980. *The ideologies of planning law*. London: Pergamon.

McConnell, S. 1981. *Theories of planning*. London: Heinemann.

McGrew, A. G. & M. J. Wilson (eds) 1982. *Decision making: approaches and analyses: a reader*. Manchester: Manchester University Press.

McGuire, M. C. & H. A. Garn 1969. The integration of equity and efficiency in public project selection. *Economic Journal* **79**, 882–93.

McHarg I. L. 1971. *Design with nature*. New York: Doubleday.

McKean R. N. 1958. *Efficiency in government through systems analysis*. New York: John Wiley.

McLoughlin, J. B. 1969. *Urban and regional planning: a systems approach*. London: Faber & Faber.

— 1973. *Control and urban planning*. London: Faber & Faber.

McMaster, J. C. & G. R. Webb (eds) 1978. *Australian project evaluation*. Sydney: Australia & New Zealand Books.

Machlup, F. 1994. Are the social sciences really inferior. In *Readings in the philosophy of social sciences*, M. Martin & L. C. McIntyre (eds), 6–19. Cambridge, Mass.: MIT Press.

Malisz, B. 1966. Urban planning theory: methods and results. In *City and regional planning in Poland*, J. C. Fisher (ed.). New York: Cornell University Press.

Markyanda, A. & J. Richardson (eds) 1992. *Environmental economics*. London: Earthscan.

Maslow, A. H. 1968. *Towards the psychology of being*. New York: Van Nostrand.

— 1974. *Motivation and personality*. New York: Harper & Row.

Massam, B. 1988. *Multi-criteria decision making*. Oxford: Pergamon.

Matthews, W. H. 1974. Objective and subjective judgements in environmental impact analysis. *Environmental conservation* **2**(2), 121–31.

Medawar, C. 1978. *The social audit consumer handbook: a guide to the social responsibilities of business to the consumer*. London: Macmillan.

Merrett, A. J. & A. Sykes 1973. *The finance and analysis of capital projects*. London: Longman.

Miles, I. 1975. *The poverty of prediction*. Farnborough, England: Saxon House.

Mill, J. S. 1974 (1863). *Utilitarianism* [edited by M. Warnock]. New York: New American Library.

Miller, D. 1976. *Social justice*. Oxford: Oxford University Press.

Millichap, D. 1995. *The effective enforcement of planning controls*. London: Butterworth.

Minister without Portfolio 1985. *Lifting the burden* [Cmd 9571]. London: HMSO.

Mishan, E. J. 1967. *The costs of economic growth*. London: Staples Press.

— 1981a. *Economic efficiency and social welfare: selected essays*. London: George Allen & Unwin.

— 1981b. *Introduction to normative economics*. New York: Oxford University Press.

— 1982. *Cost–benefit analysis: an informal introduction*. London: George Allen & Unwin.

MOWP 1942. *Expert committee on compensation and betterment* ["Uthwatt Report", Cmd 6386]. London: HMSO.

MOWP 1942. *Committee on Land Utilisation in Rural Areas* ["Scott Report"; Cmd 6378]. London: HMSO.

MOHLG (Ministry of Housing and Local Government) 1953. *Planning appeals* [Circular 61/53]. London: HMSO.

— 1955. *Green belts* [Circular 42/55]. London: HMSO.

— 1967. *Town and county planning* [Cmd 3333]. London: HMSO.

— 1970. *Development plans: a manual on form and content*. London: HMSO.

MOLGP 1951. *Progress report: town and country planning, 1943–51* [Cmd 8024]. London: HMSO.

Monopolies and Mergers Commission 1982. *Report on stage carriage services supplied by the undertak-*

ing. London: HMSO.

Morgan, P. & S. Knott 1988. *Development control: policy and practice*. London: Butterworth.

Moroni, S. 1995. Planning, assessment and utilitarianism: notes on Nathaniel Lichfield's contribution to the evaluation field. *Planning Theory* **12**, 81–107.

Morris, P. & R. Therivel (eds) 1995. *Methods of environmental impact assessment*. London: UCL Press.

Morris, P. W. G. & G. H. Hough 1987. *The anatomy of major projects*. Chichester: John Wiley.

MOTCP (Ministry of Town & Country Planning) 1944. *Control of land use* [Cmd 6537]. London: HMSO.

— 1950. *Committee on the qualification of planners* ["Schuster Report"; Cmd 8059]. London: HMSO.

— 1959. *Town and Country Planning Act 1947: planning appeals* [Circular 69]. London: HMSO.

Mynors, C. 1987. *Planning applications and appeals: a guide for architects and surveyors*. London: Architectural Press.

Myrdal, G. 1958. *Value in social theory: a selection of essays on methodology*. London: Routledge & Kegan Paul.

— 1960. *Beyond the welfare state*. New Haven, Connecticut: Yale University Press.

— 1969. *Objectivity and social research*. London: Duckworth.

Nash, C. A. 1993. Cost–benefit analysis of transport projects. In *Efficiency in the public sector: cost–benefit analysis in practice*, A. Williams (ed.), 83–105. Cheltenham: Edward Elgar.

Nash, C., D. Pearce, J. Stanley 1975a. An evaluation of cost–benefit analysis criteria. *Scottish Journal of Political Economy* **22**(21), 121–34.

— 1975b. Criteria for evaluating project evaluation techniques. *Journal of the American Institute of Planners* **41**(2), 83–9.

Nathaniel Lichfield & Associates & Imbucon/AC Management Consultants 1973. *The Oldham study; environmental planning and management*. London: HMSO.

Nathaniel Lichfield & Partners and Leslie Lintott Associates 1986. *Period building: evaluation of development conservation options: a manual for practice*. London: Henry Stewart.

Needham, B. 1971. Concrete problems, not abstract goals. *Journal of the Royal Town Planning Institute* **58**(7), 317–19.

Needham, B. & A. Faludi 1973. Planning in the public interest. *Journal of the Royal Town Planning Institute* **59**(5), 164–6.

Nelson, P. 1988. Environmental impact assessment; physical impact and hazard. *The Planner* **75**(2), 28–33.

Nijkamp, P. 1975. A multicriteria analysis for project evaluation. *Papers of Regional Science Association* **35**, 28–33.

— 1982. Soft multicriteria analysis as a tool in urban land use planning. *Environment and Planning B* **9**(2), 197–208.

Nijkamp, P., H. Leitner, N. Wrigley (eds) 1986. *Measuring the unmeasurable*. Dordrecht: Kluwer Nijhoff.

Nijkamp, P. & J. Spronk 1981. *Multiple criteria analysis: operational methods*. London: Gower.

Nijkamp, P., P. Rietvold, H. Voogd 1991. *Multiple criteria analysis in physical planning*. Amsterdam: North Holland.

Niskanen, W. A. 1973. *Bureaucracy: servant or master*. London: Institute of Economic Affairs.

Novick, D. (ed.) 1967. *Programme budgeting: program analysis and the federal budget*. Cambridge, Mass.: Harvard University Press.

O'Riordan, T. O. 1976. *Environmentalism*. London: Pion.

O'Riordan, T. & D. W. R. Sewell (eds) 1981. *Project appraisal in policy review*. Chichester: John Wiley.

Osborn, F. J. 1959. *Can man plan?*. London: Harrap.

Parry, G. & P. Morris 1974. When is a decision not a decision? In McGrew & Wilson (1982: 19–34).

Patricios, N. 1986. *International handbook on land use policy*. Westport, Connecticut: Greenwood Press.

Pearce, D. W. 1976. *Environmental economics*. London: Longman.

— (ed.) 1983a. *Dictionary of modern economics*. London: Macmillan.

— 1983b. *Cost–benefit analysis*. London: Macmillan.

Pearce, D. W. & C. A. Nash 1981. *The social appraisal of projects*. London: Macmillan.

Pearce, D., A. Markyanda, E. B. Barbier 1989. *Blueprint for a green economy*, London: Earthscan.
Pearce, D. et al. 1991. *Blueprint 2: greening the world economy*. London: Earthscan.
— 1993. *Blueprint 3: measuring sustainable development*. London: Earthscan.
Perloff, H. S. 1969. A framework for dealing with the urban environment. In *The quality of the urban environment*, H. S. Perloff, (ed.). Washington DC: Resources for the Future.
Peters, G. H. 1973. *Cost–benefit analysis in public expenditure*. London: Institute of Economic Affairs.
Phersson vs Secretary of State for the Environment 1990. Planning and Compensation Reports 266.
Phillips, L. D. 1987. Requisite decision modelling for technological projects. In *Social decision methodology for technological projects*, C. Vlek & G. Cvetkovich (eds). Amsterdam: North Holland.
Pigou, A. C. 1932. *Economics and welfare*. London: Macmillan.
Pinfield, G. 1992. Strategic environmental assessment and land use planning. *Project Appraisal* 7(3), 157–63.
PAG (Planning Advisory Group) 1965. *The future of development plans*. London: HMSO.
Prest, A. R. & R. Turvey 1965. Cost–benefit analysis: a survey, *The Economic Journal* 75(300), 683–735.
Property Advisory Group 1980. *Planning gain*, London: HMSO.
Purdue, M. 1989. Material considerations: an ever expanding concept. *Journal of Planning and Environment Law*, 145–228.

Ramsay, C. G. 1984. Assessment of hazard and risk. In Clark et al. (1984: 133–60).
Ratcliffe, J. 1978. *An introduction to urban land administration*. London: Estates Gazette.
Rau, J. G. & D. C. Wooten (eds) 1980. *Environmental impact assessment handbook*. New York: McGraw-Hill.
Ravetz, A. 1980. *Remaking cities*. London: Croom Helm.
— 1986. *The government of space: town planning in modern society*. London: Faber & Faber.
Rawls, J. 1972. *A theory of justice*. Oxford: Oxford University Press.
Reif, B. 1973. *Models in urban and regional planning*. Aylesbury: Leonard Hill.
Rickson, R. E., R. J. Burdge, A. Armour 1989. *Integrating impact assessment into the planning process: international perspectives and experience*. Impact Assessment Bulletin 8, International Association for Impact Assessment, Belhaven, North Carolina.
Rittel, H. W. J. & M. Webber 1973. Dilemmas in a general theory of planning. *Policy Sciences* 4, 155–69.
Robbins, L. 1952. *An essay on the nature and significance of economic science*. London: Macmillan.
Roberts, J. R. H. 1946. *The law relating to town and country planning*. London: Charles Knight.
Robson, B., M. Parkinson, F. Robinson 1994. *Assessing the impacts of urban policy: DoE inner cities research programme*. London: HMSO.
Rodwin, L. 1970. *Nations and cities*. Boston: Houghton & Mifflin.
Rose, R. (ed.) 1974. *The management of urban change in Britain and Germany*. London: Sage.
Roskill Commission. See Commission on the Third London Airport.
Rossi, P. H., H. E. Freeman, S. R. Wright 1979. *Evaluation: a systematic approach*. Beverley Hills: Sage.
Rosslyn Research Ltd 1990. *Planning gain: a survey for KPMG Peat Marwick*. London: KPMG.
Royal Commission on the Distribution of the Industrial Population 1940. *Report* ["Barlow Report"; Cmd 6153]. London: HMSO.
Royal Commission on Environmental Pollution 1971 *First report* [Cmd 4585]. London: HMSO.
— 1976, *Fifth report: air pollution control: an integrated approach* [Cmd 6371]. London: HMSO.
— 1988. *Twelfth report: best practicable environmental option* [Cmd 310]. London: HMSO.
RTPI (Royal Town Planning Institute) Working Party in Development Control 1978. *Development control: the present system and some proposals for the future* ["Finney Report"]. London: Royal Town Planning Institute.
RTPI 1983. *Planning gain guidelines*. London: Royal Town Planning Institute.
RTPI 1990. *General information about membership*. London: Royal Town Planning Institute.
Sabine, G. H. 1963. *Political theory*. London: Harrap.
Sadler, 1988. The evaluation of assessment: post EIS research and process. In Wathern (1988: 129–58).
Salt, A. 1991. *Planning applications: the RMJM guide*. Oxford: Blackwell Scientific.
Schaenman, P. S. 1976. *Using an impact measurement system to evaluate land development*. Washington

DC: Urban Institute.

Schaenman, P. S. & T. Muller 1974. *Measuring impacts of land development: an initial approach.* Washington DC: Urban Institute.

Schaffer, V. B. & D. C. Corbett (eds) 1966. *Decisions: case studies in Australian administration.* Melbourne: Cheshire.

Schlager, K. 1968. *The rank based expected value method of plan evaluation.* Highway Research Record 238, Highway Research Board, Washington DC. [The description is quoted from E. Boyce & N. D. Day (1969), *Metropolitan plan evaluation methodology,* Institute for Environmental Studies, University of Pennsylvania].

Schofield, J. A. 1987. *Cost–benefit analysis in urban and regional planning.* London: Allen & Unwin.

Schubert, G. A. 1961. *The public interest: a critique of the theory of a political concept.* Glencoe, Illinois: Free Press.

Schumacher, E. F. 1973. *Small is beautiful: a study of economics as if people mattered.* London: Blond & Briggs.

Schumandt, H. J. & W. Bloomberg 1969. *The quality of urban life.* Beverley Hills: Sage.

Schuster, see Committee on the Qualifications of Planners.

Schwenk, C. R. 1988. *The essence of strategic decision making.* Lexington, Massachusetts: Lexington Books.

Scott see MOWP.

Secretary of State for Employment et al. 1986. *Building business . . . not barriers* [Cmd 9794]. London: HMSO.

Secretary of State for Employment 1987. *Encouraging enterprise: a progress report on deregulation.* London: The Department.

Secretary of State for the Environment 1978. *Planning procedures: the government's response to the 8th report from the Expenditure Committee Session 1976–1977.* London: HMSO.

Secretary of State for the Environment et al. 1990. *This common inheritance: Britain's environmental strategy* [Cmd 1200]. London: HMSO

— 1991. *This common inheritance: the first year report.* London: HMSO.

— 1994. *Sustainable development, the UK strategy* [Cmd 2426]. London: HMSO.

Secretary of State for Trade and Industry 1988. *Releasing enterprise* [Cmd 512]. London: HMSO.

Seldon, A. 1994. *The state is rolling back.* London: E. & L. Books.

Select Committee on the European Communities, House of Lords 1981. *Environmental assessment of projects: draft directive concerning the assessment of the environmental effects of certain private projects.* London: HMSO.

Self, P. 1975. *Econocrats and the policy process: the politics and philosophy of cost–benefit analysis.* London: Macmillan.

Self, P. 1985. *Political theories of modern government, its role and reform.* London: Allen & Unwin.

Shankland Cox & Associates 1966. *Expansion of Ipswich – designation proposals.* London: HMSO.

Shefer, D. & H. Voogd (eds) 1990. *Evaluation methods in urban and regional plans: essays in memory of Morris (Moshe) Hill.* London: Pion.

Shelbourn C. 1989. Development control and hazardous substances. *Journal of Planning and Environment Law,* 323–30.

Simmie, J. 1974. *Citizens in conflict: the sociology of town planning.* London: Hutchinson.

Simmie, J. 1994 (ed.). *Planning London.* London: UCL Press.

Sinden, J. W. & A. C. Worrell 1979. *Unpriced values: decisions without market prices.* New York: John Wiley.

Skidmore, Owings & Merrill 1981. *Area-wide environmental impact assessment: a guidebook.* Washington DC: US Department of Housing and Urban Development.

Smart, J. J. C. & B. Williams 1973. *Utilitarianism: for and against.* Cambridge: Cambridge University Press.

Solesbury, W. 1974. *Policy in urban planning: structure plans, programmes and local plans.* Oxford: Pergamon.

Solon Consultants 1990. *Prospect Park research: public response to the revised proposals.* London: British

Airways.

Squire, L. & H. G. Van Der Tak 1975. *Economic analysis of projects: a world bank research publication.* Baltimore: Johns Hopkins University Press.

— 1979. *Framework appraisal* ["Leitch Report"]. London: HMSO.

— 1986. *Urban road appraisal* ["Williams Report"]. London: HMSO.

— 1992. *Assessing the environmental impact of road schemes* ["Wood Report"]. London: HMSO.

Stephenson, G. 1995. *Compassionate town planning.* Liverpool: Liverpool University Press.

Stone, P. A. 1980. *Building design evaluation: costs in use.* London: Spon.

Stopher, P. R. & A. H. Meyburg 1976. *Transportation systems evaluation.* Lexington, Mass.: Lexington Books.

Suddards, R. W. 1982. *Listed buildings: the law and practice.* London: Sweet & Maxwell.

Stringer vs MOHLG 1971. *All England Reports* 65.

Suchman, E. A. 1967. *Evaluative research: principles and practice in public service and social acts and programmes.* New York: Russell Sage Foundation.

Sugden, R. & A. Williams 1978. *The principles and practice of cost–benefit analysis.* Oxford: Oxford University Press.

Susskind, L., L. Bacon, M. Wheeler 1978. *Resolving environmental regulatory disputes.* Cambridge, Massachusetts: Schenkman.

Tesco Stores Ltd vs *Secretary of State for the Environment & Hounslow Borough Council* 1995. *Weekly Law Reports* 759.

Therivel, R., E. Wilson, S. Thompson, D. Heaney, D. Pritchard 1992. *Strategic environmental assessment.* London: Earthscan.

Thompson, R. 1987. Is fastest best? – the case of development control. *The Planner* **73**(9), 11–15.

Thornley, A. 1991. *Urban planning under Thatcherism: the challenge of the market.* London: Routledge.

Timmins, N. 1995. *The five giants: a biography of the Welfare State.* London: HarperCollins.

Tomlinson, P. 1986. Environmental assessment in the UK: implementation of the EEC directive. *Town Planning Review* **57**(4), 458–86.

— 1984. The issue of methods in screening and scoping. In Clark et al. (1984: 163–94).

Turvey, R. 1963. On divergence between social cost and private cost. *Economica* **30**, 309–313.

Tyson, W. J. & R. L. D. Cochrane 1977. Corporate planning and project evaluation in urban transport. *Long Range Planning* **14**, 143–61.

Underwood, J. 1981. Development control: a case study of discretion in action. In *Policy and action: essays on the implementation of public policy*, S. Barrett & C. Fudge (eds). London: Methuen.

Urban Motorways Committee 1972. *New roads in towns: report to the Secretary of State for the Environment.* London: HMSO.

Urban Motorways Project Team 1973. *Report to the Urban Motorways Committee.* London: DOE. [See chs 13 and 14 for the Travers Morgan cost–benefit matrix.]

Uthwatt   *see* Expert Committee on Compensation and Betterment.

Verheem, R. 1992. Environmental assessment at the strategic level in the Netherlands. *Project Appraisal* **7**(3), 150–56.

Von Neumann, J. & O. Morgenstern 1964. *Theory of games and economic behaviour.* New York: John Wiley.

Von Winterfeldt, D. & W. Edwards 1986. *Decision analysis and behavioural research.* Cambridge: Cambridge University Press.

Voogd, H. 1982. Multicriteria evaluation with mixed qualitative and quantitative data. *Environment and planning B* **9**(2), 221–36.

Voogd, H. 1983 *Multi-criteria evaluation for urban and regional planning.* London: Pion.

Wachs, M. (ed.) 1980. *Ethics in planning.* Centre for Public Policy Research, Rutgers University, New Jersey.

Wakeford, R. 1990. *American development control: parallels and paradoxes from an English perspective.* London: HMSO.

Wannop. U. 1985. The practice of rationality. In *Rationality in planning*, M. Breheny & A. Hooper (eds), 196–208. London: Pion.

Ward, B. & R. Dubois 1972. *Only one Earth.* Harmondsworth: Penguin.

Wathern, P. (ed.) 1988. *Environmental impact assessment: theory and practice.* London: Unwin Hyman.

Warner, M. L. & E. H. Preston 1974. *Review of environmental impact assessment methodologies.* Columbus, Ohio: Battelle.

Waterston, A. 1968. *Development planning: lessons of experience.* Baltimore: Johns Hopkins University Press.

Weiss, K. H. (ed.) 1972. *Evaluating action programmes: readings in social action and education.* Boston: Allyn & Bacon.

Whiston, T. (ed.) 1979. *The uses and abuses of forecasting.* London: Macmillan.

Wibberley, G. P. 1959. *Agriculture and urban growth: a study of the competition for rural land.* London: Michael Joseph.

Williams, A. 1972. Cost–benefit analysis: a bastard science and/or insidious poison in body politick? In *Benefit–cost and policy analysis: an Aldine annual on forecasting, decision making and evaluation*, W. A. Niskanen et al. (eds), 48–74. Chicago: Aldine.

Wilson, A. G. 1974. *Urban and regional models in geography and planning.* London: John Wiley.

Winpenny, J. T. 1991. *Values for the environment: a guide to economic appraisal.* London: HMSO.

Winterfeldt, D. V. & W. Edwards, 1986. *Decision analysis and behavioural research.* Cambridge: Cambridge University Press.

Wolf, C. P. 1983. *Social impact assessment: the state of the art.* Washington DC: Environmental Design Research Association.

Wood, C. 1988. EIA in plan making. In Wathern (1988: 98–114).

Wootton, B. 1938. *Lament for economics.* London: George Allen & Unwin.

Working Party on Economics 1995. *Economics of urban villages* ["Lichfield Report"]. London: Urban Villages Forum.

World Commission on Environment and Development 1987. *Our common future.* Brundtland Report. Oxford: Oxford University Press.

Yewlett, C. 1985. Rationality in planning: a professional perspective. In Breheny & Hooper (1985: 209–28).

Young, E. & J. Rowan-Robinson, 1985. *The Scottish planning law and procedure.* Glasgow: Hodge.

Young, G. 1991. *Hansard*, 16 May, col. 501 ff.

INDEX